AF861882

Instructor's Guide to accompany Foundations of Electronics, 5e (Electron Flow Version) and Foundations of Electronics: Circuits and Devices, 5e (Electron Flow Version) and Foundations of Electronics: Circuits and Devices, 2e (Conventional Flow Version)

Russell L. Meade

THOMSON
DELMAR LEARNING

Australia Canada Mexico Singapore Spain United Kingdom United States

THOMSON
DELMAR LEARNING

Foundations of Electronics Instructor's Guide, 5e

Russell L. Meade

Vice President, Technology and Trades ABU:
Dave Garza

Director of Learning Solutions:
Sandy Clark

Senior Acquisitions Editor:
Stephen Heiba

Senior Product Manager:
Michelle Ruelos Cannistraci

Channel Manager:
Dennis Williams

Marketing Coordinator:
Stacey Wiktorek

Production Director:
Patti Shephan

Senior Production Manager:
Larry Main

Marketing Director:
Deborah S. Yarnell

Production Editor:
Benj Gleeksman

Art/Design Coordinator:
Francis Hogan

Senior Editorial Assistant:
Dawn Daugherty

Printed in the United States of America
1 2 3 4 5 XX 06 05

For more information contact
Thomson Delmar Learning
Executive Woods
5 Maxwell Drive, PO Box 8007,
Clifton Park, NY 12065-8007

Or find us on the World Wide Web at www.delmarlearning.com

ISBN: 1418005401

DEDICATION

To my wife
Who has put up with me and encouraged me all these years.

CONTENTS

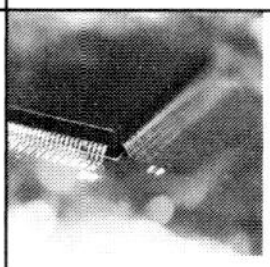

INTRODUCTION

General Information Relating to the Text

Foundations of Electronics, Fifth Edition *(Electron Flow Version), Foundations of Electronics: Circuits and Devices*, Fifth Edition *(Electron Flow Version)*, and *Foundations of Electronics: Circuits and Devices*, Second Edition *(Conventional Flow Version)* are practical theory texts. They build the fundamental of electronics knowledge and skills that every technician or engineer relies on. Critical thinking skills are emphasized so that learners acquire valuable competencies in problem solving, troubleshooting, circuit analysis, and safety consciousness.

When either of these texts is used with its companion laboratory manual, instructors gain a powerful teaching system and students gain a dynamic learning combination.

These texts use the instructional strategy of (1) present, (2) pattern, and (3) practice. That is, *information is presented* and *examples* of how to apply that information to practical situations *are shown as a pattern*. Then *the learner is asked immediately to practice* what was learned, using the pattern as an aid.

Learning material is presented through the reading assignments. Examples appear throughout the reading assignments to apply the material and give the student patterns. Finally, practice opportunities appear in several of the text features. In-Process Learning Checks and Practice Problems appear within each reading assignment, while end-of-chapter Review Questions, Problems, Analysis Questions, Using Excel, and MultiSIM Exercises verify, enhance, and reinforce learning. Students can then practice and apply their knowledge and verify what they have learned in theory by using the correlated hands-on laboratory projects.

Several special features in these texts are:

1. Computer Excel "sample" worksheet templates and *Using Excel* practice exercises at the end of many of the text chapters, which provide students with examples of how to set up worksheets for solving electronic circuit problems, and practice in doing it.
2. A new and powerful "Excel Automated formulas tool" is provided for students and instructors alike. This tool comes in the form of a complete, chronologically organized listing of all the formats in the text in a special Excel formulas template layout. These "pre-programmed" spreadsheet (Excel) formulas templates are presented in a format which makes it easy for anyone to simply insert the circuit parameters for the circuit problem of interest. The pre-programmed formulas in the last column of these templates will automatically provide the answer for the parameters entered on a given formula's row, as appropriate. (A total of 272 Excel formulas are pre-programmed and ready to go!)
3. Again, in the new editions, computerized circuit simulation exercises (using MultiSIM) are provided to encourage students in learning to use the computer for circuit simulation purposes. Also, instructors, at their discretion, can utilize this feature for additional troubleshooting exercises for their students, as well.

NOTE ➤ Answers for the Electron-Flow and Conventional-Flow Versions are shown in the Instructor's Guide. Solid-line arrows show the direction of current flow in a circuit for the electron-flow answers. Dotted-line arrows show the direction of current flow in a circuit for the conventional-flow answers. If the answers for electron-flow and conventional-flow are different, then the electron-flow answer is given first, followed by the conventional-flow answer which, will be placed in parentheses with the answer in italics.

Contents and Use of This Instructor's Guide

Contents

This Instructor's Guide has three main sections of information:

SECTION I: INSTRUCTIONAL OVERVIEW

This section provides the following instructional strategies:

- Key Topics "Quick-Find" Listing
- Course Outline and Key Topics overview
- Group Training Format and Individualized Training Format overview
- Three Training Tracks descriptors: Visual aids and explanations regarding the possible tracks that may be used in course management:
  - Track #1: "Most Comprehensive" Track (Full-length program)
  - Track #2: "Mid-Level" Track (Slightly shortened program)
  - Track #3: "Fast" Track (Shortest program)

Part Overviews with Chapter-by-Chapter Instructional Strategies (for Track #1) (NOTE: See "Sample Student Assignments for Track #2 and Track #3 Programs, and Visual Aid for Program Tracks #2 and #3 for further information on these tracks)

SECTION II: ANSWERS TO ALL CHAPTER QUESTIONS

SECTION III: ANSWERS TO LABORATORY PROJECTS

Use

Several useful graphical displays are provided in Section I (the "Instructional Overview" section). Also, once again, we are providing the Course Outline and Key Topics listings in this section, to further aid you as an instructor in planning your teaching sequence and training strategies. The new "Key Topics Quick-Find" Listing, the Track #1, Track #2, and Track #3 Resources page, and the new graphic assignment pages relating to the shortened Track #2 and Track #3 programs should be of assistance in your course planning.

Whether you are planning a schedule for standard group training methods, or for self-paced individualized instruction programs, the above-mentioned aids should help.

In the "Part Overviews" subsection, the detailed objectives listings give you insight into how and where key knowledge and skills competencies are addressed in the text.

Also, the instructional strategy samples and part overviews help you better understand how to program your sequence of learning activities.

The chapter-by-chapter and project-by-project sequentially organized answers to questions in the texts and the lab projects manual provided in Sections II and III of this guide are invaluable tools to help you grade and "upgrade" your students' practice and skill-building efforts.

Lab Approach

The laboratory projects manual is designed to aid students in understanding the basic concepts of electronics and in verifying the theory they have studied. By actually connecting circuits and making measurements in these circuits using "real-world" test equipment, the students: 1) confirm theory and 2) gain the critically important "hands-on" skills required to connect circuits from schematic diagrams and to properly connect and use test equipment. They are then taught to properly interpret and analyze their measurements.

As indicated in the student's project manual, the manual is divided into 28 parts. Each part consists of a group of related-topic projects. The manual contains 82 workbook-style projects and 20 *"Story Behind the Numbers"* projects. In both styles of projects, the students gain practical hands-on experience and analytical skills. *"The Story Behind the Numbers"* projects add the dimensions of additional graphing skills practice and technical writing and communications skills.

Each group of related projects has a set of objectives that provide a list of competencies and knowledge goals that serve as student targets of learning that students should strive to achieve during their learning activities.

Many of the questions and problems posed in the projects are designed to not only help the students in their understanding of important concepts, but also to enhance their troubleshooting knowledge and skills.

See page xiv of this introductory material for instructions on having students perform the optional troubleshooting exercises. These instructions are also found in the student's laboratory manual, but are repeated here, for your convenience.

Since the projects are closely tied to specific topics in the text, try to have the students study and/or refer to the indicated text topics suggested in the "Performance Projects Correlation Charts" prior to working on the assigned projects.

A powerful new tool for instructors is provided on the Instructor e-resource (ISBN: 1418005398) CD. This tool is comprised of "Demos" for most of the projects in the Laboratory Projects manual. If you, as an instructor, have access to using the MultiSIM circuit simulation programs (Version 2001, Version 7, Version 8, or even Version 9), the CD provides pre-programmed "initial circuit" setups for most of the 82 projects in the Projects Manual. These circuits are provided in all four versions of the MultiSIM program. This ***new tool*** gives the instructor numerous possibilities: 1) Live demonstrations and discussions of projects in a dynamic manner with students (either prior to, or after the students perform the projects), 2) use as an aid in teaching troubleshooting concepts, and 3) use to demonstrate what happens with different circuit component or parameter values.

It is the author's hope that the students will not only have valuable learning experiences while performing the "hands-on" projects, but will also find the learning process to be an enjoyable one. Further, it is the author's desire that instructors will find the training program planning and program operating tools in this instructor guide to be a great aid in planning and conducting "world class" training programs.

Prerequisites

This course assumes that learners have no previous background in electronics. Basic math skills, such as adding, subtracting, multiplying, dividing, exponents, and roots are assumed; however, any basic or special applications of these skills are always clearly presented, modeled, and reviewed right at the point where they are used. Also, use of the scientific calculator is emphasized as a tool students should use on a regular basis.

INSTRUCTIONS FOR OPTIONAL TROUBLESHOOTING EXERCISES

The Optional Troubleshooting Exercises are designed to provide hands-on troubleshooting experiences that allow students to apply the SIMPLER troubleshooting process to actual circuits, set up by their instructors.

Note that each of the "Student Log for Optional Troubleshooting Exercise" sheets is set up to lead you through the **SIMPLER** troubleshooting sequence. That is:

- In the first blank on the log sheet, you must document the starting point **S***ymptom* data.
- In the next blank you must identify the **I***nitial* "area of uncertainty" or the area of the circuit you would bracket for troubleshooting efforts.
- The third blank requires you to log the decision you have made regarding your first test (i.e., your response to the **M***ake* a decision about what type of test and where to make it step). At this point, you should **P***erform* the test and observe the results of the test.
- The next step is to use the information you have collected so far to help you log data that **L***ocates* and enables you to define the new (hopefully smaller) area to be bracketed for continued troubleshooting.
- The next blank on the log sheet requires you to **E***xamine* your available data and determine the next test you want to make. Log the type and location of the test. Now you make the test and observe results that enable you to **R***epeat* the analysis and testing steps (as indicated on the log sheet) until you find what you believe to be the problem with the circuit.

NOTE ➤ The instructor's part in this effort is to supply the student with pre-wired circuits having specific problems in them, as indicated in the Instructor's Guide. Problem circuits should be supplied to the student on a one-circuit-problem-at-a-time basis.

Instructional Overview for

Foundations of Electronics, 5e (Electron Flow Version)

Foundations of Electronics: Circuits and Devices, 5e (Electron Flow Version)

and

Foundations of Electronics: Circuits and Devices, 2e (Conventional Flow Version)

and

Laboratory Projects Manual

SECTION I

Includes:

- Three Track Program Options
- Course Outline and Key Topics (Track #1)
- Key Topic "Quick-Find" Listing
- Sample Training Approaches for Track #2 and Track #3
- General Instructional Strategies
- Part Overviews with Chapter Objectives and Chapter-by-Chapter Instructional Strategies

TRACK PROGRAM OPTIONS

Track #1—Most Comprehensive Program

General Information:

In this training track, it is appropriate to use as much of the text and as many of the hands-on laboratory projects as the program time frame will possibly allow.

In contrast to this comprehensive Track #1 approach, the sample Medium and Short Course Length programs (Tracks #2 or #3) omit a number of text chapters (e.g., Chapters 7, 8, 10, and 12). Additionally, the shorter courses may elect to use only the 14 selected *"Story Behind the Numbers"* lab projects for hands-on training. The more comprehensive Track #1 program allows students more hands-on training time than the shorter courses. This allows many of the 82 standard lab projects available in the Lab Projects manual to be used. For further enhancement of the hands-on training, the instructor may also elect to have students perform some of the *"Story Behind the Numbers"* (SBN) projects as well.

Also, for the comprehensive programs, it is good to use as many of the optional Excel, MultiSIM, and Troubleshooting Challenge Circuits exercises as is feasible. These exercises provide powerful training and reinforcement of learning. Students generally find these exercises enjoyable, as well as valuable to their training.

Depending on time constraints, the instructor may or may not wish to tailor homework assignments for their program by selecting only certain end-of-chapter Review Questions, Problems, and Analysis Questions rather than assigning every question and problem listed at the end of each chapter. This approach depends on the time constraints of students and of the course timelines.

Whether or not all the possible information in the text is used or whether your program uses the comprehensive or shorter course approaches, your students will find the text to be an invaluable "bookshelf" reference resource, not only during training, but for the rest of their technical careers.

Track #2—Medium Course Length Program

General Information:

This program is designed to be somewhat shorter than the "most-comprehensive" course program. It is shorter by virtue of omitting certain chapters and/or sections of chapters in the text, and by assigning fewer of the available hands-on lab projects, depending on time constraints.

Also, the instructor may use his or her discretion regarding the number of optional training program assignments used. For example, the instructor may not use all, but may assign only certain ones of the Excel exercises, MultiSIM exercises, and Troubleshooting Challenge Circuits provided at the end of many chapters in the text.

Ideally, students in this program will be assigned and study all of the topics listed in the "Key Topics Quick-Find" Listing, found in the Instructor's Guide.

For hands-on training, it is good, if time permits, that the 14 *"Story Behind the Numbers"* laboratory projects be performed. If time allows, some of the 82 other projects in the lab manual might be selected for additional learning reinforcement and training.

A Track #2 and Track #3 Course Length instructional matrix has been designed to aid instructors in managing their programs. This matrix provides student assignment specifics regarding:

1. The In-Chapter Practice Problems and In-Process Learning Checks, which should be used, as listed, for the Track #2 and Track #3 courses,
2. End-of-chapter Review Questions, Problems, and Analysis Questions that can be selectively assigned, in accordance with time constraints,
3. Which topics have *"Story Behind the Numbers"* hands-on projects that typically should be used in the shorter courses. Additionally, the list of available topic-appropriate Standard Lab Projects are shown, from which the instructor may also make assignments, in accordance with the course time constraints, and
4. The optional Excel, MultiSIM, and Troubleshooting Challenge Circuits work assignments, from which selections can also be made by the instructor.

For programs that use one of the shorter program options (i.e., Track #2 or Track #3 options), we highly recommend that instructors promote the idea that students keep their textbook as a <u>valuable personal "bookshelf" reference resource</u> as they continue on in their studies and in pursuing their technical careers. This idea has significant value for students who have been involved in "shortened" courses, as they will definitely need a reference resource on topics they may not have covered in classroom contact hours.

Track #3—Short Course Length Program

General Information:

This program is designed to be shorter than either the "most-comprehensive" (Track #1) course program or the "medium course length" (Track #2) program. A shorter timeline is accomplished by omitting certain chapters and/or sections of chapters in the text, and by being more selective regarding all student assignments, including hands-on lab projects assignments, homework assignments, and optional work assignments elements.

A Track #2 and Track #3 Shorter Course Length instructional matrix has been designed to aid instructors in managing their programs. This matrix provides student assignment specifics regarding:

1. In-Chapter Practice Problems and In-Process Learning Checks, which should be used, as listed, for the Track #2 and Track #3 courses,
2. End-of-chapter Review Questions, Problems, and Analysis Questions that can be selectively assigned, in accordance with time constraints,
3. Which topics have *"Story Behind the Numbers"* hands-on projects that typically should be used in the shorter courses. Additionally, the list of available topic-appropriate Standard Lab Projects are shown, from which the instructor may also make assignments, in accordance with the course time constraints, and
4. The optional Excel, MultiSIM, and Troubleshooting Challenge Circuits work assignments, from which selections can also be made by the instructor.

Optimally, the students' text work will allow them to read and study most of the topics listed in the Key Topics "Quick-Find" Listing. This means that the chapter sections listed on the "Key Topics" listing, and the In-Chapter Practice Problems and In-Process Learning Checks within those top headings, shown on the "instructional matrix," should be part of the student's study program. The areas wherein the course time frame can be significantly shortened relates to how many of the end-of-chapter questions and problems are assigned, and how many or how few of the Lab Projects and Optional Work Assignments shown on the "instructional matrix" are assigned to the students.

For programs that use one of the shorter program options (i.e., Track #2 or Track #3 options), we highly recommend that instructors promote the idea that students keep their text as a <u>valuable personal "bookshelf" reference resource</u> as they continue on in their studies and in pursuing their technical careers. This idea has significant value for students who have been involved in "shortened" courses, as they will definitely need a reference resource on topics they may not have covered in classroom contact hours.

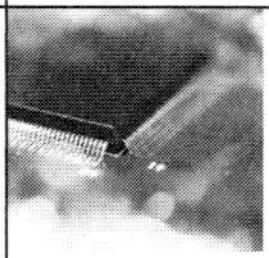

COURSE OUTLINE AND KEY TOPICS TRACK #1

PART I:	**FOUNDATIONAL CONCEPTS . . . .**	**CHAPTERS 1 THROUGH 2**
PART II:	**BASIC CIRCUIT ANALYSIS . . . .**	**CHAPTERS 3 THROUGH 8**
PART III:	**PRODUCING AND MEASURING ELECTRICAL QUANTITIES . . . .**	**CHAPTERS 9 THROUGH 12**
PART IV:	**REACTIVE COMPONENTS . . . .**	**CHAPTERS 13 THROUGH 21**
PART V:	**INTRODUCTORY DEVICES AND CIRCUITS**	**CHAPTERS 22 THROUGH 31**

INTRODUCTION

- Opportunities
- Challenges

CHAPTERS 1–2

FOUNDATIONAL CONCEPTS AND PRINCIPLES

- Basic Concepts
- Basic Units
- Basic Symbols
- Diagrams
- Resistors
- Resistor Color Code
- Measuring *V, I,* and *R*

CHAPTERS 3–8

BASIC CIRCUIT ANALYSIS

- Ohm's Law
- Series Circuits
- Parallel Circuits
- Series-Parallel Circuits
- Network Theorems
- Network Analysis Techniques

CHAPTERS 9–12

PRODUCING AND MEASURING ELECTRICAL QUANTITIES

- Sources
- Magnetism
- Measuring Instruments
- Basic AC Quantities
- The Oscilloscope

CHAPTERS 13–20

REACTIVE COMPONENTS AND CIRCUIT ANALYSIS

- Inductance (dc)
- Inductance and *RL* Circuits (ac)
- Basic Transformers
- Capacitance (dc)
- Capacitance and *RC* Circuits (ac)
- *RLC* Circuits (ac)

CHAPTER 21

RESONANT CIRCUIT CHARACTERISTICS AND ANALYSIS

- Series Resonant Circuits
- Parallel Resonant Circuits

CHAPTERS 22–24

DIODES AND POWER SUPPLY FUNDAMENTALS

- Semiconductor Materials
- Diodes
- Power Supply Circuits

CHAPTERS 25–27

INTRODUCTION TO TRANSISTORS AND TRANSISTOR AMPLIFIERS

- BJTs
- BJT Amplifier Circuits
- FETs

CHAPTERS 28–31

OP AMPS, OSCILLATORS, THYRISTORS, AND OPTOELECTRONICS

- Op Amps
- Oscillators and Multivibrators
- Thyristors
- Optoelectronics

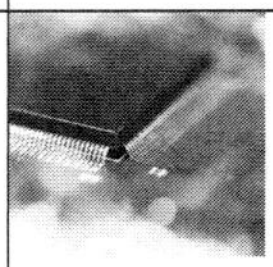

KEY TOPICS "QUICK-FIND" LISTING

Sample Student Assignments for Track # 2 and Track #3 Programs

Use In-Chapter Questions (i.e., Pr. Prob. Sets and IPLC #s) Assignments, as Shown AND

Select Desired Questions, Problems, Labs, and Optional Work Assignments from Other Listed Data

		In-Chapter Questions		Select End-of-Chap. Assign. from Listed #s			Select Lab	Optional Work Assignments		
Chap. #	Topic Heading	Pr. Prob. Set #	IPLC #	Review Q	Problems	Analysis Q	Projects **	Excel	MultiSIM	Chall. Ckts
1	Basics	-	2	13–25		7–10	-	-	-	-
2	Basics	-	1	1–25	-	1–10	(1–4)	-	-	-
3	Ohm's Law	1–4, 6–8	1 (q. 1–4) 2 (q. 1, 2)	1–3, 6, 9, 10 a, b	1–15, 17, 18, 20–25	1, 2, 6, 7	SBN or (5–8)	Excel	MultiSIM	-
4	Series Circuits	1–8	1 (q. 1–6)	1–10	1–28	1–13	SBN or (9–14)	Excel	MultiSIM	C.C 1, 2
5	Parallel Circuits	1–11, 13, 14		1–12	1–30	1–7	SBN or (15–20)	Excel	MultiSIM	C.C. 3, 4
6	Series-Parallel Circuits	1–4 & 6	1 & 2	1–10	1–3, 6–31	1, 2, 9, 10	SBN or (21–26)		MultiSIM	C.C. 5, 6
7	(Not used in this track)									
8	(Not used in this track)									
9	Magnetism	1 & 2	1 & 2	1–6	1–3	-	-	-	-	-
10	(Not used in this track)									
11	Intro to AC	1 & 3	1 & 2	5–9, 13, 14, 15, 17, 18	1–5, 7–15	2–8	-	-	-	-
12	(Not used in this track)									
13	Inductance	1 & 2	1	1–5, 11, 16, 17	1–3, 9	1, 2, 5, 7, 8	-	-	-	-
14	Inductive Reactance	1, 2, & 3	1	1–6, 8–10	1–4, 7–20	1–6, 9	SBN or (42)	Excel	-	C.C. 8
15	*RL* Circuits	1–10	1	1–20	1–10	1–12	SBN or (43, 44)	Excel	MultiSIM	C.C. 9
16	Transformers	4, 5, & 6	-	1, 4–9, 11–13	1–4, 8–15	1–6	SBN or (45, 46)	-	-	-
17	Capacitance	1, 4, & 6	1	2, 6–11, 18, 19	2–7, 10–14	-	SBN or (47, 48)	Excel	-	C.C. 10

**SBN = "Story Behind the Numbers" Projects

Sample Student Assignments for Track # 2 and Track #3 Programs

Use In-Chapter Questions (i.e., Pr. Prob. Sets and IPLC #s) Assignments, as Shown AND
Select Desired Questions, Problems, Labs, and Optional Work Assignments from Other Listed Data

		In-Chapter Questions		Select End-of-Chap. Assign. From Listed #s			Select Lab	Optional Work Assignments		
Chap. #	Topic Heading	Pr. Prob. Set #	IPLC #	Review Q	Problems	Analysis Q	Projects **	Excel	MultiSIM	Chall. Ckts
18	Capacitive Reactance	1–5	1	1–4, 9, & 14–20	1–15	1–4	SBN or (50, 51)	Excel	-	C.C. 11
19	*RC* Circuits	1–6	1 & 2	1–8, 11, 12, & 14–20	1–22	1–4	SBN or (52, 53)	Excel	MultiSIM	-
20	*RLC* Circuits	1–4	1 & 2	1–12, & 18–20	1, 2, & 6–13	1–3	-	Excel	-	C.C. 12
21	Ser. & Par. Resonance	1–3 & 5	1 & 2	1–14 & 18–20	1, 2, 5, 6, & 7–12	4–9	SBN or (54–60)	Excel	MultiSIM	C.C. 13, 14
BEGIN CIRCUIT AND DEVICES CHAPTERS (22–31)										
22	Semiconductors	-	1	1–12	1–3	1–6	-	-	-	-
23	Diodes	1 & 2	1	1–3	1–3	1, 2, & 7	61	Excel	MultiSIM	C.C. 15, 16
24	Power Supply Circuits	1–4	1 & 2 Q. 1–4	1, 3–5, 7–9, & 11	1 & 3–6	1, 2	SBN or (65, 66)	Excel	MultiSIM	C.C. 17–20
25	Transistors	1	1 & 2	1–11	1–6, 9	1–3	68	Excel	-	-
26	TransistorAmplifiers	2	1 & 2	1–5, 8, & 16–19	1–3, 7, 9, 10	-	69 & 71	Excel	-	C.C. 21, 22
27	Field-Effect Transistors	-	1, 2 & 3	1–20	1–7	1, 3, & 4	SBN or (74, 75)	-	-	-
28	Op-Amps	1 & 2	1 & 2	1–16	1–6	2	76, 77	Excel	-	C.C. 23, 24
29	Oscillators	1–3	1	1–20	1–10	1–10	79, 80	Excel	-	-
30	SCRs	-	1 & 2	1–20	-	1–4	81, 82	-	-	-
31	Optoelectronics	1 & 2	1	1–7	1–3	1 & 2	SBN	-	-	C.C. 25

**SBN = "Story Behind the Numbers" Projects

VISUAL AID FOR PROGRAM TRACKS #2 AND #3

(Showing Chapter Section Assignments to Cover "Key Topics" Listing)

BASICS Ch. 1 Sections: 9–14 Ch. 2 Sections: 1, 2, 5, 6, 7	OHM'S LAW Ch. 3 Sections: 1, 2, 4, 5, 7, 9	SERIES CIRCUITS Ch. 4 Sections: 1–7 & 11, 12	PARALLEL CIRCUITS Ch. 5 Sections: 1–9 & 13	S-P CIRCUITS Ch. 6 Sections: 1–8 & 10	MAGNETISM Ch. 9 Sections: 2, 3, 9, 10	INTRO TO AC Ch. 11 Sections: 1, 4, 5, 6, 8	INDUCT-ANCE Ch. 13 Sections: 1–4 & 10
INDUCTIVE REACTANCE Ch. 14 Sections: 2, 4, 5, 6, 7	*RL* CIRCUITS Ch. 15 Sections: 2–5	TRANS-FORMERS Ch. 16 Sections: 1, 5, 6	CAPACI-TANCE Ch. 17 Sections: 1, 3, 4, 8, 10	CAPACITIVE REACTANCE Ch. 18 Sections: 2–6 & 8	*RC* CIRCUITS Ch. 19 Sections: 2, 3, 4	*RLC* CIRCUITS Ch. 20 Sections: 2, 3, 5	SER. & PAR. RESONANCE Ch. 21 Sections: 1–4 & 6, 10, 13

SEMICON-DUCTORS Ch. 22 Sections: 1, 2	DIODES Ch. 23 Sections: 1, 2, 3	POWER-SUPPLY CKTS. Ch. 24 Sections: 1, 2, 3	TRANSIS-TORS Ch. 25 Sections: 1, 2, 3	TRANSIS-TOR AMPS. Ch. 26 Sections: 1–3 & 6	FIELD-EFFECT TRANSIS. Ch. 27 Sections: 1–4	OP-AMPS Ch. 28 Sections: 1–4	OSCILLA-TORS Ch. 29 Sections: 1–5
			SCRs Ch. 30 Sections: 1–4	OPTO-ELECTRON-ICS Ch. 31 Sections: 1–4			

TRACK #1

Resources from which instructors can create or select assignments:

<u>From any or all text chapters:</u>
1. All In-Chapter Practice Problems
2. All In-Chapter IPLC questions
3. Any or all end-of-chapter Review Questions
4. Any or all end-of-chapter Problems
5. Any or all end-of-chapter Analysis Questions
6. Any or all end-of-chapter Excel exercises
7. Any or all end-of-chapter MultiSIM exercises
8. Any or all end-of-chapter Troubleshooting Challenge Circuits

<u>From all laboratory project manual projects:</u>
1. Any or all of the 82 standard lab projects in the laboratory projects manual.
2. Any or all of the 18 *"Story Behind the Numbers"* projects in the laboratory projects manual.

<u>From the CD disk(es):</u>
1. Any or all of the MultiSIM circuits
2. Any or all of the Lab Demo circuits at the discretion of the instructor

TRACK #2

Resources from which instructors can create or select assignments:

From Track #2 and Track #3
Sample Student Assignments Matrix
<u>From Chapters 1-6; 9; 11; 13-31</u>
1. All In-Chapter Practice Problems
2. All In-Chapter IPLC questions
3. Select desired questions and problems from end-of-chapter questions and problem numbers listed in matrix for:
   Review Questions
   Problems
   Analysis Questions
4. Use SBN lab jobs listed in matrix
5. Select additional lab projects from lab project numbers listed in matrix, as appropriate
6. As time permits, add some of the Optional Work Assignments from listed Excel, MultiSIM, and Troubleshooting Challenge Circuits shown on the matrix.

TRACK #3

Resources from which instructors can create or select assignments:

From Track #2 and Track #3
Sample Student Assignments Matrix
<u>From Chapters 1-6; 9; 11; 13-31</u>
1. All In-Chapter Practice Problems
2. All In-Chapter IPLC questions
3. Only select desired questions and problems from end-of-chapter questions and problem numbers listed in matrix for:
   Review Questions
   Problems
   Analysis Questions
4. Use as many of the SBN lab projects, as time permits
5. As time permits, select some of the Optional Work Assignments from the listed Excel, MultiSIM and Troubleshooting Challenge Circuits shown on the matrix.

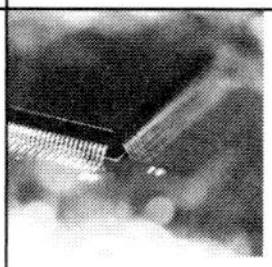

INSTRUCTIONAL STRATEGIES

Group Training Format

For traditional group training, you might have the student follow a checklist somewhat similar to the sample generic list shown here. **NOTE:** Checklists for each chapter appear later in this guide.

- ☐ Read and study the chapter assigned by the instructor. (Usually a homework assignment)
- ☐ Complete all In-Process Learning Checks and Practice Problems as you study the chapter.
- ☐ Perform Laboratory Projects assigned by the instructor. (Performed at school facilities)
- ☐ In the Laboratory Projects Manual complete Summary section pertaining to performed projects. (Done either at home or at school)
- ☐ Complete text chapter Review Questions, Problems, and Analysis Questions, as assigned. (Usually a homework assignment)
- ☐ Hand in all text and lab work assignments for grading, as appropriate.
- ☐ Attend next class session/lecture and take any quizzes or tests required. Record next assignments, as appropriate.

Individualized Instruction Training Format

For self-paced, individualized instruction (I.I.) settings, the students may use a chapter checklist to move strategically through sequenced learning activities. The general sequence followed by students in the individualized, self-paced delivery format is (1) read/study, (2) perform written and/or hands-on work, (3) get checked at strategic points via student-instructor interfaces, and (4) move on to the next learning activity.

An example of such a learning sequence is shown below.

1. Complete a brief major topic reading assignment. (Can be part or all of a chapter)
2. Perform all In-Process Checks and Practice Problems within the given study assignment.
3. Perform hands-on laboratory projects related to the study assignment.

   (**NOTE:** See the "Performance Projects Correlation Chart" at the end of each chapter in the text for information regarding logical project assignments correlating to specific text topics.)
4. Instructor checkpoints (student-instructor interfaces) are placed at strategic student progress points in each unit to assure learning is correct and progressing properly.

Note, the instructor controls how much or how little is involved in each learning activity sequence. For example, the instructor may assign just one or several laboratory projects as a learning activity to be accomplished for a given training session. In the I.I. delivery format, it can be advantageous to students to have fairly frequent instructor checkpoints to ensure that the student is learning accurately and appropriately and is progressing at a reasonable pace.

A sample chapter training/learning sequence for an I.I. setting is shown in the following chart:

CHAPTER 3: Ohm's Law—Learning Activity

CHECK AS COMPLETED

- ☐ Read Chapter 3, Sections 3–1 through 3–6.
- ☐ Perform Laboratory Projects 5 and 6.
- ☐ Have the instructor check your results on Projects 5 and 6.
- ☐ Read Chapter 3, Sections 3–7 through 3–9.
- ☐ Perform Laboratory Projects 7 and 8.
- ☐ Have the instructor check your results on Projects 7 and 8.
- ☐ Perform the "Ohm's Law: Story Behind the Numbers" project (if assigned).
- ☐ Have the instructor check your results on the "Story ..." project.
- ☐ Complete the Summary questions for the Ohm's Law series of lab projects.
- ☐ Have the instructor check your answers to the Summary questions.
- ☐ Answer the text Review and Analysis questions at the end of Chapter 3.
- ☐ Have the instructor check your work.
- ☐ Get next assignment checklist from the instructor.

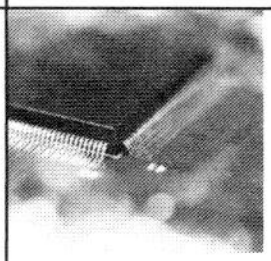

PART OVERVIEWS

with Chapter-by-Chapter Instructional Strategies

PART I: Foundational Concepts (Chapters 1–2)

Part Overview

This part provides the foundational concepts required for further study of electronics. It furnishes the underpinning necessary in the areas of basic electronic units of measure, the symbology used in electronic diagrams, and an introduction to the resistor color code. It also gives the background students need to understand and use Ohm's Law.

Knowledge and Competencies

Students should gain the knowledge and ability to:

1. Describe parts of an atom.
2. Define the differences between conductors, insulators, and semiconductors.
3. Explain the concept of polarity.
4. List and define the basic electrical units of measure.
5. Use metrics in defining units of measure.
6. Use powers of 10 in calculations.
7. Recognize and/or draw symbols for basic electrical components.
8. Interpret and use simple block, pictorial, and schematic diagrams.
9. Use meters to measure V, I, and R values.
10. Use the resistor color code.
11. Begin to use a scientific calculator for solving circuit problems.

NOTE ➤ A more detailed listing is provided in the chapter-by-chapter objectives.

Instructional Strategies

CHAPTER 1

Objectives

Students completing this chapter should be able to:

- Describe the types of tasks performed by electronics technicians, technologists, and engineers
- Define the term **matter** and list its physical and chemical states
- Describe the difference between **elements** and **compounds**
- Discuss the characteristics and structure of an **atom**, **molecule**, and **ion**
- Define the electrical characteristics of an **electron**, **proton**, and **neutron**
- Explain **valence electrons** and **free electrons**
- List the methods used to create electrical imbalances
- Describe the characteristics of **conductors**, **semiconductors**, and **insulators**

- State the law of electrical charges
- Discuss the terms **polarity** and **reference points**
- Define charge and its unit of measure, **coulomb**
- Define **potential** (emf) and give its unit of measure
- Define **current** and explain its unit of measure
- Calculate current when magnitude and rate of charge motion is known
- Define **resistance** and give its unit of measure
- List the typical elements of an electrical circuit
- Describe the difference between closed and open circuits

1. Assignment: Read Chapter 1 and perform all, or selected In-Process Learning Checks and Practice Problems as you study.
2. Complete all, or selected end-of-chapter Review Questions and Analysis Questions.
3. Lecture/Quiz/Answers Check/Next Assignment, as appropriate.

CHAPTER 2

Objectives

Students completing this chapter should be able to:

- List the units of measure for charge, potential (emf), current, resistance, and conductance and give the appropriate abbreviations and symbols for each
- Use metric system terms and abbreviations to express subunits or multiple units of the primary electrical units
- List the factors that affect the resistance of a conductor
- Recognize common types of conductors
- Use a wire table to find conductor resistance for given lengths
- Recognize and/or draw the diagrammatic representations for conductors that cross and electrically connect, and that cross and do not connect
- Define the term **superconductivity**
- Give the characteristics of several common types of resistors
- Explain the characteristics of surface-mount "chip" resistors
- Use the resistor color code
- Use other special resistor coding systems
- Explain how to connect meters to measure voltage, current, and resistance
- Recognize and/or draw the diagrammatic symbols for elemental electronic components or devices
- Interpret basic facts from block and schematic diagrams
- List key safety habits to be used in laboratory work

1. Assignment: Read Chapter 2 and complete all, or selected In-Process Learning Checks and Practice Problems as you study.
2. Complete all, or selected end-of-chapter Review Questions, Problems, and Analysis Questions.
3. Perform lab projects 1 through 4, or selected projects and complete the appropriate Summary page(s) in the lab manual.
4. Lecture/Quiz/Answers Check/Next Assignment, as appropriate.

PART II: Basic Circuit Analysis (Chapters 3–8)

Part Overview

This part provides the student with the data and the thinking processes required to analyze fundamental electronic circuit configurations. Initially, Ohm's Law and power formulas are studied. Then, students progress onto learning how to analyze series, parallel, and series-parallel circuits. Also, a basic introduction to important network theorems is provided. Additionally, this section introduces the SIMPLER troubleshooting method, which greatly enhances the students' critical and logical thinking skills. First exposure is also given to the "Chapter Challenge" problems that occur throughout the remainder of the book. These problems call upon the student to apply the SIMPLER troubleshooting technique, explained in this section.

Knowledge and Competencies

Students should gain the knowledge and ability to:

1. Calculate V, I, and R using Ohm's Law.
2. Calculate power dissipations.
3. Define key characteristics of series, parallel, and series-parallel circuits.
4. Calculate and/or determine total resistance or conductance of series, parallel, and series-parallel circuits.
5. Calculate and/or determine current(s) in series, parallel, and series-parallel circuits.
6. Calculate and/or determine voltage(s) in series, parallel, and series-parallel circuits.
7. Calculate and/or determine power dissipations in series, parallel, and series-parallel circuits.
8. Explain results of shorts or opens in series, parallel, and series-parallel circuits.
9. Troubleshoot problems in series, parallel, and series-parallel circuits.
10. Determine electrical parameters in loaded and unloaded voltage dividers.
11. Determine currents in current dividers.
12. State and use Thevenin's, Norton's, maximum power transfer, and superposition theorems.
13. Use network analysis techniques, such as loop/mesh analysis, nodal analysis, and delta-wye conversion techniques (optional requirements).
14. Use a computer spreadsheet to solve circuit problems.
15. Use a circuit simulation program (like MultiSIM) to simulate, measure, and analyze circuit parameters.

NOTE ➤ A more detailed listing is provided in the chapter-by-chapter objectives.

CHAPTER 3

Objectives

Students completing this chapter should be able to:

- Explain the relationships of current, voltage, and resistance
- Use Ohm's Law to solve for unknown circuit values
- Illustrate the direction of current flow and polarity of voltage drops on a schematic diagram

- Use metric prefixes and powers of 10 to solve Ohm's Law problems
- Use a calculator to solve circuit problems
- Use a computer spreadsheet program to solve circuit problems
- Explain power dissipation
- Use appropriate formulas to calculate values of power

1. Assignment: Read Chapter 3 and complete all, or selected In-Process Learning Checks and Practice Problems as you study.
2. Complete all, or selected end-of-chapter Review Questions, Problems, and Analysis Questions.
3. Perform lab projects 5 through 8, or selected projects and/or the "Ohm's Law: Story Behind the Numbers" project and complete the appropriate Summary page(s) in the lab manual.
4. Lecture/Quiz/Answers Check/Next Assignment, as appropriate.

CHAPTER 4

Objectives

Students completing this chapter should be able to:

- Define the term **series circuit**
- List the primary characteristics of a series circuit
- Calculate the total resistance of series circuits using two different methods
- Calculate and explain the voltage distribution characteristics of series circuits
- State and use **Kirchhoff's voltage law**
- Calculate power values in series circuits
- Explain the effects of **opens** in series circuits
- Explain the effects of **shorts** in series circuits
- List troubleshooting techniques for series circuits
- Design series circuits to specifications
- Series-connect voltage sources for desired voltages
- Analyze a **voltage divider** with reference points
- Calculate the required value of a **series-dropping resistor**
- Use the computer to solve circuit problems
- Use the SIMPLER troubleshooting sequence to solve the Chapter Challenge problems

1. Assignment: Read Chapter 4 and perform all, or selected In-Process Learning Checks and Practice Problems as you study.
2. Complete all, or selected end-of-chapter Review Questions, Problems, and Analysis Questions.
3. Perform lab projects 9 through 14, or selected projects and/or the "Series Circuits: Story Behind the Numbers" project and complete the Summary page(s) in the lab manual.
4. Lecture/Quiz/Answers Check/Next Assignment, as appropriate.

Summary/Series Circuits

In summarizing, whether you use these questions or some of your own, the instructor should be sure the students understand the following basic facts:

1. Total resistance in a series circuit is simply the sum of all the individual resistances.
2. Voltage and power distribution is directly related to the *R* distribution.

3. V_T = the sum of all V drops in series.
4. P_T = the sum of all the P dissipations.
5. The effects of opens or shorts in series circuits on R_T, I_T, P_T, and individual component parameters.

CHAPTER 5

Objectives

Students completing this chapter should be able to:

- Define the term **parallel circuit**
- List the characteristics of a parallel circuit
- Determine voltage in parallel circuits
- Calculate the total current and branch currents in parallel circuits
- Compute total resistance and branch resistance values in parallel circuits using at least three different methods
- Determine conductance values in parallel circuits
- Calculate power values in parallel circuits
- List the effects of opens in parallel circuits
- List the effects of shorts in parallel circuits
- Describe troubleshooting techniques for parallel circuits
- Use current divider formulas
- Use the computer to solve circuit problems
- Use the SIMPLER troubleshooting sequence to solve the Chapter Challenge problems

1. Assignment: Read Chapter 5 and complete all, or selected In-Process Learning Checks and Practice Problems as you study.
2. Complete all, or selected end-of-chapter Review Questions, Problems, and Analysis Questions.
3. Perform lab projects 15 through 20, or selected projects and/or the "Parallel Circuits: Story Behind the Numbers" project and complete the Summary page(s) in the lab manual.
4. Lecture/Quiz/Answers Check/Next Assignment, as appropriate.

Summary/Parallel Circuits

In summarizing, the instructor should be sure the students understand at least the following basic concepts:

1. R_e of a parallel circuit is always less than the least R branch.
2. I_T = the sum of the branch currents.
3. Each branch current is INVERSELY proportional to its branch R.
4. V is the same across all parallel branches.
5. The power distribution is INVERSE to the R distribution. That is, the smallest R dissipates the most power, the largest R dissipates the least power, and so on.
6. Total P = the sum of all the individual power dissipations (true of ANY type of circuit).
7. An open causes R_T to increase and I_T and P_T to decrease.
8. A shorted branch causes R_T to decrease drastically and I_T and P_T to increase drastically.
9. The current through the unshorted branches will drop to virtually zero.

CHAPTER 6

Objectives

Students completing this chapter should be able to:

- Define the term **series-parallel circuit**
- List the primary characteristic(s) of a series-parallel circuit
- Determine the total resistance in a series-parallel circuit
- Compute total circuit current and the current through any given portion of a series-parallel circuit
- Calculate voltages throughout a series-parallel circuit
- Determine power values throughout a series-parallel circuit
- Analyze the effects of an open in a series-parallel circuit
- Analyze the effects of a short in a series-parallel circuit
- Design a simple series-parallel circuit to specifications
- Explain the **loading effects** on a series-parallel circuit
- Calculate values relating to a **loaded voltage divider**
- Make calculations relating to **bridge circuits**
- Use the computer to solve circuit problems
- Use the SIMPLER troubleshooting sequence to solve the Chapter Challenge problems

1. Assignment: Read Chapter 6 and complete all, or selected In-Process Learning Checks and Practice Problems as you study.
2. Complete all, or selected end-of-chapter Review Questions, Problems, and Analysis Questions.
3. Perform lab projects 21 through 26, or selected projects and/or the "Series-Parallel Circuits: Story Behind the Numbers" project and complete the Summary page(s) in the lab manual.
4. Lecture/Quiz/Answers Check/Next Assignment, as appropriate.

Summary/Series-Parallel Circuits

In summarizing, the instructor should be sure the students understand at least the following basic concepts regarding series-parallel circuits:

1. R_T = sum of all the mainline components PLUS the R_es of the parallel sections of the circuit.
2. The voltage, current, and power distribution throughout the circuit follow the rules for series circuits for the series sections or components, and the rules for parallel circuits for parallel sections of the circuit.
3. A shorted element in any part of the circuit causes R_T to decrease and I_T to increase. However, the resultant V, I, or P, at any given component must be analyzed by its electrical location.
4. An open causes R_T to increase and I_T to decrease; but again, electrical location determines any parameter changes for a given component.

CHAPTER 7

Objectives

Students completing this chapter should be able to:

- State the **maximum power transfer theorem**
- Determine the R_L value needed for maximum power transfer in a given circuit

- State the **superposition theorem**
- Solve circuit parameters for a circuit having more than one source
- State **Thevenin's theorem**
- Determine V_L and I_L for various values of R_L connected across specified points in a given circuit or network using Thevenin's theorem
- State **Norton's theorem**
- Apply Norton's theorem in solving specified problems
- Convert between Norton and Thevenin equivalent parameters
- Use the computer to solve circuit problems

1. Assignment: Read Chapter 7 and complete all, or selected In-Process Learning Checks and Practice Problems as you study.
2. Complete all, or selected end-of-chapter Review Questions, Problems, and Analysis Questions.
3. Perform lab projects 27 through 29, or selected projects and/or the "Basic Network Theorems: Story Behind the Numbers" project and complete the Summary page(s) in the lab manual.
4. Lecture/Quiz/Answers Check/Next Assignment, as appropriate.

Summary/Basic Network Theorems

If time allows, give the students some problems with specific values involved. Be sure the students understand the following basic concepts:

1. According to Thevenin's theorem, an equivalent circuit comprised of a single voltage source and a single series resistor may be used to represent any two-terminal network.
2. By Norton's theorem, an equivalent circuit comprised of a single current source and a single shunt resistance may be used to represent any two-terminal network.
3. The maximum power transfer theorem says that maximum power is transferred from source to load when $R_L = R_g$.
4. Norton's theorem is often used in analyzing transistor circuits.
5. Thevenin's theorem is a powerful tool for simplifying complex network problems.
6. The maximum power transfer theorem is used in designing transmitters, determining loading factors for power plants, and for various other practical applications.

CHAPTER 8

Objectives

Students completing this chapter should be able to:

- Define the terms **mesh, loop,** and **node**
- Analyze a single-source circuit using a loop procedure
- Use the assumed mesh currents approach to find voltage and current parameters for each component in a network having two sources
- Use the nodal analysis approach to find voltage and current parameters for each component in a network having two sources
- Convert from delta (Δ) circuit configuration parameters to wye (Y) circuit configuration parameters, and vice versa
- Use the computer to solve circuit problems

1. Assignment: Read Chapter 8 and complete all, or selected In-Process Learning Checks and Practice Problems as you study.

2. Complete all, or selected end-of-chapter Review Questions, Problems, and Analysis Questions.
3. Perform lab projects 30 through 32, or selected projects and/or the "Network Analysis Techniques: Story Behind the Numbers" project and complete the Summary page(s) in the lab manual.
4. Lecture/Quiz/Answers Check/Next Assignment, as appropriate.

PART III: Producing and Measuring Electrical Quantities (Chapters 9–12)

Part Overview

Part III presents foundational knowledge in how dc and ac are produced, characteristics of magnetism and electromagnetism, and devices and methods used for measuring both dc and ac quantities. It also offers important knowledge and skills that will let students analyze the operation and characteristics of passive components in both dc and ac circuits.

Knowledge and Competencies

Students should gain the knowledge and ability to:

1. Explain and use magnetic terms and units of measure.
2. Use left-hand rules *(right hand rules)* relating to electromagnetism.
3. Interpret the hysteresis loop.
4. Define and/or calculate the important values related to ac sine-wave signals.
5. Briefly explain the uses of an oscilloscope and list its major sections.
6. List an oscilloscope's primary controls and each control's function.

NOTE ➤ A more detailed listing is provided in the chapter-by-chapter objectives.

CHAPTER 9

Objectives

Students completing this chapter should be able to:

- Define **magnetism, magnetic field**, **magnetic polarity**, and **flux**
- Draw representations of magnetic fields related to permanent magnets
- State the magnetic attraction and repulsion law
- State at least five generalizations about magnetic lines of force
- Draw representations of fields related to current-carrying conductors
- Determine the polarity of electromagnets using the **left-hand rule** *(right hand rules)*
- List and define at least five magnetic units of measure, terms, and symbols
- Draw and explain a B-H curve and its parameters
- Draw and explain a hysteresis loop and its parameters
- Explain motor action and generator action related to magnetic fields
- List the key factors related to induced emf
- Briefly explain the relationships of quantities in **Faraday's Law**
- Briefly explain **Lenz's Law**
- Use the computer to solve circuit problems

1. Assignment: Read Chapter 9 and complete all, or selected In-Process Learning Checks and Practice Problems as you study.
2. Complete all, or selected end-of-chapter Review Questions and Problems.
3. Lecture/Quiz/Answers Check/Next Assignment, as appropriate.

CHAPTER 10

Objectives

Students completing this chapter should be able to:

- List at least two key features of digital multimeters (DMMs)
- Describe at least one advantage and one disadvantage of an analog multimeter (VOM)
- Explain the meanings of the terms **autoranging** and **autopolarity**
- Describe and calculate meter loading effects for specified measurement conditions
- List at least two special-purpose measuring devices
- Define two basic methods of measuring voltage on a circuit having a ground reference
- Describe the technique for making continuity checks on a 200-foot-long cable
- Define the purpose and function of meter protection circuits

1. Assignment: Read Chapter 10 and complete all, or selected In-Process Learning Checks and Practice Problems as you study.
2. Complete all, or selected end-of-chapter Review Questions and Analysis Questions.
3. Lecture/Quiz/Answers Check/Next Assignment, as appropriate.

Summary/DC Measuring Instruments (Analog)

In summarizing, the instructor should make sure the students understand at least the following concept: To minimize changing circuit conditions when measuring electrical parameters, current meters should have as low an R as possible and voltmeters should have as high an R as possible.

CHAPTER 11

Objectives

Students completing this chapter should be able to:

- Draw a graphic illustrating an **ac** waveform
- Define **cycle**, **alternation**, **period**, **peak**, **peak-to-peak**, and **effective value (rms)**
- Compute effective, peak, and peak-to-peak values of ac voltage and current
- Explain **average** with reference to one-half cycle of sine-wave ac
- Explain **average** with reference to nonsinusoidal waves
- Define and calculate **frequency** and period
- Draw a graphic illustrating phase relationships of two **sine waves**
- Describe the phase relationships of V and I in a purely resistive ac circuit
- Label key parameters of nonsinusoidal waveforms
- Use the computer to solve circuit problems

1. Assignment: Read Chapter 11 and complete all, or selected In-Process Learning Checks and Practice Problems as you study.

2. Complete all, or selected end-of-chapter Review Questions, Problems, and Analysis Questions.
3. Lecture/Quiz/Answers Check/Next Assignment, as appropriate.

CHAPTER 12

Objectives

Students completing this chapter should be able to:

- List the key sections of the **oscilloscope**
- List precautions when using scopes
- List procedures when measuring voltage with a scope
- List procedures to display and interpret waveforms
- List procedures relating to phase measurement
- List procedures when determining frequency with a scope
- Use the computer to solve circuit problems
- Use the SIMPLER troubleshooting sequence to solve the Chapter Challenge problem

1. Assignment: Read Chapter 12 and complete all, or selected In-Process Learning Checks and Practice Problems as you study.
2. Complete all, or selected end-of-chapter Review Questions, Problems, and Analysis Questions.
3. Perform lab projects 33 through 39, or selected projects and complete the Summary page(s) in the lab manual.
4. Lecture/Quiz/Answers Check/Next Assignment, as appropriate.

Summary/The Oscilloscope

The following facts and considerations are important when using the oscilloscope to make voltage measurements.

1. Waveform deflection on the scope indicates the *peak-to-peak* value of the voltage under test.
2. For any SINE WAVE, the peak-to-peak deflection on the screen is *directly proportional* to the peak, rms, and average value of the ac applied voltage. For example, if the deflection on the CRO screen is 1″ (peak-to-peak deflection) when 1-volt rms is applied, it will be 2″ of total deflection when a 2-volt rms ac signal is applied. Therefore, it is easy to measure ac voltage values with the scope even though the deflection is peak-to-peak in nature.

PART IV: Reactive Components (Chapters 13–21)

Part Overview

This part provides the student with the data and the thinking processes required to understand and analyze reactive components. These include inductance, inductive reactance in ac, *RL* circuits, capacitance, *RC* circuits, transformer characteristics, *RLC* circuit analysis, and series and parallel resonance. This section helps students gain insight regarding basic characteristics of inductors and capacitors. Furthermore, opportunity is given to understand how all these passive components work singly or in combination with each other in both dc and ac circuits.

Knowledge and Competencies

Students should gain the knowledge and ability to:

1. Define inductance and calculate values of inductance.
2. Calculate time-constant values for *RL* circuits.
3. Calculate values of inductive reactance.
4. Analyze ac circuit parameters for inductive circuits.
5. Describe transformer action and calculate important transformer ratio parameters.
6. Define capacitance and calculate values of capacitance.
7. Calculate time-constant values for *RC* circuits.
8. Calculate values of capacitive reactance.
9. Analyze ac circuit parameters for capacitive circuits.
10. Explain the characteristics of series and parallel resonant circuits.
11. Calculate electrical quantities in or related to series resonant and parallel resonant circuits.
12. Use a scientific calculator to perform a wide variety of circuit analysis calculations.
13. Use a computer spreadsheet program (like Excel) to perform calculations and plot graphs.
14. Use a computer circuit simulation program (like MultiSIM) to simulate, measure, and analyze circuit parameters.

NOTE ➤ A more detailed listing is provided in the chapter-by-chapter objectives.

CHAPTER 13

Objectives

Students completing this chapter should be able to:

- Define **inductance** and **self-inductance**
- Explain **Faraday's** and **Lenz's laws**
- Calculate induced **cemf** values for specified circuit conditions
- Calculate inductance values from specified parameters
- Calculate inductance in series and parallel
- Determine energy stored in a magnetic field
- Draw and explain time-constant graphs
- Calculate time constants for specified circuit conditions
- Use ϵ to aid in finding circuit quantities in dc *RL* circuits
- List common problems of inductors
- Use the computer to solve circuit problems

1. Assignment: Read Chapter 13 and perform all, or selected In-Process Learning Checks and Practice Problems as you study.
2. Complete all, or selected end-of-chapter Review Questions, Problems, and Analysis Questions.
3. Perform lab project 40 and/or the "Inductance: Story Behind the Numbers" project and complete the Summary page(s) in the lab manual.
4. Lecture/Quiz/Answers Check/Next Assignment, as appropriate.

CHAPTER 14

Objectives

Students completing this chapter should be able to:

- Illustrate V-I relationships for a purely resistive ac circuit
- Illustrate V-I relationships for a purely inductive ac circuit
- Explain the concept of **inductive reactance**
- Write and explain the formula for inductive reactance
- Use Ohm's Law to solve for X_L
- Use the X_L formula to solve for inductive reactance at different frequencies and with various inductance values
- Use the X_L formula to solve for unknown L or f values
- Determine X_L, I_L, and V_L values for series- and parallel-connected inductances
- Use the computer to solve circuit problems
- Use the SIMPLER troubleshooting sequence to solve the Chapter Challenge problem

1. Assignment: Read Chapter 14 and complete all, or selected In-Process Learning Checks and Practice Problems as you study.
2. Complete all, or selected end-of-chapter Review Questions, Problems, and Analysis Questions.
3. Perform lab projects 41 and 42, or selected projects and/or the "Inductive Reactance: Story Behind the Numbers" project and complete the Summary page(s) in the lab manual.
4. Lecture/Quiz/Answers Check/Next Assignment, as appropriate.

CHAPTER 15

Objectives

Students completing this chapter should be able to:

- Use **vectors** to determine magnitude and direction
- Determine circuit impedance using the **Pythagorean theorem**
- Determine V_T and I_T using the Pythagorean theorem
- Determine ac circuit parameters using trigonometry
- Calculate ac electrical parameters for series RL circuits
- Calculate ac electrical parameters for parallel RL circuits
- List at least three practical applications of inductive circuits
- Use the computer to solve circuit problems
- Use the SIMPLER troubleshooting sequence to solve the Chapter Challenge problem

1. Assignment: Read Chapter 15 and complete all, or selected In-Process Learning Checks and Practice Problems as you study.
2. Complete all, or selected end-of-chapter Review Questions, Problems, and Analysis Questions.
3. Perform lab projects 43 and 44, or selected projects and/or the "RL Circuits in AC: Story Behind the Numbers" project and complete the Summary page(s) in the lab manual.
4. Lecture/Quiz/Answers Check/Next Assignment, as appropriate.

Summary/*RL* Circuits in AC

The instructor may wish to further test the students' knowledge by giving *RL* circuit problems with specific given values, and requiring quantitative results. The key items to be sure the students understand are:

1. Circuit parameters cannot be solved by simple arithmetic sums but must be solved vectorially.
2. In a "perfect" inductance, V leads I by 90 degrees.
3. In a resistor, V and I are in phase.
4. In a circuit composed of both resistance and inductance, the circuit V and I are out of phase by an angle between 0 and 90 degrees.
5. The more inductive the circuit, the larger the phase angle.
6. The more resistive the circuit, the smaller the phase angle.
7. For series *RL* circuits, the larger the X_L, the more inductive the circuit (for any given R).
8. For parallel *RL* circuits, the smaller the X_L, the more inductive the circuit (for any given R).

CHAPTER 16

Objectives

Students completing this chapter should be able to:

- Define **mutual inductance**
- Calculate mutual inductance values
- Calculate **coefficient of coupling** values
- Calculate **turns**, **voltage**, **current**, and **impedance ratios**
- List, draw, or explain physical, magnetic, electrical, and schematic characteristics of various **transformers**
- List common transformer color codes
- Define at least two types of **core losses**
- List common problems found in transformers
- List troubleshooting procedures
- Use the computer to solve circuit problems

1. Assignment: Read Chapter 16 and complete all, or selected In-Process Learning Checks and Practice Problems as you study.
2. Complete all, or selected end-of-chapter Review Questions, Problems, and Analysis Questions.
3. Perform lab projects 45 and 46, or selected projects and/or the "Basic Transformer Characteristics: Story Behind the Numbers" project and complete the Summary page(s) in the lab manual.
4. Lecture/Quiz/Answers Check/Next Assignment, as appropriate.

CHAPTER 17

Objectives

Students completing this chapter should be able to:

- Define **capacitor**, **capacitance**, **dielectric**, **dielectric constant**, **electric field**, farad, ***RC*** **time constant**, and leakage resistance

- Describe **capacitor charging** action and **discharging** action
- Calculate charge, voltage, capacitance, and stored energy, using the appropriate formulas
- Determine total capacitance in circuits with more than one capacitor (series and parallel)
- Calculate circuit voltages using appropriate *RC* time-constant formulas
- List and describe the physical and electrical features of at least four types of capacitors
- List typical capacitor problems and describe troubleshooting techniques
- Use the computer to solve circuit problems
- Use the SIMPLER troubleshooting sequence to solve the Chapter Challenge problem

1. Assignment: Read Chapter 17 and complete all, or selected In-Process Learning Checks and Practice Problems as you study.
2. Complete all, or selected end-of-chapter Review Questions, Problems, and Analysis Questions.
3. Perform lab projects 47 and 48, or selected projects and/or the "Capacitance in DC: Story Behind the Numbers" project and complete the Summary page(s) in the lab manual.
4. Lecture/Quiz/Answers Check/Next Assignment, as appropriate.

Summary/Capacitance (DC Characteristics)

The instructors may wish to include facts relating to the $Q = CV$ formula in their summary discussion. At any rate, you should be sure the students understand the concepts of:

1. Charge and discharge action
2. V_C (charged capacitor voltage) being series-opposing to the source
3. One $TC = R \times C$
4. Five time constants are needed to fully charge or discharge a capacitor through a resistance
5. Series capacitors adding like parallel resistances and parallel capacitors adding like series resistances.

For example, C_T of series capacitors = the product-over-the-sum (two capacitors) and C_T for parallel capacitors = $C_1 + C_2 + \ldots C_n$.

CHAPTER 18

Objectives

Students completing this chapter should be able to:

- Illustrate *V-I* relationships for purely resistive and purely capacitive circuits
- Explain **capacitive reactance**
- Use Ohm's Law to solve for X_C value(s)
- Use the capacitive reactance formula to solve for X_C value(s)
- Use the X_C formula to solve for unknown *C* and *f* values
- Use Ohm's Law and **reactance** formulas to determine circuit reactances, voltages, and currents for series- and parallel-connected capacitors
- List two practical applications for X_C
- Use the computer to solve circuit problems
- Use the SIMPLER troubleshooting sequence to solve the Chapter Challenge problem

1. Assignment: Read Chapter 18 and complete all, or selected In-Process Learning Checks and Practice Problems as you read.
2. Complete all, or selected end-of-chapter Review Questions, Problems, and Analysis Questions.

3. Perform lab projects 49 through 51, or selected projects and/or the "Capacitive Reactance in AC: Story Behind the Numbers" project and complete the Summary page(s) in the lab manual.
4. Lecture/Quiz/Answers Check/Next Assignment, as appropriate.

Summary/Capacitive Reactance in AC

The instructor may wish to teach the students how to use the X_C formula to solve for an unknown C, when X_C and f are known, or how to solve for an unknown f, when C and X_C are known $C = \frac{1}{2\pi f X_C}$; and $f = \frac{1}{2\pi C X_C}$.

Key points to cover in summarizing, are:

1. The inverse relationship of X_C to both C and f
2. X_Cs in series add like Rs in series
3. X_Cs in parallel add like parallel Rs

It is important to emphasize that capacitance does the opposite of the X_C in this regard.

CHAPTER 19

Objectives

Students completing this chapter should be able to:

- Draw or describe operation of simple R and C circuits
- Analyze appropriate series and parallel RC circuit parameters using the Pythagorean theorem
- Use vector analysis to analyze series and parallel RC circuit parameters
- List differences between RC and RL circuits
- Predict output waveform(s) from a waveshaping network
- List two practical applications for RC circuits
- Apply troubleshooting hints to help identify problems
- Use the computer to solve circuit problems

1. Assignment: Read Chapter 19 and complete all, or selected In-Process Learning Checks and Practice Problems as you study.
2. Complete all, or selected end-of-chapter Review Questions, Problems, and Analysis Questions.
3. Perform lab projects 52 and 53, or selected projects and/or the "*RC* Circuits in AC: Story Behind the Numbers" project and complete the Summary page(s) in the lab manual.
4. Lecture/Quiz/Answers Check/Next Assignment, as appropriate.

Summary/*RC* Circuits in AC

The instructor may wish to quiz the students by giving problems with specific values and having them draw V-I vector diagrams for the problems.

In summarizing, the instructor should be sure to stress the following key points:

1. I leads V by 90 degrees for a capacitor.
2. For series RC circuits, the higher the X_C compared to the R, the more capacitive the circuit acts.
3. For parallel RC circuits, the higher the X_C, the less capacitive the circuit.
4. For either series or parallel RC circuits, the higher the X_C, the higher the circuit impedance (Z).

5. Total voltage in series *RC* circuits equals the vector resultant of V_C and V_R.
6. Total current in parallel *RC* circuits equals the vector resultant of branch currents I_C and I_R.

Finally, the students should clearly understand the inverse relationship of X_C to BOTH capacitance (C) and frequency (f).

CHAPTER 20

Objectives

Students completing this chapter should be able to:

- Solve *RLC* circuit problems using the Pythagorean approach and trig functions
- Define and illustrate ac circuit parameters using both **rectangular** and **polar form notation**
- Define **real numbers** and **imaginary numbers**
- Define real power, **apparent power**, **power factor**, and **voltampere-reactive**
- Calculate values of real power, apparent power, and power factor, and draw the power triangle
- Analyze *RLC* circuits and state results in rectangular and polar forms
- Use the computer to solve circuit problems
- Use the SIMPLER troubleshooting sequence to solve the Chapter Challenge problem

1. Assignment: Read Chapter 20 and complete all, or selected In-Process Learning Checks and Practice Problems as you study.
2. Complete all, or selected end-of-chapter Review Questions, Problems, and Analysis Questions.
3. Lecture/Quiz/Answers Check/Next Assignment, as appropriate.

CHAPTER 21

Objectives

Students completing this chapter should be able to:

- List the key characteristics of series and parallel resonant circuits
- Calculate the resonant frequency of circuits
- Calculate $\boldsymbol{L}$ or $\boldsymbol{C}$ values needed for **resonance** at a given f_r
- Calculate $\boldsymbol{Q}$ factor for series and parallel resonant circuits
- Determine **bandwidth** and **bandpass** of resonant circuits
- Draw circuit diagrams for three types of **filters**
- Use the computer to solve circuit problems
- Use the SIMPLER troubleshooting sequence to solve the Chapter Challenge problems

1. Assignment: Read Chapter 21 and complete all, or selected In-Process Learning Checks and Practice Problems as you study.
2. Complete all, or selected end-of-chapter Review Questions, Problems, and Analysis Questions.
3. Perform lab projects 54 through 60, or selected projects and/or the "Series Resonance and Parallel Resonance: Story Behind the Numbers" projects and complete the appropriate Summary page(s) in the lab manual.
4. Lecture/Quiz/Answers Check/Next Assignment, as appropriate.

Summary/Series Resonance

The instructor should be sure in summarizing that the students understand the following:

1. For a series resonant circuit, I is maximum, Z is minimum, phase angle is zero, and voltage across the reactive components = $Q \times V_A$.
2. Bandwidth = f_r/Q.
3. The circuit acts "capacitive" below resonance and "inductive" above resonance.
4. A high Q means sharp tuning and narrow bandwidth, while a low Q means broad tuning and wide bandwidth.

Summary/Parallel Resonance

The instructor may give additional summary problems with specific values, demanding quantitative answers if more practice is desirable. The instructor may also mention some key parameters generally associated with parallel resonance: zero phase angle; $P.F. = 1$; maximum total impedance and minimum line current; $X_L = X_C$. Point out that all these events DO NOT occur *precisely* at the same frequency in practical parallel LC circuits. The higher the circuit Q, the closer to the same frequency these events occur. The term *antiresonance* is sometimes used for the frequency at which circuit Z is maximum.

In summarizing, be sure the students understand the following key points:

1. At resonance, I_T is minimum.
2. Z_T is maximum.
3. Phase angle = zero degrees.
4. Current through the reactive branches = $Q \times I_T$.
5. The parallel LC circuit acts capacitively above resonance and inductively below resonance.
6. $B.W. = f_r/Q$.
7. High Q means sharp tuning and narrow bandwidth, while low Q means broad tuning and wide bandwidth.

PART V: Circuits and Devices (Chapters 22–31)

Part Overview

This part introduces the student to diodes, power supply circuits, bipolar junction transistors and BJT amplifier circuits, field-effect transistors and circuits, operational amplifiers, oscillators and multivibrators, thyristors, and optoelectronic-related devices. The student will learn how these semiconductor devices work and will have an opportunity to examine a number of practical circuit applications for each of them.

Knowledge and Competencies

Students should gain the knowledge and ability to:

1. Describe the basic characteristics of various types of semiconductor diodes, including rectifier, switching, zener, optoelectric, and other types of diodes.
2. Explain forward and reverse biasing of semiconductor devices.

3. Draw circuits for, and explain the operation of, half-wave, full-wave, and bridge rectifier circuits, with and without filters.
4. Determine dc voltage values for power supply circuits without filtering and approximate voltage output values with filtering.
5. Describe the elements and junctions involved in NPN and PNP transistors.
6. Explain appropriate biasing for each transistor junction.
7. List the key characteristics of the three basic configurations of transistor amplifiers (CE, CB, and CC).
8. Perform transistor amplifier circuit calculations.
9. List common troubles that can occur in transistors and semiconductors, and some of the means used for preventing problems.
10. Explain the operation of bipolar junction transistors.
11. Explain the operation of FETs and MOSFETs.
12. Describe the basic characteristics of operational amplifiers.
13. Identify several specific applications for op-amps.
14. Define the four requirements for creating/sustaining oscillations in an electronic oscillator circuit.
15. Identify and draw the circuits for four basic types of *LC* oscillator circuits.
16. Identify and draw the circuits for at least two types of *RC*-controlled oscillators.
17. Calculate the operating frequency of *LC* and *RC* oscillators.
18. Explain the operational characteristics of thyristors, such as SCRs, diacs, and triacs.
19. Describe at least four optoelectronic devices.

CHAPTER 22

Objectives

Students completing this chapter should be able to:

- Describe the difference between **valence electrons** and **conduction-band electrons**
- Describe the main difference between **n-type** semiconductor materials and **p-type** semiconductor materials
- Draw a diagram of a **P-N junction**, including the **depletion region**
- Draw a P-N junction that shows the polarity of applied voltage for **forward biasing** the junction
- Draw a P-N junction that shows the polarity of applied voltage for **reverse biasing** the junction
- Explain the difference between the **barrier potential** and **reverse breakdown voltage** for a P-N junction
- Sketch the ***I-V* curve** for a typical P-N junction, showing both the forward and reverse bias parts of the curve

1. Assignment: Read Chapter 22 and perform all, or selected In-Process Learning Checks and Practice Problems as you study.
2. Complete all, or selected end-of-chapter Review Questions, Problems, and Analysis Questions.
3. Lecture/Quiz/Answers Check/Next Assignment, as appropriate.

CHAPTER 23

Objectives

Students completing this chapter should be able to:

- Describe how to connect a dc source to a **junction diode** for forward bias and for reverse bias
- Determine the distribution of voltages in a series circuit consisting of a diode and resistor
- Sketch the waveforms found in an ac circuit consisting of a diode and resistor
- Explain the function of diode **clamping** and **clipper circuits**
- Understand the meaning and importance of diode specifications
- Describe the operation and specifications for **zener diodes**
- Explain the function of a simple zener diode circuit
- Describe the operation of **LEDs**
- Determine the value of a resistor to be placed in series with an LED for proper operation

1. Assignment: Read Chapter 23 and complete all, or selected In-Process Learning Checks and Practice Problems as you study.
2. Complete all, or selected end-of-chapter Review Questions, Problems, Analysis Questions, Excel and MultiSIM Exercises, and the Troubleshooting Challenge.
3. Perform lab projects 61 through 64, or selected projects and complete the Summary page(s) in the lab manual.
4. Lecture/Quiz/Answers Check/Next Assignment, as appropriate.

Summary/The Semiconductor Diode

The instructor also may wish to discuss the point-contact germanium diode, which was not the type used in our project. The difference in typical forward-bias voltage drop (typical for germanium = 0.3 V) and in general application should be mentioned. In summarizing, be sure the students understand:

1. the diode symbol
2. the meaning of forward and reverse bias
3. the term *rectification*
4. how reversing the diode reverses polarity of output
5. front-to-back ratio
6. terms such as *reverse-bias, leakage current,* and *breakover* or *avalanche voltage*
7. how to calculate dc resistance of the diode
8. how to calculate ac resistance of the diode

CHAPTER 24

Objectives

Students completing this chapter should be able to:

- List the basic elements of a power supply system
- Draw the three basic types of rectifier circuits: **half-wave**, **center-tapped full-wave**, and **bridge circuits**
- Explain the paths for current flow through the three basic types of rectifier circuits

- Describe the waveforms found across the diode(s) and at the output of the three basic types of rectifier circuits
- Determine the unfiltered dc output voltage of specified rectifier circuits
- Briefly describe power supply **filter action**
- Identify power supply filter configurations
- Explain the purpose of a power supply **voltage regulator**
- Recognize two **voltage multiplier** circuits
- Briefly describe the most common types of power supply troubles

1. Assignment: Read Chapter 24 and complete all, or selected In-Process Learning Checks and Practice Problems as you study.
2. Complete all, or selected end-of-chapter Review Questions, Problems, Analysis Questions, Excel and MultiSIM Exercises, and Troubleshooting Challenges.
3. Perform lab projects 65 through 67, or selected projects and/or the "Voltage Regulator: Story Behind the Numbers" project, and complete the Summary page(s) in the lab manual.
4. Lecture/Quiz/Answers Check/Next Assignment, as appropriate.

Summary/Half-Wave and Bridge Rectifier

In summarizing, the instructor may wish to include a comparison of the half-wave rectifier and the bridge rectifier system relating to: PIV, ripple frequency, ripple amplitude, average dc output per given ac input, etc. The instructor may also wish to bring the standard full-wave rectifier system into the comparisons. Basic facts to review regarding the bridge rectifier are:

1. Unfiltered dc output equals $0.9 \times V_{\text{eff}}$ of ac input.
2. Ripple frequency = 2 × ac input frequency.
3. Two diodes conduct during each alternation.

If time permits, this is also a good time to discuss component ratings, such as PIV, current ratings, and so forth.

CHAPTER 25

Objectives

Students completing this chapter should be able to:

- Draw the symbols and identify the **emitter**, **base**, and **collector** leads for NPN and PNP transistors
- Draw the symbols for **NPN** and **PNP transistors** and show the proper **voltage polarities** for the base-emitter terminals and for the collector-base terminals
- Explain the meaning and cite the mathematical symbols for **emitter current**, **base current**, **collector current**, **base-emitter voltage**, and **collector-emitter voltage**
- Describe how increasing the **forward-bias base current** in a **BJT amplifier** decreases the voltage between the emitter and collector
- Describe the operation of a BJT when applied as a switch
- Explain the meaning of the curves shown on a family of **collector characteristic curves**
- Describe the meaning of the **maximum voltage**, **current**, and **power ratings** listed in BJT data sheets

1. Assignment: Read Chapter 25 and complete all, or selected In-Process Learning Checks and Practice Problems as you study.
2. Complete all, or selected end-of-chapter Review Questions, Problems, Analysis Questions, and Excel Problems.

3. Perform lab project 68, or selected projects, and/or the "BJT Transistor Characteristics: Story Behind the Numbers" project, and complete the Summary page(s) in the lab manual.
4. Lecture/Quiz/Answers Check/Next Assignment, as appropriate.

CHAPTER 26

Objectives

Students completing this chapter should be able to:

- Explain the basic **transistor amplification process**
- Describe the input and output characteristics of **common transistor amplifier stages**
- List the advantages of each common type of transistor amplifier stage
- Describe the difference between **small-signal** and **power amplifier circuits**
- Classify amplifiers by **class of operation**
- Describe the classification of amplifiers and their operation from their **load lines**
- List typical applications for each classification of amplifier
- Perform a basic **analysis** of a common-emitter, Class A BJT that uses voltage-divider biasing

1. Assignment: Read Chapter 26 and perform all, or selected In-Process Learning Checks and Practice Problems as you study.
2. Complete all, or selected end-of-chapter Review Questions, Problems, Analysis Questions, Excel and MultiSIM Exercises, and the Troubleshooting Challenge.
3. Perform lab projects 69 through 73, or selected projects and complete the Summary page(s) in the lab manual.
4. Lecture/Quiz/Answers Check/Next Assignment, as appropriate.

CHAPTER 27

Objectives

Students completing this chapter should be able to:

- Describe the **semiconductor structure** and identify the schematic symbols for N- and P-channel **JFETs**, **D-MOSFETs**, and **E-MOSFETs**
- Determine the proper voltage polarities for operating **N-** and **P-channel FETs**
- Explain the difference between depletion and enhancement modes of operation for FETs
- Identify and explain the operation of **common-source**, **common-drain**, and **common-gate FET amplifier** circuits
- Name some common practices for storing and handling MOSFET devices to ensure that they are not destroyed by static electricity

1. Assignment: Read Chapter 27 and perform all, or selected In-Process Learning Checks and Practice Problems as you study.
2. Complete all, or selected end-of-chapter Review Questions, Problems, and Analysis Questions.
3. Perform lab projects 74 and 75, or selected projects, and/or the "JFET Characteristics: Story Behind the Numbers" project, and complete the Summary page(s) in the lab manual.
4. Lecture/Quiz/Answers Check/Next Assignment, as appropriate.

CHAPTER 28

Objectives

Students completing this chapter should be able to:

- Explain the derivation of the term *operational amplifier* **(op-amp)**
- Draw op-amp symbol(s)
- Define the term **differential amplifier**
- Draw a block diagram of typical circuits used in op-amps
- List the key characteristics of an ideal op-amp
- Identify linear and nonlinear applications circuits for op-amps
- Distinguish between **inverting** and **noninverting op-amp circuits**
- Perform voltage gain and resistance calculations for standard inverting and noninverting op-amp circuits
- Describe the operation of op-amps in voltage amplifiers, voltage followers, comparators, and Schmitt-trigger amplifiers
- Describe the function of op-amps in circuits originally designed for analog computers: summing amplifiers, differential amplifiers, differentiators, and integrators

1. Assignment: Read Chapter 28 and perform all, or selected In-Process Learning Checks and Practice Problems as you study.
2. Complete all, or selected end-of-chapter Review Questions, Problems, and Analysis Questions.
3. Perform lab projects 76 through 78, or selected projects and complete the Summary page(s) in the lab manual.
4. Lecture/Quiz/Answers Check/Next Assignment, as appropriate.

CHAPTER 29

Objectives

Students completing this chapter should be able to:

- Identify from schematic diagrams the BJT, FET, and op-amp versions of the **Hartley**, **Colpitts**, and **Clapp oscillators**
- Identify the tuning components and describe the procedure for determining the oscillating frequency of the Hartley, Colpitts, and Clapp oscillators
- Explain the operation of a **crystal oscillator**
- Identify from schematic diagrams the **phase-shift** and **Wien-bridge oscillators**
- Identify the tuning components and describe the procedure for determining the oscillating frequency of the phase-shift and Wien-bridge oscillators
- Define the operation of a **monostable multivibrator** and calculate the duration of the output pulse
- Define the operation of an **astable multivibrator** and determine the operating frequency for both symmetrical and nonsymmetrical output waveforms

1. Assignment: Read Chapter 29 and perform all, or selected In-Process Learning Checks and Practice Problems as you study.
2. Complete all, or selected end-of-chapter Review Questions, Problems, and Analysis Questions.
3. Perform lab projects 79 and 80, or selected projects and complete the Summary page(s) in the lab manual.
4. Lecture/Quiz/Answers Check/Next Assignment, as appropriate.

CHAPTER 30

Objectives

Students completing this chapter should be able to:

- Describe what a **thyristor** is
- Describe in detail the way an **SCR** can be switched on and off
- Explain the operation of simple SCR circuits including a power "on/off " push-button control circuit and an electronic "crowbar"
- Identify the symbols for, and describe the operation of, the gate-controlled switch, silicon-controlled switch, and light-activated VSCR
- Identify and explain the purpose of thyristors connected in inverse parallel, or back-to-back
- Identify the schematic symbol and explain the operation of a **diac**
- Explain the details for starting and ending the conduction of a **triac**
- Identify phase-control power circuits that use thyristors
- Describe basic troubleshooting procedures for thyristors

1. Assignment: Read Chapter 30 and perform all, or selected In-Process Learning Checks and Practice Problems as you study.
2. Complete all, or selected end-of-chapter Review Questions and Analysis Questions.
3. Perform lab projects 81 and 82, or selected projects and complete the Summary page(s) in the lab manual.
4. Lecture/Quiz/Answers Check/Next Assignment, as appropriate.

CHAPTER 31

Objectives

Students completing this chapter should be able to:

- Describe the operation of LEDs and photodiodes
- Determine the value of a resistor to be placed in series with an LED for proper operation
- Describe the purpose of laser diodes
- Understand the operation of seven-segment displays
- Understand the operation of optocouplers
- Understand fiber-optic cables
- Understand photoemitters and photodetectors

1. Assignment: Read Chapter 31 and perform all, or selected In-Process Learning Checks and Practice Problems as you study.
2. Complete all, or selected end-of-chapter Review Questions, Analysis Questions, and Challenge Circuit Troubleshooting.
3. Perform the "Fiber-Optic System Characteristics: Story Behind the Numbers" project.
4. Lecture/Quiz/Answers Check/Next Assignment, as appropriate.

Answers to Questions in Foundations of Electronics, 5e (Electron Flow Version) and Foundations of Electronics: Circuits and Devices, 5e (Electron Flow Version) and Foundations of Electronics: Circuits and Devices, 2e (Conventional Flow Version)

Includes answers for in-chapter:

- In-Process Learning Checks
- Practice Problems

And answers for end-of-chapter:

- Review Questions
- Problems
- Analysis Questions
- Using Excel Problems
- MultiSIM Exercises

SECTION II

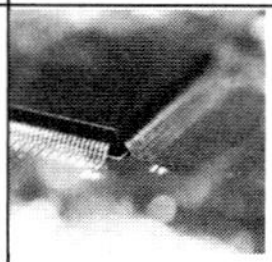

ANSWERS

CHAPTER 1

IN-PROCESS LEARNING CHECK 1

1. Matter is anything that has *weight* and occupies *space*.
2. Three physical states of matter are *solid*, *liquid*, and *gas*.
3. Three chemical states of matter are *elements*, *compounds*, and *mixtures*.
4. The smallest particle that a compound can be divided into and still retain its physical characteristics is the *molecule*.
5. The smallest particle that an element can be divided into and still retain its physical characteristics is the *atom*.
6. The three parts of an atom that interest electronics students are the *electron*, *proton*, and *neutron*.
7. The atomic particle having a negative charge is the *electron*.
8. The atomic particle having a positive charge is the *proton*.
9. The atomic particle having a neutral charge is the *neutron*.
10. The *protons* and *neutrons* are found in the atom's nucleus.
11. The particle that orbits the nucleus of the atom is the *electron*.

IN-PROCESS LEARNING CHECK 2

1. An electrical system (circuit) consists of a source, a way to transport the electrical energy, and a *load*.
2. Static electricity is usually associated with *nonconductor*-type materials.
3. The basic electrical law is that *unlike* charges attract each other and *like* charges repel each other.
4. A positive sign or a negative sign often shows electrical *polarity*.
5. If two quantities are "directly" related, as one increases the other will *increase*.
6. If two quantities are "inversely" related, as one increases the other will *decrease*.
7. The unit of charge is the *coulomb*.
8. The unit of current is the *ampere*.
9. An ampere is an electron flow of one *coulomb* per second.
10. If two points have different electrical charge levels, there is a difference of *potential* between them.
11. The volt is the unit of *electromotive* force, or *potential* difference.

REVIEW QUESTIONS

1. Matter is anything that has weight and occupies space whether it is a solid, liquid, or gas.
2. Three physical states of matter are solid, liquid, and gas. Examples are iron (solid), water (liquid), and oxygen (gas).
3. The three chemical states of matter are elements, compounds, and mixtures. Copper is an element, sugar is a compound, and sand and gold dust are mixtures.

4. An element is a form of matter that cannot be chemically divided into simpler substances and that has only one type of atom. Examples are gold and silver.
5. A compound is a form of matter that can be chemically divided into simpler substances and that has two or more types of atoms. Examples are water (hydrogen and oxygen) and sugar (carbon, hydrogen, and oxygen).
6. Sketch of a hydrogen atom with particles identified:

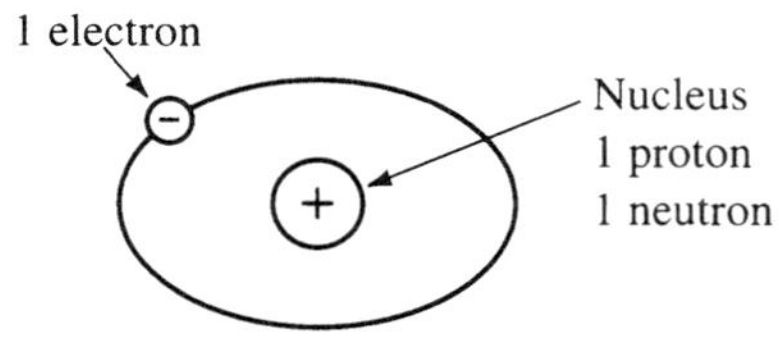

7. Identification of charges on particles
8. a. Proton
   b. The proton is about 1,836 times heavier than the electron.
9. Free electrons are electrons in the atomic structure that can be easily moved or removed from their original atom. These are outer-ring or valence electrons in conductor materials.
10. a. An ion is an atom that has lost or gained electron(s). Therefore, the atom is no longer electrically neutral.
    b. A positive ion is an atom that has lost electron(s). *Note:* A negative ion is an atom that has gained excess electron(s).
11. Chemical, mechanical, light, and heat
12. a. Less than four
    b. Four
    c. More than four
13. Unlike charges attract and like charges repel each other.
14. The force of attraction or repulsion between two charges is directly related to the product of the two charges and inversely related to the square of the distance between them.
15. Force of attraction or repulsion will be one-ninth the original force.
16. Two points with a difference of potential have a difference in charge levels (measured in volts) between them. This potential difference, or voltage, has the ability to move electrons between the two points if a current path exists.
17. The unit of measure of electromotive force (emf) is the volt.
18. Current flow indicates the progressive movement of electrons, (*or charges*) in (or through) a circuit.
19. a. The unit of measure for current is the ampere.
    b. 4 amperes of current flow
20. Resistance is opposition to electrical current flow. The unit of resistance is the ohm. One ohm of resistance limits the current to one ampere when one volt is applied.
21. A closed circuit has an unbroken path for current flow. An open circuit has discontinuity or a broken path; therefore, the circuit does not provide a continuous path for current.
22. A source, a means to transport electrons, (*or charges*) from one point to another, and a load, which uses electrical energy
23. A minus (−) sign
24. Negative; Positive (*Positive, Negative*)
25. Point B represents the + (positive) (− *negative*) side of the source.

ANALYSIS QUESTIONS

1. Drawing
2. Labeling of drawing:

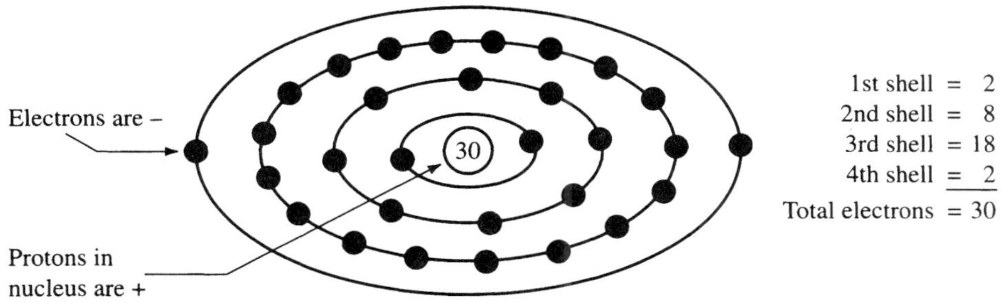

3. Zinc
4. Conductor
5. 30
6. Polarity expresses a difference in charge condition between two points in a circuit. For electrical polarity, the point having the most electrons is the negative point, and the point with fewer electrons is the positive point.
7. Circuit diagram:

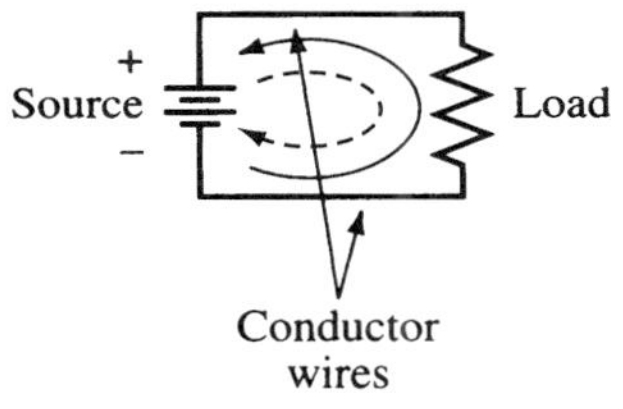

8. — Broken conductor or broken contact with conductor
   — Open switch contacts
   — Bad resistor or component
9. A short circuit is (usually) an undesired very low-resistance path around part or all of a circuit.
10. Consequences of a short circuit may include melted wires, fire, and damage to other components and/or to the source.

CHAPTER 2

IN-PROCESS LEARNING CHECK 1

1. Charge is represented by the letter "Q." The unit of measure is the *coulomb*, and the abbreviation for coulomb is C.
2. The unit of potential difference is the *volt*. The symbol is V.
3. The abbreviation for current is *I*. The unit of measure for current is the *ampere* and the symbol for this unit is A.
4. The abbreviation for resistance is *R*. The unit of measure is the *ohm*, and the symbol for this unit is Ω (Greek letter omega).
5. Conductance is the *ease* with which current can flow through a component or circuit. The abbreviation is *G*. The unit of measure for conductance is the *siemens*. The symbol for this unit is S.

6. 0.0000022 amperes = *2.2 microamperes* (μA). Could be expressed as 2.2×10^{-6}.
7. One-thousandth is represented by the metric prefix *milli*. When expressed as a power of 10, it is 10^{-3}.
8. 10,000 ohms might be expressed as *10 kilohms* (kΩ). Expressed as a whole number times a power of 10 = 10×10^{3} *ohms*.

PRACTICE PROBLEMS 1

The circular-mil cross-sectional area = 2.5^2 = 6.25 CM.

PRACTICE PROBLEMS 2

300 feet of copper wire having a cross-sectional area of 2,048 CM = 1.523 Ω.

PRACTICE PROBLEMS 3

1. 500 feet of #12 wire has a resistance of approximately 0.81 Ω.
2. The gauge of wire having a cross-sectional area of 10,380 CM is #10 wire.
3. #10 wire has an approximate resistance of 1.018 Ω per thousand feet of length.

IN-PROCESS LEARNING CHECK 2

1. The greater the resistivity of a given conductor, the *higher* its resistance will be per given length.
2. The circular-mil area of a conductor is equal to the square of its *diameter* in *mils* (thousandths of an inch).
3. The smaller the diameter of a conductor, the *higher* its resistance will be per given length.
4. The higher the temperature in which a typical metal conductor must operate, the *higher* will be its resistance.
5. Wire tables show the AWG wire *size*, the cross-sectional area of wires in *circular mils*, the resistance per given length of wire, and the operating *temperature* at which these parameters hold true.

PRACTICE PROBLEMS 4

1. 39 kΩ, 10% tolerance
2. A resistor color-coded brown, gray, red, gold, and brown is a precision five-band color-coded resistor having a value of 18.2 Ω, a tolerance of 1%.

IN-PROCESS LEARNING CHECK 3

1. Meter negative lead connected to Point *B*. Meter positive lead connected to Point *A*.
2. Meter negative lead connected to Point *B*. Meter positive lead connected to Point *C*.
3. First thing to do is to remove the source voltage from the circuit.

PRACTICE PROBLEMS 5

1. It is a TV receiver system.
2. Signal flows from left to right, as illustrated.
3. The receiver circuitry receives signals from the TV antenna.
4. Two types of outputs include picture information and sound information.

PRACTICE PROBLEMS 6

1. There are three kinds of components (resistor, switch, battery).
2. The highest-value resistor is 100 kΩ.
3. An SPST switch is used.
4. The voltage source is a battery.

PRACTICE PROBLEMS 7

1. There are five resistors.
2. The lowest R value resistor is 1 kΩ (R_3).
3. There are two switches in the circuit.
4. The circuit applied voltage is 100 volts.
5. S_2 is opened, and S_1 is closed.

REVIEW QUESTIONS

1. 0.003 V
2. 12 μA = 12×10^{-6} = 0.000012 amperes
3. A micro unit is one-thousandth of a milli unit.
4. 1,000 kΩ
5. Schematic diagram of battery, two resistors, and an SPST switch:

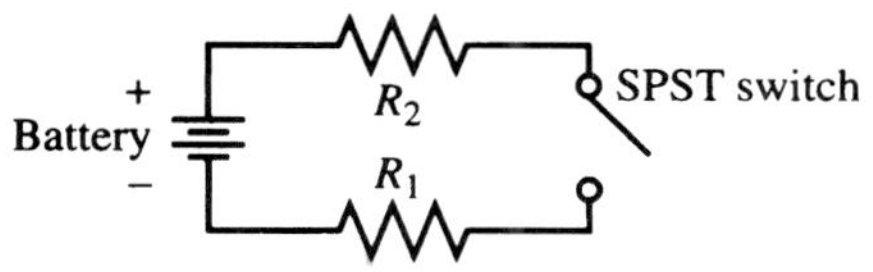

6. Resistor
7. Resistor
8. Voltmeter
9. Battery
10. a. 0.015 A
    b. 15×10^{-3} A
    c. 5 kV
    d. 5×10^{3} V
    e. 500,000 Ω
    f. 0.1 MΩ
    g. 500 mA
11. Carbon or composition resistors and wirewound resistors.

    Carbon/composition resistors are composed of carbon or graphite mixed with powdered insulating material.

    Wirewound resistors are constructed with special resistance wire wrapped around a ceramic or insulator core.
12. a. 470 kΩ, 10% tolerance
    b. 2.2 MΩ, 10% tolerance
    c. 33 Ω, 5% tolerance
    d. 910 Ω, 5% tolerance
    e. 1.2 Ω, 5% tolerance
13. Value, power rating, tolerance, and physical size
14. The fifth color band designates the tolerance (for precision resistors).
15. 22,200 Ω ± 2%
16. ± 20%
17. Wirewound resistor
18. 44,000 Ω
19. 5th band; 0.01%
20. Turn power off first
21. Use only one hand, insulate yourself from ground, and use only insulated handle tools.
22. Discharge the capacitors before working on circuitry.
23. Electrical shock and physical injury

24. Properly locate soldering iron, wear safety glasses, and be careful when wiping the iron.
25. Part of the body is providing a current path between points having a potential difference.

ANALYSIS QUESTIONS

1. The resistor color code is necessary because many resistors are too small to enable printing their values on them. Also, color code provides a "universal" method, readable by anyone in the world, no matter their language.
2. If the operating temperature for a conductor increases, (a) resistance increases and (b) current-carrying capacity decreases.
3. Generally, reliability refers to the ability of the component to perform as it is expected to, for the length of time intended, under normal operating conditions. Resistor reliability ratings typically indicate the percentage of failure for 1,000 hours of operation.
4. a. To measure dc voltage: turn off power to circuit; make sure meter is set on high enough range and is in the voltage-measuring mode; connect meter leads "across" component, using correct polarity; turn on power to circuit; read meter.
   b. To measure dc current: turn off power to circuit; break circuit where meter will be inserted; make sure meter is set on high enough range and is in the current-measuring mode; connect meter leads *into* the circuit, making sure to observe proper polarity; turn on power to circuit; read meter.
   c. To measure a resistor's value: turn off power and/or disconnect power source from circuit; electrically isolate resistor to be measured by disconnecting one end of resistor from circuit; make sure meter is on the ohms mode and set on an appropriate range if not an "auto-ranging" meter; connect test leads across the resistor; read the meter.
5. Block diagram of a simple power distribution system:

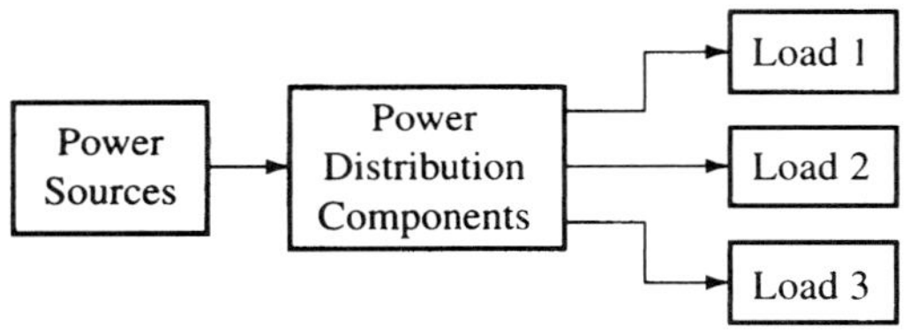

6. Schematic diagram of two resistors, one meter, two switches, and a battery:

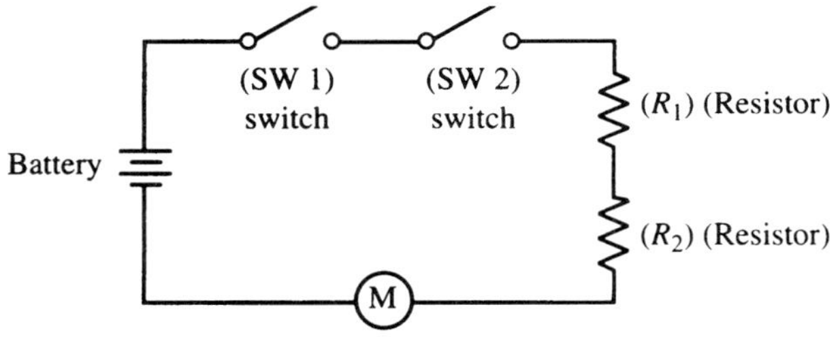

7. a. Brown, Black, Red, Silver
   b. Red, Violet, Orange
   c. Brown, Black, Green, Silver
   d. Brown, Black, Black, Gold
8. Power off, correct meter mode, proper range, polarity (if dc measurement)
9. Two connection points
10. Three connection points

CHAPTER 3

PRACTICE PROBLEMS 1

1. $I = V/R = 150\ \text{V}/10\ \Omega = 15\ \text{A}$
2. 7.5 A
3. a. Remain the same
   b. $\dfrac{300\ \text{V}}{20\ \Omega} = 15\ \text{A}$

PRACTICE PROBLEMS 2

1. 2.5×10^4 (or 25×10^3)
2. 1.2335×10^1
3. 1.0×10^{-1}
4. 1.5×10^{-3}
5. 23,500
6. 1,000
7. 10,000
8. 0.015

PRACTICE PROBLEMS 3

1. 1,456,000
2. 0.0333
3. 1.5
4. 1,000,000

IN-PROCESS LEARNING CHECK 1

1. $I = 1.5$ A or 1,500 mA. ($I = V/R$; $I = 15\ \text{V}/10\ \Omega = 1.5\ \text{A}$)
2. $V = 30$ V. ($V = I \times R$; $V = 2\ \text{mA} \times 15\ \text{k}\Omega = 30\ \text{V}$)
3. $R = 27{,}000\ \Omega$ or 27 kΩ. ($R = V/I$; $R = 135\ \text{V}/5\ \text{mA} = 27\ \text{k}\Omega$)
4. a. Current would *decrease*.
   b. Voltage applied would be *higher*.
   c. Circuit resistance must have *halved*.

PRACTICE PROBLEMS 4

1. P = energy/time = 25 J/2 sec = 12.5 watts
2. P = energy/time = 1 J/100 sec = .01 watts or 10 milliwatts

PRACTICE PROBLEMS 5

1. Energy = Power × Time = 250 W × 8 hrs = 2,000 Wh or 2 kWh

PRACTICE PROBLEMS 6

1. $P = V \times I = 150\ \text{V} \times 10\ \text{mA} = 1{,}500$ mW, or 1.5 W
2. a. I would decrease to half the original value
   b. Power dissipation = $V \times I = 150 \times 5\ \text{mA} = 750$ mW, or 0.75 W

IN-PROCESS LEARNING CHECK 2

1. Electron current flow is considered to flow from the *negative* side of the source, through the external circuit, back to the *positive* side of the source.
2. "Conventional" current is considered to flow from the *positive* side of the source, through the external circuit, back to the *negative* side of the source.
3. This text uses the *electron, (conventional)* current flow approach.
4. One of the basic differences between dc and ac is that current in a dc circuit always flows in *one* direction, whereas current flow in an ac circuit *changes* direction for each alternation of the ac voltage.
5. Work is the *expenditure* of energy; energy is the ability to do *work*, or that which is expended in doing *work*. Power is the rate of using *energy*.

6. Power in watts is equal to energy in *joules* divided by time in *seconds*.
7. Energy = power (in watts) × time. A common measure of energy consumption used by power companies for billing customers is the *kilowatt-hour*.

PRACTICE PROBLEMS 7

1. $V_A = I \times R = 1.5\text{ mA} \times 4.7\text{ k}\Omega = 7.05\text{ V}$
2. $I = V/R = 100\text{ V}/27\text{ k}\Omega = 3.704\text{ mA}$
3. $R = V/I = 100\text{ V}/19.6\text{ mA} = 5.102\text{ k}\Omega$

PRACTICE PROBLEMS 8

1. $P = 40\text{ W}$; $I = 2\text{ A}$
2. $I = 1\text{ mA}$; $R = 100\text{ k}\Omega$
3. $I = 12\text{ mA}$; $P = 1.44\text{ W}$ (or 1,440 mW)

USING EXCEL

1. 2 A; 3 A; 1 A (see Figure 3-1)
2. $P = 40\text{ W}$; $I = 2\text{ A}$ (see Practice Problems 8, Circuit 1)

REVIEW QUESTIONS

1. a. $I = V/R$; $V = IR$; $R = V/I$
   b. $P = V \times I$; $P = I^2R$; $P = V^2/R$
2. a. 3.3456×10^4
   b. 2.5×10^1
   c. 1.055×10^6
   d. 1.0×10^1
3. a. 5.1×10^3
   b. 47×10^3
   c. 0.000001×10^3
   d. 0.039×10^3
4. Work is the expenditure of energy.
5. Energy is the ability to do work.
6. Power is the rate of using energy to perform work.
7. Work is measured in foot-poundals, or foot-pounds, or horsepower. (**NOTE:** 32.1 foot-poundals = 1 foot-pound, at sea level.)
8. Energy is measured in joules.
9. Power (electrical) is measured in watts.
10. a. For a given *R*: power is related to current squared.
    b. For a given *R*: voltage is directly related to current.
    c. For a given time: energy is directly related to power.

PROBLEMS

1. The new current will be four times the original current.
2. Power will be nine times the original power.
3. a. Schematic:
   b. $R = V/I = 30\text{ V}/2\text{ mA} = 15\text{ k}\Omega$;
      $P = V \times I = 30\text{ V} \times 2\text{ mA} = 60\text{ mW}$
   c. See diagram

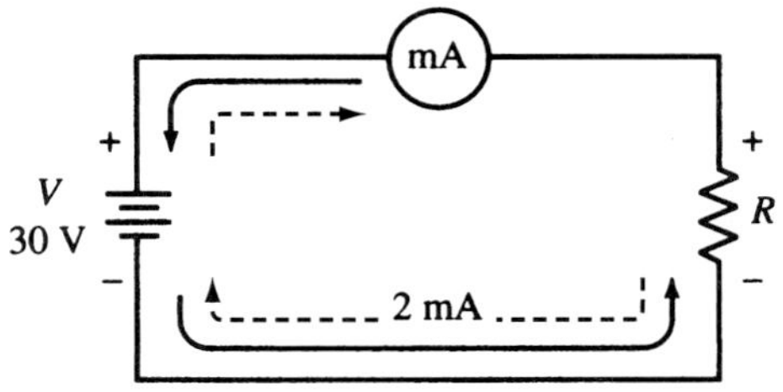

4. $I = V/R = 50\text{ V}/33\text{ k}\Omega = 1.515\text{ mA}$
5. $V = I \times R = 3\text{ mA} \times 12\text{ k}\Omega = 36\text{ V}$

6. a. V has increased
   b. R will remain the same
   c. I will increase
7. $R = V/I = 41\text{ V}/50\text{ mA} = 820\ \Omega$
8. $V = 8\text{ V}$
9. $I = 3.162\text{ A}$
10. $P = 20\text{ W}$
11. $R = V/I = 100\text{ V}/8.5\text{ mA} = 11.765\text{ k}\Omega$
12. Circuit current will decrease (to 2.66 mA from 4 mA).
13. Power dissipated will increase ($P = V \times I$ and I remained the same while V doubled).
14. 18.75 mA
15. 2.8125 W
16. 8.4375 Wh
17. 20.83 mA
18. 5.2075 W
19. 33.85 Wh
20. 200 kΩ
21. 31.6 V
22. Schematic diagram of source, resistor, and current meter:
    (Solving for I):
    $P = I^2R$;
    $1{,}000\text{ W} = I^2 \times 10\ \Omega$
    $I^2 = P/R$;
    $I^2 = 1{,}000/10$
    $I = \sqrt{P/R}$
    $I = \sqrt{1{,}000/10} = \sqrt{100} = 10\text{ A}$

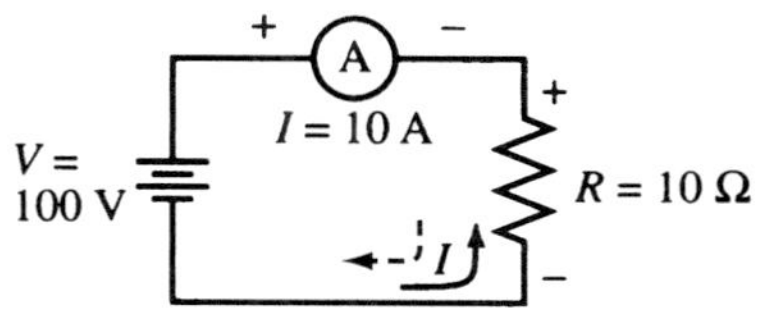

23. $V_A = I \times R$;
    $= 10\text{ A} \times 10\ \Omega = 100\text{ V}$
24. Labeled diagram
25. a. 0.01
    b. 12th power (10^{12})
    c. 8.570×10^4 (or 85,700)
    d. 6.1664×10^3 (or 6.17×10^3)

ANALYSIS QUESTIONS

1. The EE (exponent entry) key allows entering numbers with exponents, such as numbers raised to some power of 10.
2. 222^{04} (**NOTE:** If you follow these strokes with an equal, the display will then show 2220000.)
3. Current only flows in one direction with the dc source. It alternates direction of flow with ac applied.
4. DC = one polarity, all the time.
   AC = alternating polarity, every half-cycle.
5. The direction a + (positive) charge would move.
6. $P = 746 \times 3.5 = 2{,}611$ watts
7. V has doubled

CHAPTER 4

PRACTICE PROBLEMS 1

1. Total resistance = 74 kΩ
2. Total resistance = 16 kΩ

PRACTICE PROBLEMS 2

1. $R_1 = 47\ \Omega$; $R_T = 100\ \Omega$
2. New $R_T = 10.5$ kΩ
3. $R_2 = 3.9$ kΩ

PRACTICE PROBLEMS 3

1. $V_A = 411$ V; $I_T = 3$ mA; $V_{R_2} = 81$ V; $V_{R_1} = 300$ V
2. $V_A = 342.5$ V; $I_T = 2.5$ mA; $V_{R_2} = 67.5$ V; $V_{R_1} = 250$ V

PRACTICE PROBLEMS 4

1. $V_X = (R_X/R_T) \times V_T = (2.7\ \text{k}\Omega/17.4\ \text{k}\Omega) \times 50\ \text{V} = 0.155 \times 50 = 7.75\ \text{V}\ (V_{R_1})$
2. $(4.7\ \text{k}\Omega/17.4\ \text{k}\Omega) \times 50 = 13.5\ \text{V}\ (V_{R_2})$
3. $(10\ \text{k}\Omega/17.4\ \text{k}\Omega) \times 50\ \text{V} = 0.575 \times 50 = 28.75\ \text{V}\ (V_{R_3})$
4. 7.75 + 13.5 + 28.75 = 50 V (Yes, they add up correctly)

PRACTICE PROBLEMS 5

$V_{R_2} = 30$ V; $V_A = 75$ V

IN-PROCESS LEARNING CHECK 1

1. The primary identifying characteristic of a series circuit is that the *current* is the same throughout the circuit.
2. The total resistance in a series circuit must be greater than any one *resistance* in the circuit.
3. In a series circuit, the highest value voltage is dropped by the *largest* or *highest* value resistance, and the lowest value voltage is dropped by the *smallest* or *lowest* value resistance.
4. V dropped by the other resistor must be *100* V.
5. Total resistance *increases*. Total current *decreases*. Adjacent resistor's voltage drop *decreases*.
6. Applied voltage is *130* V.

PRACTICE PROBLEMS 6

P_{R_1} = 90 W	I_T = 3 A	V_{R_3} = 30 V
P_{R_2} = 90 W	V_{R_1} = 30 V	R_3 = 33% of R_T
P_{R_3} = 90 W	V_{R_2} = 30 V	P_{R_3} = 33% of P_T
P_T = 270 W		

PRACTICE PROBLEMS 7

1. a. Increase
   b. Decrease
   c. Decrease
   d. Increase
   e. Decrease
2. Yes

PRACTICE PROBLEMS 8

1. With the 47-kΩ resistor shorted, the voltage drops are:

10-kΩ resistor	= 14.6 V
27-kΩ resistor	= 39.4 V
47-kΩ resistor	= 0 V
100-kΩ resistor	= 146 V

2. Voltage drops across the other resistors increase.

PRACTICE PROBLEMS 9

Givens: $R_1 = 20$ kΩ; $V_1 = 2/5 \times V_A$; Both V_2 and V_3 equal $1.5/5 \times V_A$; $V_A = 50$ V; $R_2 = 15$ kΩ; $R_3 = 15$ kΩ; $V_{R_1} = 20$ V; $V_{R_2} = 15$ V; $V_{R_3} = 15$ V; $P_T = 50$ mW; $P_{R_1} = 20$ mW; $P_{R_2} = 15$ mW; $P_{R_3} = 15$ mW; $I = 1$ mA.

Circuit:

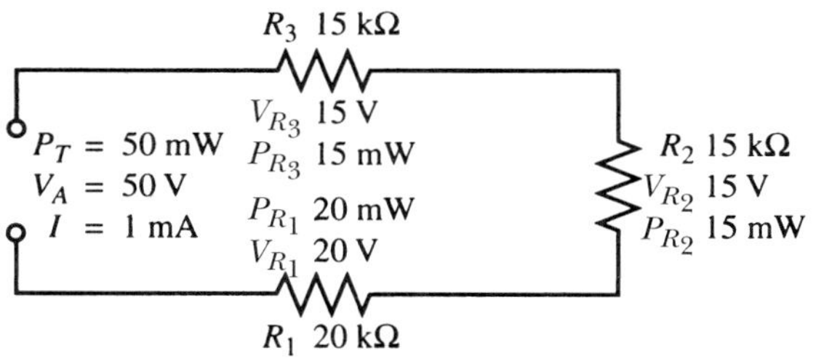

PRACTICE PROBLEMS 10

1. $V_{R_1} = 4.5$ V
2. $V_{R_2} = 1.5$ V

PRACTICE PROBLEMS 11

D to C = 94 volts (D being negative with respect to C)

C to B = 54 volts (C being negative with respect to B)

B to A = 40 volts (B being negative with respect to A)

D to B = 148 volts (D being negative with respect to B)

PRACTICE PROBLEMS 12

1. Voltage across $R_2 = 27.49$ V (or approximately 27.5 V)
2. Voltage at top of R_2 is +27.5 V with respect to point C
3. Voltage across $R_1 = 9.995$ V (or approximately 10 V)
4. Voltage at point C with respect to point A = −37.5 V

PRACTICE PROBLEMS 13

1. Minimum power rating should be 2 times (60 V × 30 mA), or 2 × 1,800 mW = 3,600 mW (3.6 W)
2. *R* dropping = 110 V/30 mA = 3.66 kΩ
   Minimum power rating should be 2 times 110 V × 30 mA, or 2 × 3,300 mW = 6,600 mW (6.6 W)

USING EXCEL

1. $V_{R_2} = 13.5$ V
2. $V_{R_1} = 7.75$ V

REVIEW QUESTIONS

1. Current is the same throughout the circuit because there is only one path for current.
2. (a)
3. R_T equals the sum of all the resistor values in series.
4. (b)
5. (b)
6. (b)
7. (a)
8. (b)
9. (c)
10. (b)

PROBLEMS

1. $R_T = 10/5 = 2\ \Omega$

   Each resistor = 1 Ω
2. $I_T = V_T/R_T$

   $I_T = 40\ \text{V}/60\ \text{k}\Omega = 0.67\ \text{mA}$
3. Diagram, labeling, and calculations for a series circuit:

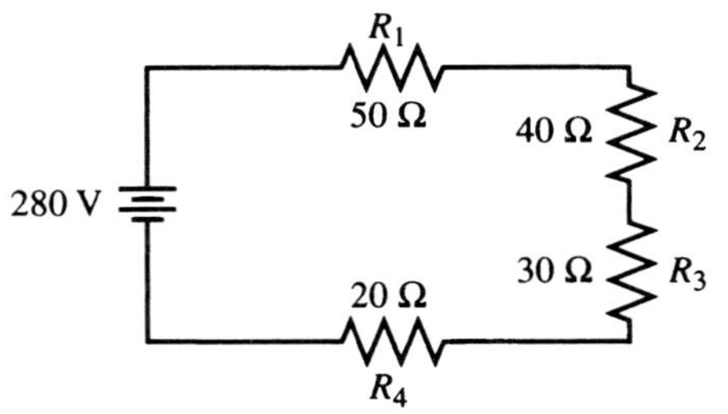

   a. $R_T = 50 + 40 + 30 + 20 = 140\ \Omega$

   b. V applied $= IR = 2\ \text{A} \times 140\ \Omega = 280\ \text{V}$

   c. $V_{R_1} = 100\ \text{V}$ $\quad V_{R_3} = 60\ \text{V}$

   $V_{R_2} = 80\ \text{V}$ $\quad V_{R_4} = 40\ \text{V}$

   d. $P_T = I_T \times V_T = 2\ \text{A} \times 280\ \text{V} = 560\ \text{W}$

   e. $P_{R_2} = 2\ \text{A} \times 80\ \text{V} = 160\ \text{W}$

   $P_{R_4} = 2\ \text{A} \times 40\ \text{V} = 80\ \text{W}$

   f. one-seventh V_A dropped by R_4

   g. (1) Total resistance increases.

   (2) Total current decreases.

   (3) V_{R_1} decreases; V_{R_2} decreases; V_{R_4} decreases.

   (4) P_T decreases since I decreases and V remains the same.

   h. (1) Total resistance decreases.

   (2) Total current increases.

   (3) V_{R_1} increases; V_{R_3} increases; V_{R_4} increases (due to increased I with Rs remaining the same).
4. V applied = sum of V drops = 110 V

   Algebraic sum = zero (counting the source)
5. Diagram of three sources to acquire 60 V, if sources are 100 V, 40 V, and 120 V:

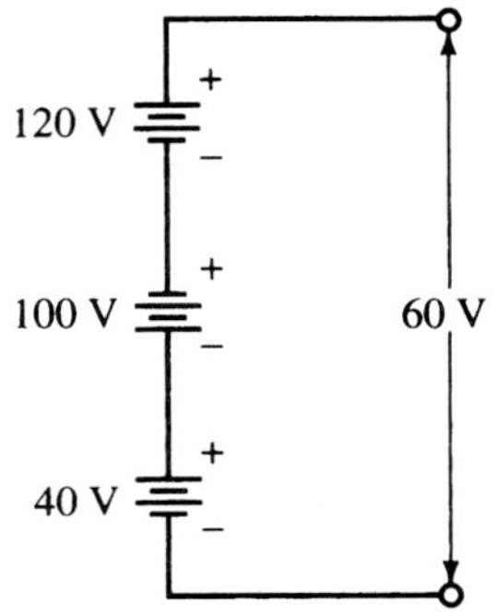

6. a. $P_T = I^2R = (3 \times 10^{-3})^2 \times 74 \times 10^3 = 0.666$ W or 666 mW
$P_{R_1} = I^2R_1 = (3 \times 10^{-3})^2 \times 10 \times 10^3 = 0.09$ W or 90 mW
b. $V_T = I_T \times R_T = 3$ mA $\times$ 74 kΩ = 222 V
c. $R_T = R_1 + R_2 + R_3 + R_4 = 10$ kΩ + 10 kΩ + 27 kΩ + 27 kΩ = 74 kΩ
d. P_{R_4} is 2.7 times greater than P_{R_1}.
e. P_T becomes 1.049 W. P_{R_1} becomes 222 mW (assuming V was still 222 V).
7. a. $V_A = 34$ V (4 V + 20 V + 10 V)
b. $R_3 = 2.5$ kΩ (to drop 5 V of the 10 V dropped by R_3 and R_4)
c. $R_2 = 10$ kΩ (to drop 20 V with 2 mA of current)
8. Diagram of a three-resistor voltage divider:

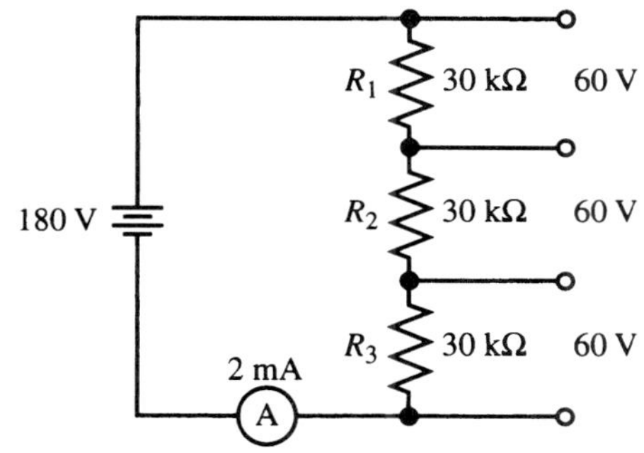

9. 17.22 V
10. 1.435 mA (approx. 1.44 mA)
11. $V_{R_2} = 2.15$ V
12. $V_{R_3} = 6.74$ V
13. $V_{R_4} = 3.87$ V
14. With R_3 shorted, voltmeter reading = 22.2 V
15. Most power = R_1; Least power = R_2
16. Percentage would stay the same, since R is the same % of R_T as before.
17. $V_3 = 70$ V
18. V applied = 125 V
19. a. R_1 is set at 3.81 kΩ
b. $I_T = 4.2$ mA
c. R_1 dissipates 16% of total power
d. $I_T = 4$ mA with R_1 set at middle of its range
e. Voltmeter would indicate 50 V
20. a. $R_T = 91$ kΩ, $V_T = 182$ V
b. M_1 will indicate 94 V
c. M_3 will indicate 2 V
d. $I_T = 2$ mA
e. New $V_T = 257.4$ V; New $I_T = 2.828$ mA
21. a. $P_{R_2} = 423$ mW
b. $P_T = 819$ mW
c. $I_T = 3$ mA
22. Circuit diagram where $V_{R_3} = 3V_{R_1}$; $V_{R_2} = 2V_{R_1}$; and $V_A = 60$ V:
a. $R_1 = 20$ kΩ
b. $V_{R_1} = 10$ V
$V_{R_2} = 20$ V
$V_{R_3} = 30$ V
c. $I = 0.5$ mA; $P_T = 30$ mW

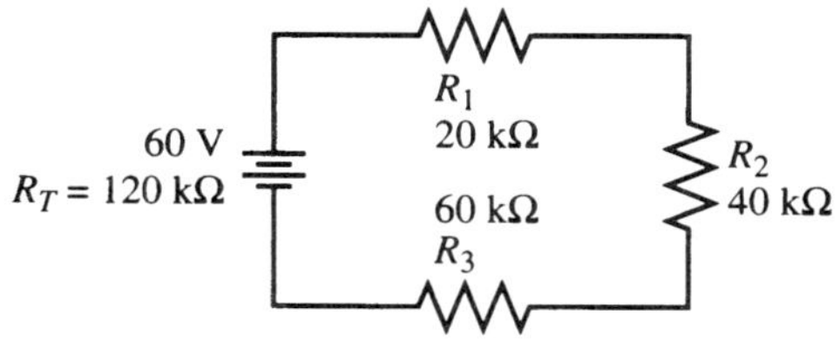

23. Circuit diagram with series dropping resistor:

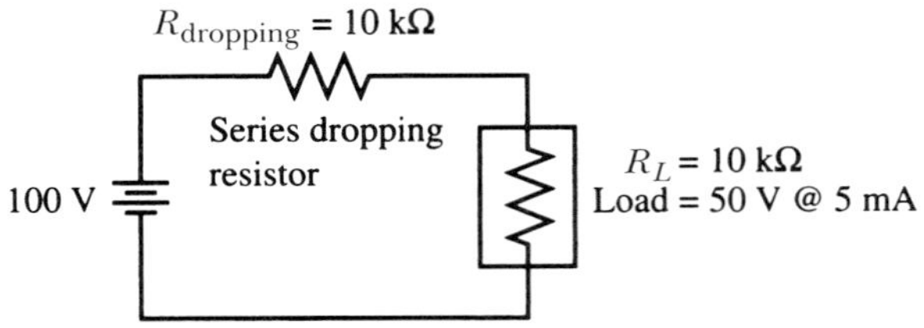

24. Increase
25. Decrease
26. Remain the same
27. $V_{applied} = 182$ V
28. V has doubled, which causes I to double also, since R remained the same.

ANALYSIS QUESTIONS

1. In Scientific mode—the results always indicate a number between 1 and 10 times some power of 10.

   In Engineering mode—the results show a number times a power of 10 with an exponent, which is some multiple or submultiple of 3 (taking advantage of the metric system).
2. To provide different voltage levels to different loads.

   To provide both a + and – voltage output with respect to a given reference point.
3. The "divide-and-conquer" or "split" rule indicates that where there is a linear, or series flow, it is often good to make the first check at the middle of the questionable area. This enables eliminating half the questionable area with just one test.
4. Three tools to help might include: information from user; looking at easy-to-check indicators; and comparing known "norms" to actual present parameters.
5. To identify "initial suspect area": analyze all symptom information; determine all sections or components in the system that could cause the symptoms; then "bracket" the area of uncertainty, or suspect area in which to make tests and diagnoses.
6. Two decisions: where to test and what kind of test to make.
7. Bracketing: enclosing the area of uncertainty or "suspect area" (either mentally or on paper) in parentheses or brackets to highlight the area in which you will troubleshoot.
8. Used where there are several components, devices, or subsystems in series, forming a linear flow path, all of which are in the area of uncertainty.
9. Useful documentation includes: block diagrams, schematic diagrams, operator manuals, and the like.
10. Precautions include turning the power off and using resistance measurements to troubleshoot rather than voltage measurements. This is to prevent further damage to the circuit and/or a possible fire hazard from excessive current due to the short condition.
11. If current drops to zero, an open condition has occurred somewhere in the series path.
12. d. (R_2 and R_3 have shorted out.)
13. The suggested method to use is the "divide-and-conquer" (half-split) technique.

MULTISIM EXERCISE

2. a. +30 V (Point C is + with respect to Point D)

   b. −60 V (Point C is – with respect to Point A)

   c. −30 V (Point C is – with respect to Point B)
3. Yes

CHAPTER 5

PRACTICE PROBLEMS 1

$V_{R_1} = V_{R_2} = 2 \text{ mA} \times 27 \text{ k}\Omega = 54 \text{ V}$

$V_A = V_{R_1} = V_{R_2} = 54 \text{ V}$

PRACTICE PROBLEMS 2

$I_T = I_1 + I_2 + I_3 + I_4 + I_5$

The lowest R value branch passes the most current.

The highest R value branch passes the least current.

PRACTICE PROBLEMS 3

$V_{R_2} = 50$ V; $V_{R_1} = 50$ V; $V_A = 50$ V; $I_1 = 5$ A; $I_T = 6$ A

PRACTICE PROBLEMS 4

1. $R_1 = 15$ kΩ; $I_2 = 2.5$ mA; $I_3 = 2.5$ mA; $R_2 = 30$ kΩ
2. $R_1 = 27$ kΩ; $I_2 = 3.78$ mA; $I_3 = 2.23$ mA; $R_2 = 33$ kΩ; $R_T = 11.75$ kΩ

PRACTICE PROBLEMS 5

1. $V_{R_1} = 30$ V; $V_T = 30$ V
2. $R_T = 1.67$ kΩ

PRACTICE PROBLEMS 6

1. $R_T = 5.35$ kΩ
2. $R_T = 0.76$ kΩ or 760 Ω

PRACTICE PROBLEMS 7

1. The calculations are correct. $R_T = 2.5$ Ω
2. $R_T = 25$ Ω

PRACTICE PROBLEMS 8

1. $R_T = 12$ Ω
2. $R_T = 7.67$ kΩ

PRACTICE PROBLEMS 9

$R_T = 27.15$ Ω

PRACTICE PROBLEMS 10

$R_2 = 100$ Ω

PRACTICE PROBLEMS 11

$P_T = 2{,}250$ mW or 2.25 W

PRACTICE PROBLEMS 12

Labeled diagram:

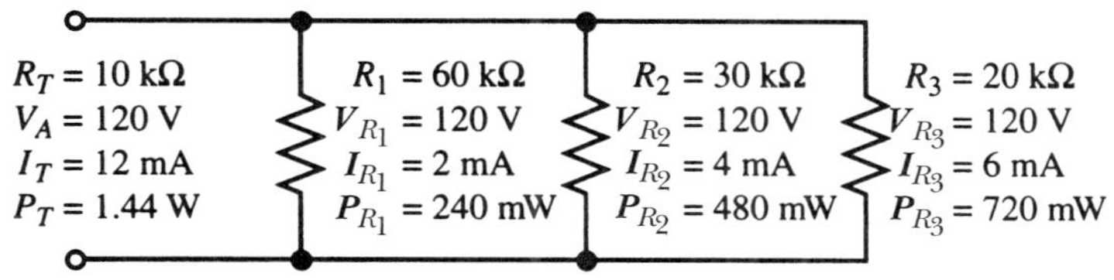

PRACTICE PROBLEMS 13

1. $I_1 = 1.85$ mA (through 27-kΩ resistor)
   $I_2 = 1.07$ mA (through 47-kΩ resistor)
   $I_3 = 0.50$ mA (through 100-kΩ resistor)
2. a. $R_T = 1.54$ kΩ
   b. $I_1 = 0.616$ mA (through 10-kΩ resistor),
   $I_2 = 1.1$ mA (through 5.6-kΩ resistor),
   $I_3 = 2.28$ mA (through 2.7-kΩ resistor).

PRACTICE PROBLEMS 14

1. $I_1 = 1.65$ mA
   $I_2 = 0.35$ mA
2. $I_1 = 16.4$ mA
   $I_2 = 3.6$ mA

USING EXCEL

1. $R_T = 760\ \Omega$
2. $I_1 = 15.38$ mA

REVIEW QUESTIONS

1. (b)
2. (b)
3. (c)
4. (c)
5. (a)
6. $R_T = V_T/I_T$
7. $1/R_T = 1/R_1 + 1/R_2 + \ldots\ 1/R_n$ and/or "Reciprocal of the reciprocals":
   $$\frac{1}{1/R_1 + 1/R_2 + \ldots 1/R_n}$$
8. $R_T = 1/G_T$
9. $R_T = R_1 \times R_2/R_1 + R_2$
10. First step: Assume a voltage applied that is easily divided by the various branch resistance values; usually, some multiple of these values. (This makes it easy to calculate what each branch current would be.)
    Second step: Solve for each branch current based on the assumed voltage.
    Third step: Add the branch currents to get the total circuit current.
    Fourth step: Divide the assumed voltage by the calculated total current to get the R_T value.
11. $R_u = R_k \times R_e/R_k - R_e$
12. (b)

PROBLEMS

1. $R_T = 17.15$ kΩ; product-over-the-sum method
2. $R_T = 12.5$ kΩ; repeatedly use equal-resistor branches technique, where two 100s = 50, two 50s = 25, and two 25s equal 12.5 kΩ.
3. $I_2 = 30$ mA; Kirchhoff's current law
4. $R_T = 25$ kΩ (150/6)
   $I_T = 300$ V/25 kΩ = 12 mA
   $P_T = 3.6$ W (300 V × 12 mA)
   meter 1 reads 12 mA
   meter 2 reads 6 mA

5. a. I_1 will RTS
I_2 will D
I_3 will RTS
I_T will D
P_T will D
V_{R_1} will RTS
V_T will RTS

b. I_1 will D
I_2 will D
I_3 will I
I_T will I

6. $R_1 = 6$ kΩ (Using $R_u = R_k \times R_e / R_k - R_e$)
7. $I_1 = 12$ mA
$I_2 = 6$ mA (since I branch is inverse to R branch)
8. $R_1 = 12$ kΩ (since R_T must equal 4 kΩ to have 10 mA of I_T).
9. $P_{R_1} = 0.5 \text{ A} \times 50 \text{ V} = 25$ W
10. $R_T = 1$ kΩ (since $V_A = 30$ V and sum of branch Is = 30 mA).
$I_T = 30$ mA (sum of branch currents)
$P_T = 900$ mW ($V_T \times I_T$)
$I_{R_1} = 15$ mA (30 V/2 kΩ)
$I_{R_2} = 5$ mA (30 V/6 kΩ)
$I_{R_3} = 6$ mA (30 V/5 kΩ)
$I_{R_4} = 2$ mA (30 V/15 kΩ)
$I_{R_5} = 2$ mA (30 V/15 kΩ)
$P_{R_1} = 450$ mW (30 V × 15 mA)
$P_{R_2} = 150$ mW (30 V × 5 mA)
$P_{R_3} = 180$ mW (30 V × 6 mA)
$P_{R_4} = 60$ mW (30 V × 2 mA)
$P_{R_5} = 60$ mW (30 V × 2 mA)
11. a. $M_1 = 2.13$ mA
b. $M_2 = 37$ mA
c. $M_3 = 39.13$ mA
d. $M_4 = 10$ mA
e. $M_5 = 49.13$ mA
f. $P_{R_1} = 213$ mW
g. $P_{R_2} = 3.7$ W
h. $P_{R_3} = 1$ W
i. $P_T = 4.91$ W
j. New $P_T = 1.21$ W
12. $R_T = 14.64$ kΩ
13. $R_X = 31.59$ kΩ
14. R_T would increase from 10 kΩ to 10.93 kΩ. No.
15. a. R_4
b. $P_{R_4} = 71.7$ mW
16. a. R_2
b. $P_{R_2} = 19.36$ mW
17. $V_T = 18$ V
18. $R_1 = 1.5$ kΩ
19. $I_{R_2} = 5.45$ mA
20. $R_2 = 3.3$ kΩ
21. R_1 dissipates the most power.
22. R_2 dissipates the least power.
23. I_T would be 34 mA if R_2 were changed to a 1.5-kΩ resistor value.
24. Total power would decrease if R_1 were to increase in value.
25. Power dissipated by R_3 would not change as long as V_T and R_3 were the same values.
26. $R_2 = 27$ kΩ; $R_4 = 47$ kΩ
27. 120 V
28. P_T would be 1,741 mW, or 1.741 W
29. $R_e = 5.45$ kΩ
30. R to be added = 15 kΩ

ANALYSIS QUESTIONS

1. The purpose of performing mental approximations is to provide a rapid check as to whether you might have made a mistake along the way in solving a problem. If your approximation comes out completely different from your other calculations, chances are you have a simple mathematical error that needs correcting.
2. One possible application of the current-divider circuit is illustrated in meter circuits, with current meter shunts.
3. a. R_T = 4.864864865 03
   b. R_T = 4.864864865 03
4. For rough approximation, you could have considered the circuit as approximating three 15-kΩ resistors in parallel, or 15 kΩ/3 = approximately 5 kΩ total resistance.
5. Circuit diagram of a three-branch parallel circuit:

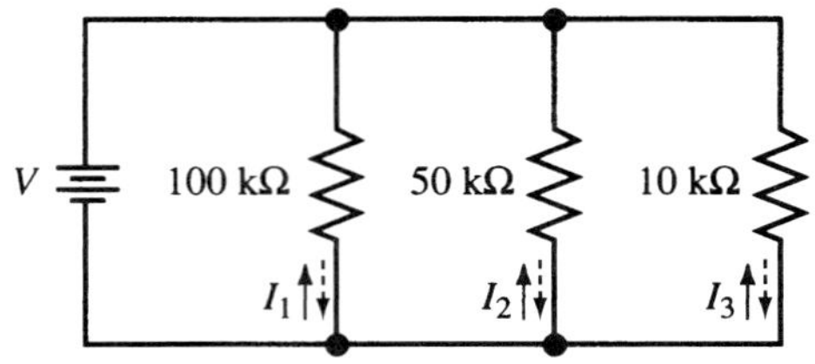

6. R needed = $R_k R_e / R_k - R_e$ = 50 kΩ × 10 kΩ/40 kΩ = 500 kΩ/40 kΩ = 12.5 kΩ
7. Isolate by lifting one end of the branch from the circuit to electrically isolate it. (This may mean unsoldering for tests and resoldering after tests.)

MULTISIM EXERCISE

2. a. 30 V
   b. 30 mA, 1 kΩ
   c. 6 mA, 5 kΩ
3. Yes
4. Yes

CHAPTER 6

PRACTICE PROBLEMS 1

1. The only component carrying total current is R_7.
2. The only single components in parallel with each other are R_2 and R_3.
3. The only single components that pass the same current are R_1, R_4, and R_5.

PRACTICE PROBLEMS 2

1. R_T for Figure 6–11 = 12.1 kΩ
2. R_T for Figure 6–12 = 7.63 kΩ

IN-PROCESS LEARNING CHECK 1

1. For Figure 6–13a: R_T = 10 Ω
   For Figure 6–13b: R_T = 120 Ω

PRACTICE PROBLEMS 3

1. $I_T = 2$ A; $I_1 = 1.33$ A; $I_2 = 0.67$ A; $I_3 = 0.33$ A
2. $I_T = 6.17$ mA; $I_1 = 2.36$ mA; $I_2 = 1.38$ mA; $I_3 = 2.43$ mA

PRACTICE PROBLEMS 4

1. *Voltage across*: $R_1 = 30$ V; $R_2 = 100$ V; $R_3 = 25$ V; $R_4 = 50$ V; $R_5 = 50$ V; $R_6 = 25$ V
   Current through: $R_1 = 3$ A; $R_2 = 0.5$ A; $R_3 = 2.5$ A; $R_4 = 0.5$ A; $R_5 = 2$ A; $R_6 = 2.5$ A
2. *Voltage across*: $R_1 = 50$ V; $R_2 = 10.7$ V; $R_3 = 8.9$ V; $R_4 = 15.2$ V; $R_5 = 4.45$ V; $R_6 = 0.72$ V; $R_7 = 1.94$ V; $R_8 = 2.0$ V; $R_9 = 2.0$ V
   Current through: $R_1 = 5$ mA; $R_2 = 2.28$ mA; $R_3 = 2.28$ mA; $R_4 = 2.72$ mA; $R_5 = 2.0$ mA; $R_6 = 0.72$ mA; $R_7 = 0.72$ mA; $R_8 = 0.36$ mA; $R_9 = 0.36$ mA

IN-PROCESS LEARNING CHECK 2

$I_1 = 0.4$ mA; $I_T = 1$ mA; $V_T = 85$ V; $R_T = 85$ kΩ; $P_{R_3} = 36$ mW
The 100-kΩ resistor (R_3) dissipates the most power.

"SPECIAL THINKING EXERCISE"

If R_2 opened:

V_T remains the same.
V_{R_1} decreases due to lower total current through R_1.
V_{R_2} increases because V_T is the same, and V_1 and V_4 decrease due to lower total current through them.
V_{R_3} increases for the same reason V_2 increases.
V_{R_4} decreases due to lower I_T caused by higher R_T.
I_T decreases due to higher R_T with same applied voltage.
I through R_1 decreases due to higher circuit total R.
I through R_2 decreases to zero due to its opening.
I through R_3 increases due to higher V across it.
I through R_4 decreases due to lower I_T through it.
P_T decreases due to lower I_T with same V_T.
P_{R_3} increases due to higher I through same R value.
P_{R_4} decreases due to lower current through same R value.

IN-PROCESS LEARNING CHECK 3

1. In Figure 6–25, the defective component is R_5. It has drastically increased in value from 3.3 kΩ to 17.3 kΩ.
2. In Figure 6–26, total circuit current should be 2 mA, meaning R_3 should drop 94 volts (2 mA × 47 kΩ volts). Instead, it is dropping 121 − 17.6 = 103.4 volts, which is higher than it should be.
   Possible bad components could include R_1 or R_2 decreased in value, or R_3 increased in value.
3. In Figure 6–27, if R_2 increases in value, V_{R_1} increases and V_{R_3} decreases. (This is due to total circuit resistance increasing, which causes a decrease in the total current through R_3, resulting in R_3 dropping less voltage. Kirchhoff's Voltage Law would indicate, therefore, that the voltage across parallel resistors R_1 and R_2 has to increase so the loop voltages equal V applied.)

PRACTICE PROBLEMS 5

Circuit diagram:

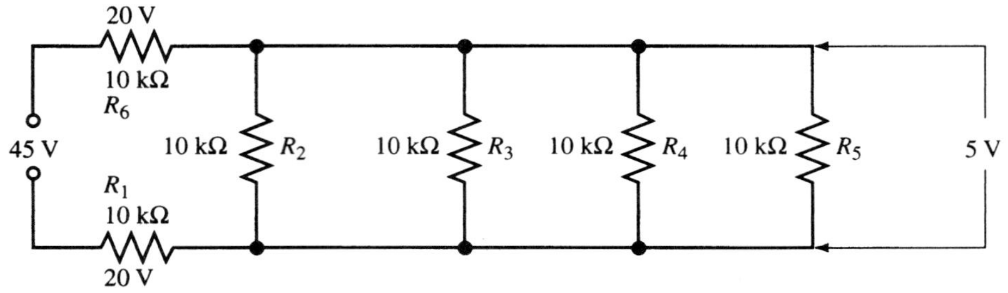

PRACTICE PROBLEMS 6

1. Circuit diagram:

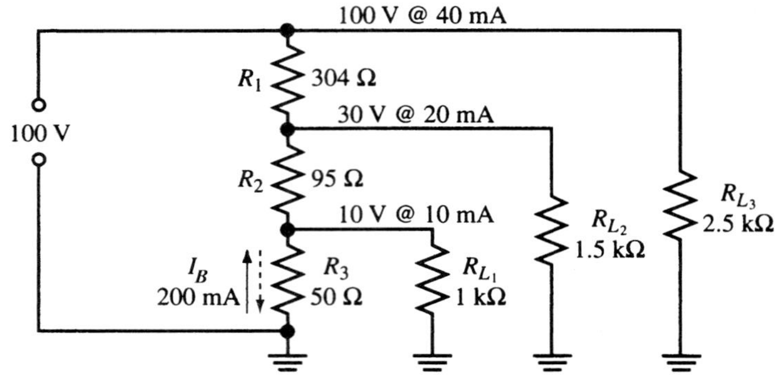

2. Load #1 = 1 kΩ — R_1 divider resistor = 304 Ω
   Load #2 = 1.5 kΩ — R_2 divider resistor = 95 Ω
   Load #3 = 2.5 kΩ — R_3 divider resistor = 50 Ω
3. See labeling of diagram in answer 1.

PRACTICE PROBLEMS 7

1. $V_{R_L} = 150$ V; $P_{R_L} = 300$ mW
2. $R_L = 75$ kΩ
3. $R_2 = 50$ kΩ
4. The change, as specified, means R_2 has decreased.
5. The change, as specified, means R_1 has increased.
6. V_{R_2} equals 166.66 V
7. If R_1 opens, V_{R_L} drops to zero volts.
   If R_2 shorts, V_{R_L} drops to zero volts.

USING EXCEL

$R_X = 100\ \Omega$

REVIEW QUESTIONS

1. A series-parallel circuit has portions that are connected in series and portions that are connected in parallel. The series portions must be analyzed using series circuit principles, and the parallel portions must be analyzed using parallel circuit principles. The results of the analysis of each type of circuit portion must be combined with results from other connected portions in order to solve for the total circuit quantities.

2. Parallel portions or branches are identified by virtue of current "splits" and by virtue of having the same (common) voltage across their extremities.
3. Series portions of series-parallel circuits are identified as those through which the same (common) current passes through the components, and by the fact that the components are in "tandem."
4. A general approach for solving for total resistance is to start at the furthest point from the power source and work backward through the circuit toward the power source, solving series and parallel combinations as you progress.
5. A general approach for solving for a series-parallel circuit's current distribution is to start at the source end of the circuit and work out.
6. (c)
7. (a)
8. (d)
9. (d)
10. (b)

PROBLEMS

1. $R_T = 22\ \Omega$
2. $I_T = 1$ A
   $V_{R_2} = 10$ V
   $V_{R_3} = 55$ V
   $V_{R_5} = 23.75$ V
   Current through R_2 (I_{R_2}) = 1 A
   Current through R_3 (I_{R_3}) = 0.5 A
3. R_T will D
   R_4 will RTS
   V_{R_1} will I
   V_{R_3} will D
   P_T will I
4. 20 mA
5. Minimum voltage is 13 V. Maximum voltage is 18 V.
6. $R_1 = 2.5\ \Omega$ (since R_T must = 10 Ω to have $I_T = 10$ A).
   $P_{R_2} = 375$ W (since V_{R_2} must equal 75 V and its $I = 5$ A).
7. $R_4 = 6$ kΩ (so that branches voltage equals 80 V).
   $I_2 = 10$ mA since $I_T = 20$ mA and I_1 can be calculated as 10 mA, due to V across the 8-kΩ resistor having to be 80 V (the 100 V applied minus the 20-V drop across the 1-kΩ resistor that has I_T passing through it).
8. 24 kΩ
9. The light gets dimmer due to the additional current that passes through the resistor in series with the source. This drops the voltage across the remaining sections of the circuit.
10. Applied voltage = 60 V. Ohm's Law will find the current through the 20-kΩ resistor with the meter reading 1 mA. Likewise, voltage across the series 18-kΩ resistor and 2-kΩ resistor must equal the same 20-V value; hence, 1 mA through that branch. Current through the 20-kΩ resistor in series with the source must be 2 mA; hence its $V = 40$ V. Using Kirchhoff's Voltage Law, V applied must equal 40 V + 20 V = 60 V.
11. a. $R_T = 169$ kΩ
    b. $I_T = 2$ mA
    c. $V_{R_1} = 14$ V
    d. $V_{R_2} = 14$ V
    e. $V_{R_3} = 14$ V
    f. $V_{R_4} = 14$ V
    g. $V_{R_5} = 62$ V
    h. $V_{R_6} = 62$ V
    i. $V_{R_7} = 138$ V
    j. $V_{R_8} = 200$ V

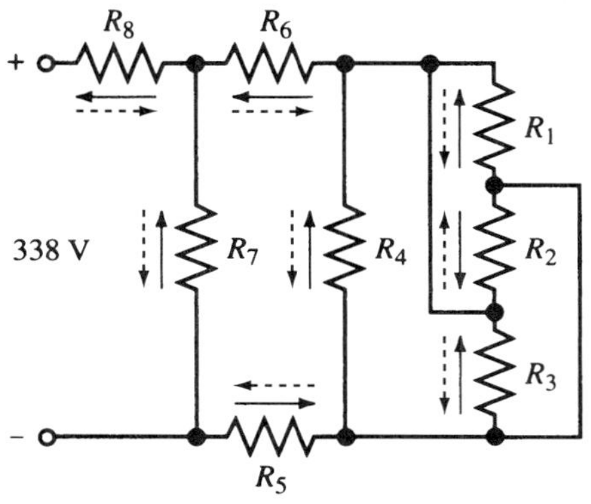

12. a. I through R_5 increases
    b. V_{R_9} will increase
    c. V_{R_2} decreases; V_{R_8} increases
    d. $R_T = 27.85$ kΩ
    e. $V_{R_1} \cong 14.25$ V
    f. $P_{R_9} = 250$ mW
    g. R_6 and R_9 (most power)
    h. R_1 and R_2 (least power)
    i. R_6 and $R_9 = 4$ W; R_1 and $R_2 = 81.23$ mW
13. Prior to connecting voltmeter, $V_{R_1} = 5$ V.
14. Prior to connecting voltmeter, $V_{R_2} = 5$ V.
15. Approximately 4 volts
16. Approximately 6 volts
17. Yes, connecting the meter changes R_T and I_T. A higher I_T will pass through R_1, causing its voltage to increase from 5 V to 6 V. Less voltage will be dropped across R_2 than was present before the meter was connected.
18. If the meter were connected across R_1 rather than R_2, the reverse conditions would occur, as described in problem 17. That is, R_2 would drop about 6 volts, and R_1 would drop about 4 volts. However, the total current would increase by the same amount in either of the two positions described for meter connection.
19. Total current with the meter across one resistor = 60 mA.
20. Without the meter connected, the $I_T = 50$ mA.
21. If the meter were connected from the bottom of R_2 to the top of R_1, the voltages across the two resistors would not be affected.
22. $R_{L_1} = 100$ kΩ; $R_{L_2} = 150$ kΩ; $R_{L_3} = 20$ kΩ
23. $R_3 = 33.3$ kΩ; $R_2 = 12.5$ kΩ; $R_1 = 10$ kΩ
24. Loaded
25. R_1 would have to be decreased in value.
26. R_1's new value would be 3.57 kΩ.
27. Bleeder current would become 5.08 mA.
28. Minimum rating should be 500 mW.
29. Minimum rating should exceed 400 mW (R_2).
30. Minimum rating should exceed 600 mW.
31. P_T being supplied is 1.5 mA × 200 V = 3,000 mW or 3 W.

ANALYSIS QUESTIONS

1. If R_2 decreased greatly in value:
   a. Voltage across R_3 would I (increase).
   b. Voltage across R_1 would I (increase).
   c. Current through load connected across R_4 would I (increase).
2. If R_2 opened:
   a. Voltage across load connected to R_1 would increase.
   b. The voltage across R_1 would decrease (since R_T has increased and I_T has decreased).
3. If the sensor is connected across a balanced bridge circuit, when the bridge is unbalanced by a change in conditions the sensor's output can then be fed to control elements, which respond to the changed parameter across the sensor, and make appropriate system corrections, based on this feedback.
4. $R_1 = 0.5$ kΩ
   $R_2 = 0.5$ kΩ
   $R_3 = 0.75$ kΩ
   $R_4 = 6$ kΩ

(**NOTE:** Current through R_4 = 10 mA; through R_3 = 20 mA; through R_2 = 30 mA; through R_1 = 60 mA. R_1 drops 30 V with 60 mA through it, therefore = 0.5 kΩ; R_2 drops 15 V w/30 mA through it, therefore = 0.5 kΩ; R_3 drops 15 V w/20 mA through it, therefore = 0.75 kΩ; and R_4 drops 60 V w/10 mA through it, therefore = 6 kΩ.)

5. R_X = 200 kΩ
6. R_X = 57.45 kΩ
7. 0.0555:1 (R_1:R_2) or 18:1 (R_2:R_1)
8. R_2 has the greater voltage drop.
9. First, set the calculator to the engineering mode. Next, put in one resistor's value, press the $1/x$ key, then the plus (+) key. Then, put in the second resistor's value, press the $1/x$ key, then the + key. Then, input the third resistor's value, press the $1/x$ key, then press the equal (=) key. Press the $1/x$ key, one more time, for the answer.
10. The voltmeter places a resistance value less than infinity in parallel with the portion of the circuit across which it is connected. This lowers the equivalent resistance of that portion of the circuit. This lowering of resistance, then causes other parameters in the circuit to change and respond accordingly.

MULTISIM EXERCISE 1

24 kΩ (see Problems question 8)

MULTISIM EXERCISE 2

2. 15 mA
3. 20 mA
4. 20 mA; point A; R_T = minimum under this condition
5. Yes

CHAPTER 7

PRACTICE PROBLEMS 1

$I = 4$ A

$V_L = 16$ V

$P_T = 80$ W

$P_L = 64$ W

Efficiency = 80%

PRACTICE PROBLEMS 2

1. I through R_1 = 0 mA
   V across R_1 with respect to A = 0 V
   I through R_2 = 7.5 mA
   V across R_2 with respect to A = +75 V
   I through R_3 = 7.5 mA
   V across R_3 with respect to A = −75 V
2. I through R_1 = 10 mA
   V across R_1 with respect to A = +100 V
   I through R_2 = 12.5 mA
   V across R_2 with respect to A = −125 V
   I through R_3 = 2.5 mA
   V across R_3 with respect to A = +25 V

IN-PROCESS LEARNING CHECK 1

1. For maximum power transfer to occur between source and load, the load resistance should *equal* the source resistance.

2. The higher the efficiency of power transfer from source to load, the *greater* the percentage of total power that is dissipated by the load.
3. Maximum power transfer occurs at *50%* efficiency.
4. If the load resistance is less than the source resistance, efficiency is *less* than the efficiency at maximum power transfer.
5. To analyze a circuit having two sources, the superposition theorem indicates that Ohm's Law *can* be used.
6. What are two key observations needed in using the superposition theorem to analyze a circuit with more than one source? *Noting the direction of current flow and polarity of voltage drops.*
7. Using the superposition theorem, if the sources are considered "voltage" sources, are these sources considered shorted or opened during the analysis process? *Shorted.*
8. When using the superposition theorem when determining the final result of your analysis, the calculated parameters are combined, or superimposed, *algebraically*.

PRACTICE PROBLEMS 3

1. If $R_L = 175\ \Omega$, $I_L = 0.25$ A, and $V_L = 43.75$ V (Figure 7–10).
2. If $I_L = 125$ mA, then $R_L = 375\ \Omega$.
3. $I_L = 1.16$ mA; $V_{R_L} = 0.55$ V
4. In Figure 7–13, $R_{TH} = 4$ kΩ and $V_{TH} = 50$ V.
   If $R_L = 16$ kΩ, then $V_L = 40$ V, and $I_L = 2.5$ mA.

PRACTICE PROBLEMS 4

If R_L is 60 Ω, $I_L = 0.588$ A, and $V_L = 35.28$ V.

PRACTICE PROBLEMS 5

1. If $R_L = 100\ \Omega$ in Figure 7–17, $I_L = 0.4$ A, and $V_L = 40$ V.
2. If R_L changes from 25 Ω to 50 Ω, I_L decreases (from 1 A to 0.66 A), and V_L increases (from 25 V to 33 V).

USING EXCEL

1. $I_L = 0.694$ A
2. $V_L = 32.6$ V

REVIEW QUESTIONS

1. Matching the impedance of an antenna system to the impedance of a transmitter output stage, and matching the output impedance of an audio amplifier to the impedance of a speaker.
2. 50% efficiency
3. $P_{out}/P_{in} \times 100$
4. A bilateral linear network is one that has equal R in either direction, which does not change the way it behaves with changing values of V or I.
5. To "Thevenize" a circuit:
   a. Open or remove R_L.
   b. Determine the open circuit V at points where R_L is to be connected. (This value is V_{TH}.)
   c. Determine the resistance looking toward the source from the R_L connection points, assuming the source is shorted. (This value is R_{TH}.)
   d. Draw the Thevenin equivalent circuit with V_{TH} as the source and R_{TH} in series with the source.
   e. Calculate I_L and V_L for given values of R_L, using a two-resistor series circuit analysis technique.

6. To "Nortonize" a circuit:
   a. Determine the I_N value by assuming R_L = zero ohms.
   b. Find the value of R_N as being the R, looking back from open R_L terminals toward the source, assuming the source is shorted.
   c. Draw the Norton equivalent circuit as a current source = to I_N and a resistance in parallel with the source = to R_N.
   d. Complete R_L parameters using parallel circuit analysis techniques.
7. The most important application of the superposition theorem is for analyzing circuits having more than one source.
8. (a)
9. (c)
10. (b)
11. (d)
12. (a)

PROBLEMS

1. Zero volts
2. $V_{R_1} = 20$ V
   $V_{R_2} = 20$ V
3. $V_{TH} = 40.5$ V (voltage at points A and B with R_L removed)
   $R_{TH} = 14.85$ kΩ (R at points A and B with source shorted)
   $I_L = 1.63$ mA (V_L/R_L in Thevenin equivalent circuit)
   $V_L = 16.3$ V (from simple series Thevenin equivalent circuit analysis)
4. $V_{TH} = 40.5$ V; $I_L = 0.968$ mA
   $R_{TH} = 14.85$ kΩ; $V_L = 26.14$ V
5. The Norton equivalent circuit is a constant current source with a parallel resistance, called R_N, and the load (R_L) in parallel with R_N and the source. To convert parameters:
   $R_N = R_{TH} = 14.85$ kΩ
   I_N (constant current source) = $V_{TH}/R_{TH} = 40.5$ V/14.85 kΩ = 2.73 mA
   The circuit is shown.

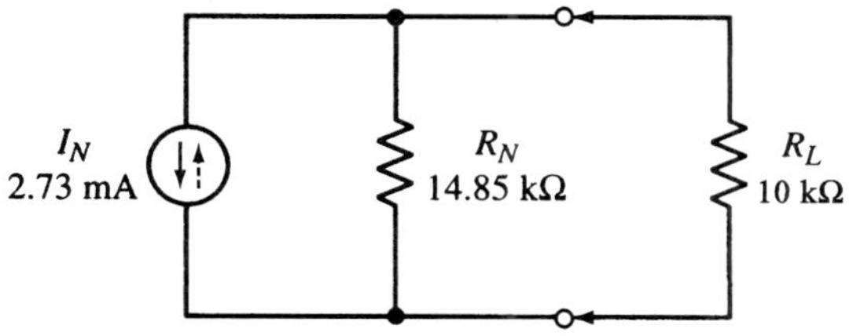

6.

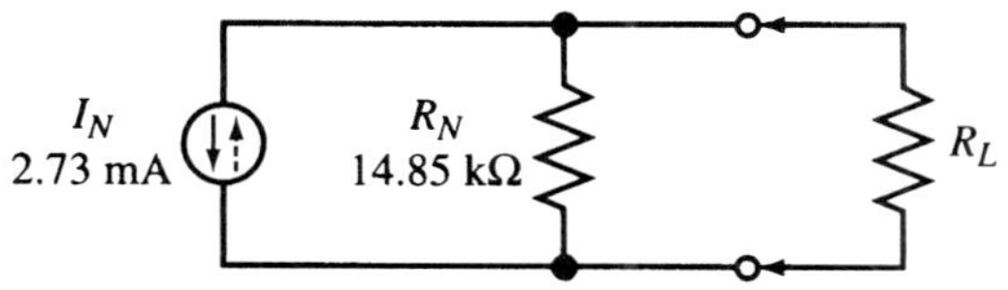

7. The maximum power that can be delivered to the load in Figure 7–22 is 1.25 W.
8. 1 Ω
9. $R_{TH} = 29.49$ Ω; $V_{TH} = 34.46$ V
10. $R_N = 3.67$ Ω; $I_N = 10$ A
11. $R_{TH} = 10$ Ω; $V_{TH} = I_N \times R_N = 3\text{ A} \times 10\ \Omega = 30$ V

12. $R_N = 15\ \Omega$; $I_N = \frac{V_{TH}}{R_{TH}} = \frac{25\text{ V}}{15\ \Omega} = 1.67\text{ A}$

13. 83.2%
14. Yes. No.
15. 5 Ω

ANALYSIS QUESTIONS

1. It is important for a voltage source to have as low an internal resistance (r_{int}) as possible in order that source output voltage changes will be minimized as current demands for the connected circuit change. Ideally, a voltage source would have zero internal resistance and, thus, would not have any internal voltage drop for any current demand level.
2. This is to enable maximum transfer of power from source to load.
3. Load resistance should be decreased.
4. Thevenin's theorem
5. Norton's theorem
6. a. The reason for carefully observing the direction of current flow when using the superposition theorem is to provide proper assignment of polarities. This allows for correct algebraic addition of the parameter results when combining the circuit actions of each source that is considered independently.
   b. Components common to both sources
7. R_{TH} is the fixed-value, Thevenin equivalent-circuit series resistor considered to be in series with whatever the R_L value is, for calculation purposes. Finding this value enables using a simple two-resistor series circuit analysis in calculating circuit parameters for various values of R_L. R_{TH} is determined by looking back into the network at the points R_L will be connected and considering the voltage source to be shorted (or zero ohms).
8. V_{TH} is the open-terminal voltage felt across the two load terminals (with R_L removed). To find V_{TH}, Ohm's Law or the voltage-divider rule is used.
9. R_N is the fixed-value, Norton equivalent-circuit parallel resistor considered to be in parallel with the "constant current" source and R_L. R_N (like R_{TH}) is also found by looking back into the network at the points R_L is to be connected and considering the voltage source as being replaced by its internal R (r_{int}), or zero ohms, in this case. Knowing the R_N value (along with the equivalent circuit I_N value) enables reducing circuit analysis to a simple two-resistor, parallel circuit, current-division approach.
10. I_N is the current available from the Norton equivalent "current source." I_N is found by assuming that R_L was shorted and determining the current that would flow through the zero-ohm R_L portion of the circuit.
11. a. Various answers possible (e.g., some IC circuits).
    b. Superposition theorem
12. For the circuit of Figure 7–23:

    To find R_{TH}:

    Remove the power supply, and replace it with a jumper wire short between the points to which the power supply was previously connected. Use an ohmmeter and read the R value between points A and B. This = R_{TH}.

    To find V_{TH}:

    Reconnect the source, as appropriate. Use a voltmeter and measure the voltage between points A and B. This = V_{TH}.
13. Yes, it is possible to measure the circuit R_N. Since this is the same circuit (Figure 7–23) and R_N is determined from the same points, under the same conditions, you would use the same technique as described to measure R_{TH} in the previous question.

14. To determine the r_{int} of a voltage source, measure the no-load voltage and the full-load voltage from the source. Then, use the formula:

$$r_{int} = \frac{V\text{ (no load)} - V\text{ (loaded)}}{I\text{ (load)}}$$

15. Battery resistance would equal: 12 V – 11 V/50 A; 1 V/50 A = 0.02 Ω

MULTISIM EXERCISE 1

2. $V_{R_1} = 40$ V; $I_{R_2} = 5$ mA
3. Yes

MULTISIM EXERCISE 2

2. $V_{TH} = 50$ V; $R_{TH} = 25\ \Omega$
3. Yes
4. $V_L = 45.7$ V; $I_L = 169$ mA

CHAPTER 8

PRACTICE PROBLEMS 1

1. $I_A = 0.55$ A
2. $I_B = 0.304$ A
3. $V_{R_1} = 9.6$ V
4. $V_{R_2} = 8.25$ V
5. $V_{R_3} = 12.1$ V
6. $V_{R_4} = 8.25$ V
7. $V_{R_5} = 11.75$ V
8. V_{A-B} = close to 0 V

Considered from points *A* and *B* with respect to the opposite ends of resistors R_2 and R_4, respectively: Point *A* is +8.25 V with respect to the common connection points of R_2 and R_4, and point *B* is +8.25 V with respect to the common connection points of R_2 and R_4. Thus, point *A* and point *B* are at the same potential.

Considered from points *A* and *B* with respect to the opposite ends of R_3 and R_5, respectively: Point *A* is –12.1 V with respect to the common connection points of R_3 and R_5, and point *B* is –11.75 V with respect to the common connection points of R_3 and R_5. Therefore, point *A* is –0.6 V with respect to point *B*, or point *B* is +0.6 V with respect to point *A*.

(**NOTE:** Variance is noticeable because the math was not carried out to a large number of decimal places.)

PRACTICE PROBLEMS 2

1. $I_A = 0.714$ A
2. $I_B = 0.143$ A
3. $V_{R_1} = 7.14$ V (current through $R_1 = 0.714$ A)
4. $V_{R_2} = 12.85$ V (current through $R_2 = 0.857$ A)
5. $V_{R_3} = 2.145$ V (current through $R_3 = 0.143$ A)

IN-PROCESS LEARNING CHECK 1

1. A loop is considered to be a closed, or complete *path* within a circuit.
2. In the loop or mesh approach, it is important to label the assumed *mesh* currents on the diagram you use for analysis.

3. When assigning polarities to sources in the Kirchhoff equations, a source voltage is considered *positive* if the assumed mesh current enters the positive terminal of the source.
4. When writing Kirchhoff equations for each loop's voltage drops, the component voltage is considered *positive* when caused by its own mesh current, and *negative* when caused by an adjacent mesh's current flow.

PRACTICE PROBLEMS 3

1. $V_{R_1} = 18.5$ V
2. $V_{R_2} = 41.5$ V
3. $V_{R_3} = 6.5$ V
4. Current through $R_1 = 0.925$ A
5. Current through $R_2 = 4.15$ A
6. Current through $R_3 = 3.25$ A

IN-PROCESS LEARNING CHECK 2

1. A node is a *current* junction point for two or more components in a circuit.
2. A node where three or more components are joined is sometimes called a *major node.*
3. For nodal analysis, one node is chosen as the *reference node.*
4. In nodal analysis, a Kirchhoff *current* law equation is written for each major node, with the exception of the reference point node.

PRACTICE PROBLEMS 4

1. $R_A = 31.82\ \Omega$; $R_B = 70\ \Omega$; $R_C = 46.67\ \Omega$
2. $R_1 = 22\ \Omega$; $R_2 = 10\ \Omega$; $R_3 = 15\ \Omega$
3. Yes

IN-PROCESS LEARNING CHECK 3

1. Because of their visual shapes and electrical connections:
   a. The wye circuit configuration is sometimes also called the *T (tee)* configuration.
   b. The delta circuit configuration is sometimes also called the *π (pi)* configuration.
2. Delta and wye circuit configurations are often found in *AC (or electrical)* power applications.
3. The value in knowing how to determine equivalent circuits, or convert from delta-to-wye and wye-to-delta circuits is that the equivalent circuits will present the same resistance and draw the same *current* from the source feeding the circuit.
4. The voltage between connection points in wye and delta configurations *is not* the same.

PRACTICE PROBLEMS 5

($R_1 = 7\ \Omega$; $R_2 = 3.5\ \Omega$; $R_3 = 2.33\ \Omega$)

1. $R_T = (R_D + R_2) \parallel (R_E + R_3) + R_1$

   $R_T = 13.44\ \Omega$
2. $I_T = V_T/R_T = 84\ \text{V}/13.44\ \Omega = 6.25$ A

USING EXCEL

1. $R_A = 51.6\ \Omega$
2. $R_1 = 3.33\ \Omega$

REVIEW QUESTIONS

1. A loop is a complete path in a circuit, from one side of the source, through a series of circuit components, back to the other side of the source.
2. A mesh is an unbroken, closed loop within a network.

3. Mesh currents are "assumed" currents (and current directions) used in conjunction with Kirchhoff's voltage law to help in analyzing networks. Mesh currents are different from actual currents in that they are assumed not to branch.

4. (b)
5. (c)
6. (a)
7. (a)
8. (b)
9. (a)
10. (b)
11. (b)
12. (c)

PROBLEMS

1. $I_A = 0.8$ A
2. $I_B = 0.4$ A
3. $V_{R_1} = 8$ V
4. $V_{R_2} = 12$ V
5. $V_{R_3} = 4$ V
6. $I_{R_1} = 0.8$ A
7. $I_{R_2} = 1.2$ A
8. $I_{R_3} = 0.4$ A

Solution:

Equation 1: $20\,I_A - 10\,I_B = 20$

Equation 2: $-10\,I_A + 20\,I_B = 16$

Simultaneous equations:

$$\begin{array}{lll} 40\,I_A & -20\,I_B & = 40 \\ -10\,I_A & +20\,I_B & = 16 \\ \hline 30\,I_A & & = 24 \\ I_A & & = 0.8\text{A} \\ & I_B & = 0.4 \end{array}$$

$V_{R_1} = 0.8 \times 10 = 8$ V

$V_{R_2} = (0.8 + 0.4) \times 10 = 12$ V

$V_{R_3} = 0.4 \times 10 = 4$ V

Verification:

$V_{R_1} + V_{R_2} = 8\text{ V} + 12\text{ V} = 20$ V

$V_{R_2} + V_{R_3} = 12\text{ V} + 4\text{ V} = 16$ V

9. I_1 (current through R_1) = 1.98 A
10. I_2 (current through R_3) = 0.498 A
11. I_3 (current through R_2) = 2.48 A
12. $V_{R_1} = 9.34$ V
13. $V_{R_2} = 11.66$ V
14. $V_{R_3} = 2.34$ V

Solution:

$(V_{R_1}/R_1) + (V_{R_3}/R_3) = V_{R_2}/R_2$

$V_{R_1} + V_{R_2} = 21$ V

$V_{R_1} = 21 - V_{R_2}$

$V_{R_3} = 14 - V_{R_2}$

$(21 - V_{R_2}/4.7) + (14 - V_{R_2}/4.7) = V_{R_2}/4.7$

Multiplying each term by 4.7 to remove the denominators:

$21 - V_{R_2} + 14 - V_{R_2} = V_{R_2}$

$35 = 3\ V_{R_2}$

$V_{R_2} = 11.66$ V

$V_{R_1} = 21 - 11.66 = 9.34$ V

$V_{R_3} = 14 - 11.66 = 2.34$ V

I_1 (current through $R_1 = 9.34/4.7 = 1.98$ A)

I_2 (current through $R_3 = 0.498$ A)

I_3 (current through $R_2 = 2.48$ A)

Verification:

$V_{R_1} + V_{R_2} = V_{S_1}$

$9.34 + 11.66 = 21$

$V_{R_2} + V_{R_3} = V_{S_2}$

$11.66 + 2.34 = 14$

15. $R_A = 45\ \Omega$
16. $R_B = 37.5\ \Omega$
17. $R_C = 30\ \Omega$
18. $R_1 = R_2 = R_3 = 50\ \Omega$ (for symmetrical same-value R's wye R's $= \frac{1}{3}$ delta R values)

ANALYSIS QUESTIONS

1. Mesh or nodal analysis techniques are used in situations where basic series and parallel analysis techniques might be cumbersome or very difficult.
2. Typically, one equation is written for each mesh; therefore for a three-mesh circuit there would be three equations.
3. In industrial settings, you might find delta and wye circuit configurations related to three-phase power, three-phase motors, etc.
4. It is very common to find pi circuit configurations in the power supply circuitry of many electronic devices and systems.

MULTISIM EXERCISE

2. a. $V_{R_1} = 4.62$ V
   b. $V_{R_2} = 10.38$ V
   c. $V_{R_3} = 1.62$ V
3. Yes

CHAPTER 9

IN-PROCESS LEARNING CHECK 1

1. Small magnets that are suspended and free to move align in a *North* and *South* direction.
2. Materials that lose their magnetism after the magnetizing force is removed are called *temporary magnets*. Materials that retain their magnetism after the magnetizing force is removed are called *permanent magnets*.
3. A wire carrying dc current *does* establish a magnetic field. Once a magnetic field is established, and if the current level doesn't change, the field is *stationary*.
4. A law of magnetism is that like poles *repel* each other and unlike poles *attract* each other.
5. A maxwell represents *one* line of force.
6. A weber represents *10^8* lines of force.
7. Lines of force are *continuous*.
8. Lines of force related to magnets exit the *North* pole of the magnet and enter the *South* pole.

9. Nonmagnetic materials do *not* stop the flow of magnetic flux through themselves.
10. Yes, magnetism *can* be "induced" from one magnetic object to another object, if it is a magnetic material.

PRACTICE PROBLEMS 1

1. Drawing:
2. See arrows on drawing.

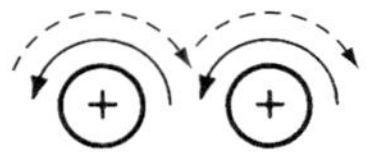

PRACTICE PROBLEMS 2

Figure 9–13a = Move away from each other.

Figure 9–13b = Move toward one another.

Figure 9–13c = Move away from each other.

PRACTICE PROBLEMS 3

Figure 9–17a = N at right end, S at left end.

Figure 9–17b = N at left end, S at right end.

Figure 9–17c = N at left end, S at right end.

IN-PROCESS LEARNING CHECK 2

1. For a given coil dimension and core material, what two factors primarily affect the strength of an electromagnet? *Number of turns* and *amount of current.*
2. The left-hand rule (*right-hand rule*) for determining the polarity of electromagnets states that when the fingers of your left (*right*) hand point *in the same direction* as the current passing through the coil, the thumb points toward the *North* pole of the electromagnet.
3. Adjacent current-carrying conductors which are carrying current in the same direction tend to *attract each other.*
4. If you grasp a current-carrying conductor so that the thumb of your left hand (*right hand*) is in the direction of current through the conductor, the fingers "curled around the conductor" *will indicate* the direction of the magnetic field around the conductor.
5. When representing an end view of a current-carrying conductor pictorially, it is common to show current coming out of the paper via a *dot*.

IN-PROCESS LEARNING CHECK 3

1. A *B-H* curve is also known as a *magnetization* curve.
2. *B* stands for flux *density*.
3. *H* stands for magnetizing *intensity*.
4. The point where increasing current through a coil causes no further significant increase in flux density is called *saturation*.
5. The larger the area inside a "hysteresis loop" the *larger* the magnetic losses represented.

USING EXCEL

$V_{\text{ind}} = 625$ V

REVIEW QUESTIONS

1. (b)
2. (a)
3. (b)
4. (d)
5. (d)

6. –Lines of force are continuous.
   –Lines of force are assumed to flow from N to S, external to the magnet.
   –Lines of force attempt to take the shortest or easiest path.
   –Lines of force in the same direction repel each other, and in opposite directions attract (or cancel) each other.
   –Lines of force penetrate nonmagnetic materials.
   –Lines of force do not cross each other.
7. Permeability
8. Reluctance
9. ϕ, weber
10. B, webers per square meter, or tesla
11. mmf, ampere-turns (NI)
12. H, ampere-turns per meter
13. 10^8
14. The magnetic path cannot contain any more lines of flux.
15. The shield simply diverts lines of flux through itself, preventing flux from reaching the items inside the shield enclosure.
16. A Gaussmeter measures flux density.
17. Drawing of a hysteresis loop, with callouts for saturation, residual magnetism, and coercive force:

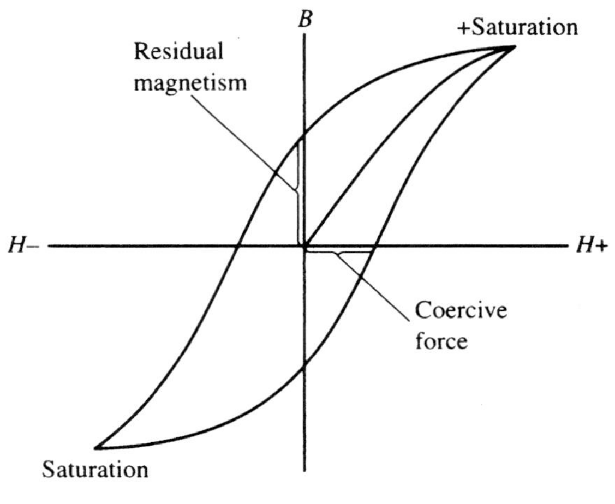

18. (d)
19. (b)
20. (c)
21. The direction of an induced voltage or current is such that it tends to oppose the change causing it.
22. The amount of induced emf depends on the rate of cutting flux lines (i.e., the number of flux lines cut per unit of time).
23. Motor effect in a generator relates to the induced current causing an opposition to the motion that induced it.
24. Generator effect in a motor relates to the induced current being in the opposite direction from the current causing the motor's rotation.
25. Toroidal coil forms are efficient conductors of magnetic flux due to the high-permeability materials used in them, and due to their geometry not allowing much leakage flux.

PROBLEMS

1. Field patterns are as shown.

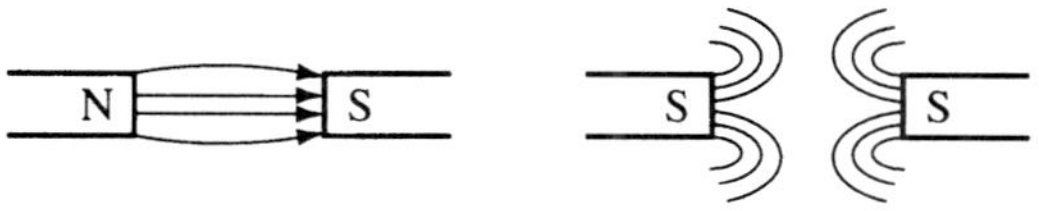

2. Field lines around current-carrying conductors are shown.

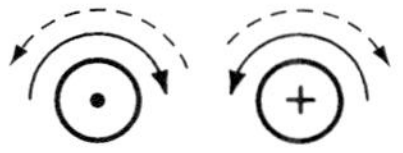

3. North and South poles are indicated.

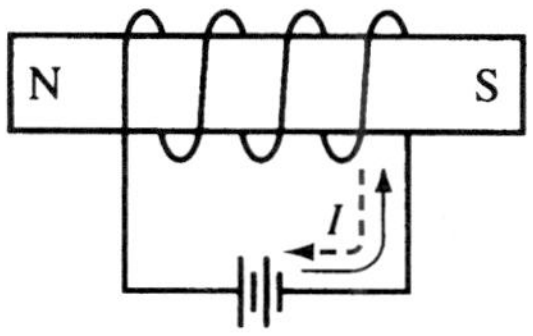

4. a. $10^8/10^6 = 100$ lines of flux

   b. Teslas = webers per square meter. Since there are 100 lines in 0.005 m^2, there are $1/0.005 \times 100$ lines in a square meter, or lines $= 200 \times 100 = 20{,}000$.

   Number of webers per square meter: teslas $= 20{,}000/100{,}000{,}000 = 0.0002$ T

5. $V_{ind} = \Delta$ flux (Wb)/Δ t (seconds) $\times$ number of turns

   $V_{ind} = 10 \times 10^{-6}/1 \times 500 = 5{,}000\ \mu V$

6. $H = 3 \times 1{,}500/1 = 4{,}500$ ampere-turns per meter

7. Lines $= 4 \times 10^8 = 400{,}000{,}000$

CHAPTER 10

REVIEW QUESTIONS

1. Two advantages of DMMs are that they are easy to read, and they produce less meter "loading effect" on the circuit under test.
2. (b)
3. (c)
4. Power off, break circuit, insert meter (observing polarity), power on, read meter.
5. Do not have to break circuit to make measurement.
6. For certain adjustments and tuning procedures on tuned circuits.
7. (c)
8. (b)
9. (c)

ANALYSIS QUESTIONS

1. A current meter should have low resistance so as to minimize changing the circuit under the test's resistance value, since the current meter is placed in series with the circuit to be tested.

2. Drawing of a three-resistor parallel circuit with dc source, showing how current meters should be connected to measure I_T and the current through the middle branch:

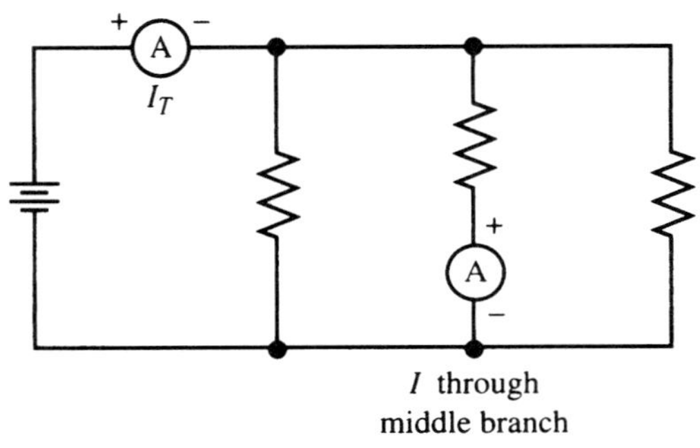

3. A voltmeter should have high resistance so as to not "load" the circuit under test and change its parameters. Since the meter is connected in parallel with the circuit being tested, the higher the meter's *R*, the better.

CHAPTER 11

IN-PROCESS LEARNING CHECK 1

1. The basic difference is that *dc* (direct current) *is in one direction and of one polarity*. Alternating current (*ac*) *periodically changes direction and polarity.*
2. The reference, or zero-degree position, is *horizontally to the right* when describing angular motion.
3. The y-axis is the *vertical axis* in the coordinate system.
4. The second quadrant in the coordinate system is that quadrant between *90 and 180 degrees.*
5. A *vector* represents a given quantity's magnitude and direction with respect to location in space.
6. A *phasor* is a rotating vector representation showing relative position with respect to time.

PRACTICE PROBLEMS 1

1. The sine of 35 degrees = 0.5736.
2. Cycle 1 ends and cycle 2 begins at point *G*.
3. Maximum rate of change for the waveform shown occurs at points *J, N, R,* and *T.*

IN-PROCESS LEARNING CHECK 2

1. 10 kHz (10,000 Hz)
2. 0.0025 seconds, or 2.5 msec
3. 50 µsec
4. Amplitude
5. 45 and 135 degrees; 225 and 315 degrees
6. 1 and 2
7. 0.067 µsec
8. 40 Hz
9. As frequency increases, *T decreases.*
10. The longer a given signal's period, the *longer* the time for each alternation.

PRACTICE PROBLEMS 2

1. 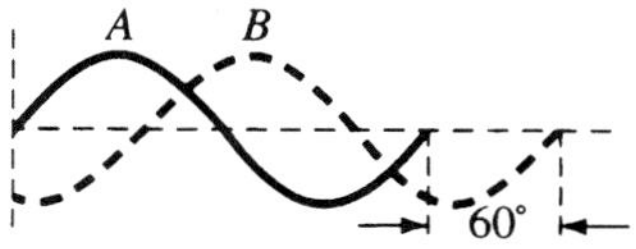

2. 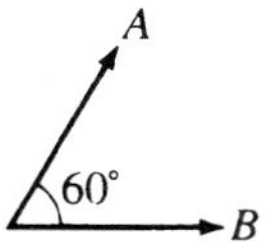

PRACTICE PROBLEMS 3

1. 10 V
2. 9 V
3. 90%
4. No
5. Yes
6. a. 120.8 V
   b. 171 V
   c. 120.8 V
   d. 171 V
   e. 108.9 V

PRACTICE PROBLEMS 4

1. a. V_{R_1} = 50 V
   b. V_{R_2} = 50 V
   c. I_p = 7.07 mA
   d. P_p = 999.7 mW
   e. P_{R_1} = 250 mW
2. a. I_1 = 10 mA
   b. I_2 = 10 mA
   c. I_p = 14.14 mA
   d. $V_{p\text{-}p}$ = 282.8 V
3. a. 66.7 V
   b. 33.3 V
   c. 33.3 V
   d. 444 mW

PRACTICE PROBLEMS 5

The duty cycle = Pulse width/Time for one period times 100

Duty cycle = (1 μs/15 μs) × 100 = *6.67% duty cycle*

PRACTICE PROBLEMS 6

1. V_{avg} = Baseline value + (duty cycle × pk-pk amplitude)
   V_{avg} = 0 + (pulse width × pk-pk amplitude)
   $V_{avg} = 0 + [(2 \text{ ms}/30 \text{ ms}) \times 6 \text{ V}] = 0 + 0.066 \times 6 \cong 0.4 \text{ V}$
2. −2.604 V

USING EXCEL

1. f = 400 Hz
2. V_{inst} = 128.56 V

REVIEW QUESTIONS

1. (d)
2. (b)
3. (b)
4. (b)
5. (b)
6. (c)
7. (c)
8. (b)
9. (b)
10. (c)
11. (a)
12. (b)
13. (c)
14. (c)
15. (c)

16. Fundamental means the frequency of the primary signal. Harmonic means a multiple of the fundamental frequency. That is, a second harmonic of the fundamental signal would be two times the fundamental frequency; a third harmonic would be three times, and so on.
17. A periodic wave is one that repeats itself in time and form.
18. (a)
19. (b)
20. (c)

PROBLEMS

1. $T = 1/f = 1/2{,}000 = 0.0005$ seconds, or 0.5 msec
2. $f = 1/T = 1/0.2$ μsec = 5 MHz
3. One one-hundred-twentieth second
4. 120 V
5. 400 Hz
6. Pk-to-pk voltage is approximately 63 volts.
($P = V^2/R$; 50 mW = $V^2/10$ kΩ; 500 = V^2; 22.36 = rms V; rms × 2.828 = pk-to-pk V; 22.36 × 2.828 = 63.24 V)
7. If the rms value of an ac voltage doubles, the peak-to-peak value must *double*.
8. Power will quadruple.
9. Frequency has been reduced to one-third the original frequency.
10. If rms $V = 75$ volts, peak value = 1.414 × 75 = 106 V. Thus, peak-to-peak value equals 212 V.
11. Time is one-half the time of a period, where $T = 1/f = 1/500 = 0.002$ seconds. Therefore, one alternation equals 0.001 seconds, or 1 millisecond.
12. The frequency = 1,000 Hz. Two periods = 2,000 μs.
13. Diagram:

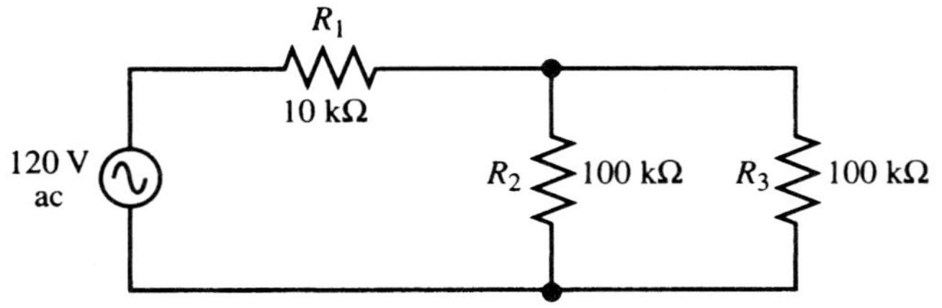

a. $V_{R_1} = 20$ V
b. $I_{R_1} = 2$ mA
c. $P_{R_1} = 40$ mW
d. $I_{R_2} = 1$ mA
e. $V_{R_3} = 100$ V
f. θ between V applied and I total = 0°.

14. λ = 300 meters
15. λ will increase to three times the original.

ANALYSIS QUESTIONS

1. Three differences between ac and dc are: (1) ac changes polarity periodically, dc does not; (2) ac continuously varies in amplitude, dc does not; (3) ac can be stepped up or down via transformer action, dc cannot.

2. Drawing of waveform with descriptors:

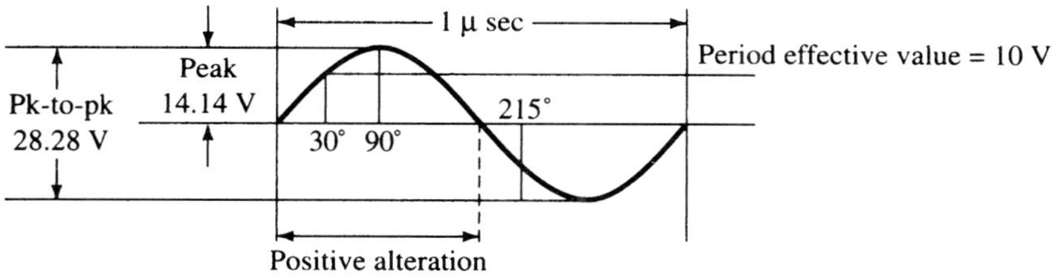

3. If rms value = 100 V, then peak value = 141 V and peak-to-peak value = 282 V.
4. B
5. 70.7 V
6. 14.14 V
7. Since A = $1/0.002 \times 10^{-3}$ = 500 kHz, then B = 250 kHz.
8. 21.2 V
9. a. Graphic drawing of rectangular wave:

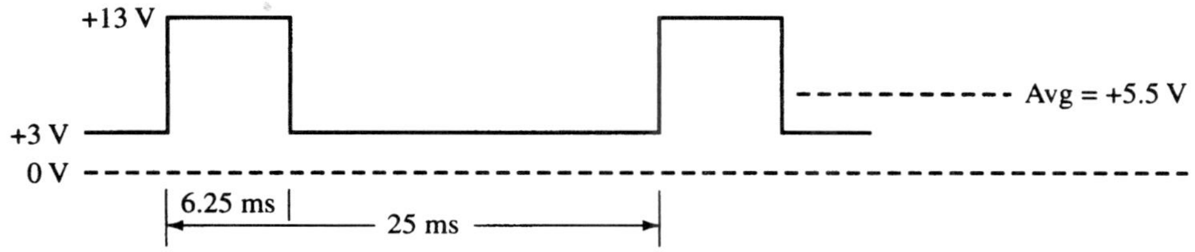

   b. The V_{avg} = Baseline value + (duty cycle × pk-pk amplitude)
   V_{avg} = 3 + (pulse width/period × pk-pk amplitude)
   $V_{avg} = 3 + [(6.25 \text{ ms}/25 \text{ ms}) \times 10 \text{ V}] = 3 + 0.25 \times 10 = 5.5 \text{ V}$
10. Diagram of sine wave and cosine wave:

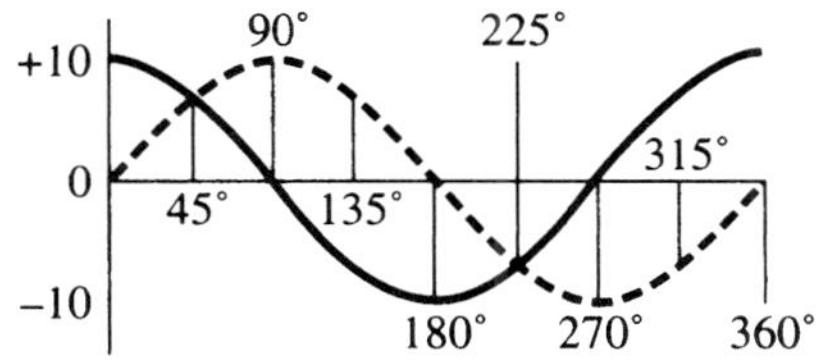

   a. See diagram.
   b. Amplitude of sine wave at 45 degrees = +7.07 units; at 90 degrees = +10 units
   c. Amplitude of cosine wave at 0° = +10 units; at 45° = +7.07 units; at 90° = 0 units
   d. Cosine wave is leading sine wave by 90°

CHAPTER 12

IN-PROCESS LEARNING CHECK 1

1. The part of the oscilloscope producing the visual display is the *cathode-ray tube.*
2. The scope control that influences the brightness of the display is the *intensity* control.
3. A waveform is moved up or down on the screen by using the *vertical position* control.
4. A waveform is moved left or right on the screen by using the *horizontal position* control.

5. For a signal fed to the scope's vertical input, the controls that adjust the number of cycles seen on the screen are the horizontal *time/div* (or *sec/div*) control and the horizontal *time/var* (or *Var*) control.
6. The control(s) that help keep the waveform from moving or jiggling on the display are associated with the *synchronization* or *trigger* circuitry.

PRACTICE PROBLEMS 1

When the vertical frequency is four times the horizontal frequency, the waveform shows four cycles of the given waveform.

PRACTICE PROBLEMS 2

1. 30 V (pk-to-pk); 10.6 V (rms)
2. 1.76 V/div (rms)
3. Use the 10 position setting of the vertical volts/div control.

PRACTICE PROBLEMS 3

1. 1.5 V (pk-to-pk); 0.53 V (rms)
2. 3 V (pk-to-pk); 1.06 V (rms)

PRACTICE PROBLEMS 4

72 degrees per major division

PRACTICE PROBLEMS 5

1. a. Period (T) = 0.25 s (seconds)
   b. Frequency = 4 Hz
2. a. Period (T) = 50 μs
   b. Frequency = 20 kHz

REVIEW QUESTIONS

1. The name is cathode-ray tube and abbreviation is CRT.
2. a. Show waveform(s)
   b. Measure voltage
   c. Determine frequency
   d. Determine phase differences between signals
3. To protect and add life to the coating on the CRT screen
4. Variable time/div and variable time control(s)
5. Vertical input terminal(s)
6. Horizontal input jack ("Ext X input")
7. "Ext Trig" terminal
8. Vertical (or y-axis)
9. Settings of the vertical gain control and vertical input switch should remain unchanged after calibration has been done to accurately measure voltages. If these controls are changed, the original calibration is no longer true.
10. An attenuator probe is a scope test lead probe used to attenuate or reduce the signal into the scope by a selected known factor. A typical attenuation would be a factor of 10. This would mean that you would multiply by 10 times what you are reading on the scope to get the actual amplitude of the signal being tested.

PROBLEMS

1. $4 \times 0.1 = 0.4$ ms/cycle; $f = 1/T = 1/0.4$ ms = 2,500 Hz
2. 50 Hz
3. 2 divisions
4. 30 V
5. $T = 0.004$ s
6. Horizontal time/cm (or time/div) and/or time variable
7. 0.0004
8. rms = 10.6 V; pk = 15 V; pk-pk = 30 V
9. pk-pk = 5 V; rms $V = 1.77$ V; $I = 1.77$ V/4.7 kΩ = 0.3765 mA
10. 2 cycles

ANALYSIS QUESTIONS

1. List of control names should be based on the specific oscilloscope name that students can observe on your equipment in your training setting:
   a. Control for trace focus ______________________
   b. Control for trace intensity ______________________
   c. Control(s) for vertical position ______________________
   d. Control(s) for horizontal position ______________________
   e. Vertical attenuator/amplifier controls ______________________
   ____
   f. Horizontal sweep frequency controls ______________________
   ____
2. List of switches and controls and their settings should be based on the specific oscilloscope that students can observe on your equipment in your training setting:
   a. Maximum possible vertical input signal sensitivity setting ______________________.
   b. Minimum possible vertical input signal sensitivity setting ______________________.
3. Diagram B: vertical variable and horizontal position controls
   Diagram C: vertical variable control
   Diagram D: horizontal sweep frequency controls (horizontal time/div and time variable)
   Diagram E: vertical variable control
   Diagram F: horizontal sweep frequency controls (horizontal time/div and time variable)
   Diagram G: vertical variable control

CHAPTER 13

PRACTICE PROBLEMS 1

Since 1 weber = 1^8 line and cutting 1 weber per second induces 1 volt; then, cutting 2×10^8 lines in 0.5 seconds will induce 4 volts.

PRACTICE PROBLEMS 2

$V_L = L(di/dt)$

$V_L = 5(3/1) = 15$ V

IN-PROCESS LEARNING CHECK 1

1. All other factors remaining the same, when an inductor's number of turns increases four times, the inductance value *increases* by a factor of *16 times*.
2. All other factors remaining the same, when an inductor's diameter triples, the inductance value *increases* by a factor of *nine times*. (**NOTE:** Area is directly related to diameter squared.)
3. All other factors remaining the same, when an inductor's length increases, the inductance value *decreases*.
4. When the core of an air-core inductor is replaced by material having a permeability of 10, the inductance value *increases* by a factor of *10 times*.

PRACTICE PROBLEMS 3

10 mH + 6 mH + 100 μH = 16.1 mH total inductance

PRACTICE PROBLEMS 4

1. Three equal-value parallel inductors (15 H each) have a total inductance of one-third of one of them. $L_T = 5$ H
2. L_T of 10-mH, 12-mH, 15-mH, and 20-mH inductors in parallel is 3.33 mH.

PRACTICE PROBLEMS 5

a. DC current through coil = 5 A

b. Energy = $(LI^2)/2 = 25/2 = 12.5$ J (joules)

PRACTICE PROBLEMS 6

2 seconds ≈ 2.4 time constants (τ)

After 2.4 τ, V_R is about 91% of V applied, or 9.1 volts.
After 2.4 τ, V_L is about 9% of V applied, or 0.9 volts.

PRACTICE PROBLEMS 7

0.92 mA

USING EXCEL

1. $V_L = 15$ V
2. $I_{inst} = 0.92$ mA

REVIEW QUESTIONS

1. Inductance (or self-inductance) is the property of a circuit that opposes a *change* in current flow.
2. A magnetic field
3. a. Number of turns on the coil
   b. Length of the coil
   c. Cross-sectional area of the coil
   d. Relative permeability of the core material
4. Induced current will always set up a field that opposes the change that originally caused it.
5. The amount of emf induced in a circuit is related to the number of magnetic flux lines cut or linked per unit time.
6. For inductors, one time constant $\tau = L/R$, where τ is the time for one time constant, L equals inductance in henrys, and R equals resistance in ohms.

7. (c)
8. (b
9. (c)
10. (a)
11. (b)
12. (c)
13. (a)
14. (b)
15. (c)
16. (b)
17. Be careful regarding the "high-voltage kick" produced when the test leads break connection with the inductor just tested. Use appropriate personal safety measures.
18. Know the normal value of R for the inductor being tested in order to properly interpret the test result.
19. The 63.2% indicates the percentage of change that will occur from the present level toward the final (stable) level, or value during the period of one time constant.
20. Yes.

PROBLEMS

1. Inductance will increase to 400 mH, (L varies as the square of the turns).
   Expressed in μH = 400,000 μH, or 0.4 henrys.
2. $L = V_L/(di/dt) = 30$ mV/(100 mA/sec) = 0.3 H
3. New inductance is 500 × 100 μH = 0.05 H
4. Energy (joules) = $1/2 \times LI^2 = 15 \times (2)^2/2 = 30$ J
5. $L_T = L_1 + L_2 + L_3 = 0.1$ H + 0.2 H + 0.001 H = 0.301 H, or 301 mH
6. $L_T = L_1L_2/L_1 + L_2 = 10 \times 15/25 = 150/25 = 6$ H
7. 1 time constant = $L/R = 250 \times 10^{-3}/100 = 0.0025$ s, or 2.5 ms.
   It will take 5 times 2.5 ms, or 12.5 ms for current to complete the change from one level to another.
8. After 4 time constants, the voltage across the resistor is approximately 98.5% of V applied, and the voltage across the inductor is 100% – 98.5%, or approximately 1.5% of V applied.
9. Tapping a coil at midpoint yields an inductance of one-fourth the value of the total inductance of the coil. In the case of the 0.4 H coil, L equals about 0.1 H at the 250 turn tap.
10. V across the resistor in a series RL circuit will achieve 86.5% of V applied after two time constants. For our case, 1 time constant = 5/10 or 0.5 sec., and 2 time constants = 1 second.
11. V_L after 0.6 μs ≈ 6.76 volts
12. It will take about 1 time constant, or 1 millisecond for V_R to arrive at a value of 38 volts.

ANALYSIS QUESTIONS

1. Inductance will decrease.
2. There will not be an induced back-emf (Faraday's Law).
3. Total inductance will increase.
4. Total inductance will decrease.
5. Since the inductance change is proportional to the square of the turns,
   $L_1/L_2 = (N_1)^2/(N_2)^2$; $4/6 = (1{,}000)^2/(N_2)^2$; $N_2 = 1{,}224.74$ turns
6. Drawing of graph showing V_R and V_L voltages:
   @ 1τ: $V_L = 7.4$ V, $V_R = 12.6$ V
   @ 2τ: $V_L = 2.7$ V, $V_R = 17.3$ V
   @ 3τ: $V_L = 1$ V, $V_R = 19$ V
   @ 4τ: $V_L = 0.36$ V, $V_R = 19.64$ V
   @ 5τ: V_L = close to 0 V, V_R = close to 20 V

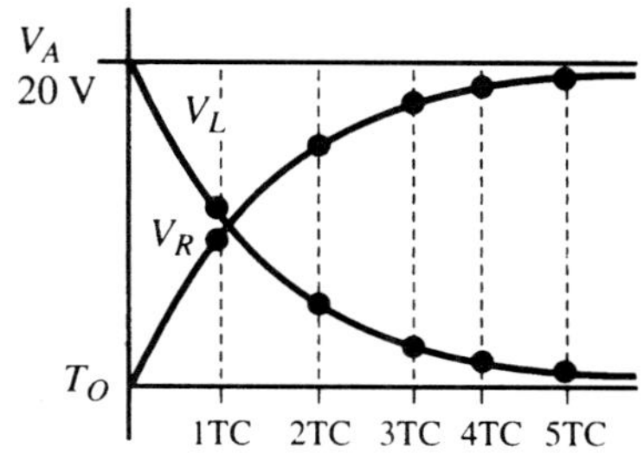

7. Because the voltage-per-turn is less than the varnish insulation breakdown voltage. (The voltage-per-turn in this case is approximately 2 V/turn).
8. The resistance in a coil limits the current as per Ohm's Law, and is not frequency-conscious. The property of inductance limits current based upon the amount of induced back-emf, and is frequency-conscious.
9. The joules of energy stored by an inductor is directly related to the amount of inductance the inductor has because the strength of the magnetic field built up by a given value of current through the coil is directly related to the inductance value. The energy is stored in the form of the magnetic field.
10. The direct relationship between current through the coil and the energy stored by the coil is based upon the fact that the strength of the magnetic field produced by the coil is directly related to the current through it. Since the energy is stored in the magnetic field, the stronger the field, the greater the energy stored in the field.

CHAPTER 14

PRACTICE PROBLEMS 1

1. 3 kΩ
2. 333.3 Ω

IN-PROCESS LEARNING CHECK 1

1. When current increases through an inductor, the cemf *hinders* the current increase.
2. When current decreases through an inductor, the cemf *hinders* the current decrease.
3. In a pure inductor, the *voltage* leads the *current* by 90 degrees.
4. The memory aid that can help you remember the relationship described in question 3 is "*Eli the ice man.*"
5. The opposition that an inductor shows to ac is termed *inductive reactance.*
6. The opposition that an inductor shows to ac *increases* as inductance increases.
7. The opposition that an inductor shows to ac *decreases* as frequency decreases.
8. X_L is *directly* related to inductance value.
9. X_L is *directly* related to frequency.

PRACTICE PROBLEMS 2

1. 6,280 Ω
2. 79.6 mH, or approximately 80 mH
3. 238.85 Hz, or approximately 239 Hz

PRACTICE PROBLEMS 3

1. 5 kΩ
2. $X_L = 5{,}024\ \Omega$, so, $IX_L = 3\text{ mA} \times 5{,}024\ \Omega$ or approximately 15 V
3. 40 mA

USING EXCEL

1. $X_L = 6{,}280\ \Omega$
2. $L = 79.6\text{ mH}$

REVIEW QUESTIONS

1. (c)
2. (a)
3. (c)
4. (b)
5. (b)
6. (b)
7. (d)
8. (c)
9. (b)
10. (b)

PROBLEMS

1. $L = X_L/2\pi f = 376.8/6.28 \times 100 = 376.8/628 = 0.6$ H; at 450 Hz
   $X_L = 4.5 \times 376.8\ \Omega = 1695.6\ \Omega$
2. $X_L = 2\pi fL = 6.28 \times 120 \times 5 = 3768\ \Omega$
   At 600 Hz, X_L of the 5-H inductor equals 18,840 Ω.
3. Total inductance = 2 mH × 6 mH/2 mH + 6 mH = 12/8 = 1.5 mH
   Frequency = $X_L/2\pi L$ = 39.8 kHz
4. L_T = 15 H (assuming zero mutual inductance)
   X_{L_T} will equal 5,000 ohms at approximately 53 Hz.
5. $Q = X_L/R = 6.28 \times 20$ MHz × 10 μH/10 Ω = 125.6/10 = 12.56
   Q will increase, since $Q = X_L/R$ and X_L triples while R only doubles.
   Q is not an appropriate consideration for inductors in dc circuits. X_L is reactance to ac due to the changing current inducing back-emf.
6. Waveform showing circuit V and I for a simple one inductor circuit with an X_L of 1 kΩ and a 10-V (rms) source is shown. (Peak values shown.)

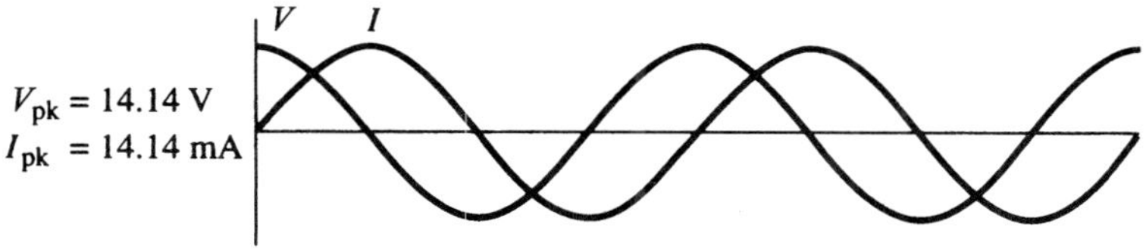

7. V_T = 45 V
   V_{L_1} = 15 V
   I_T = 0.75 mA
8. Current through L_2 is approximately 4.5 mA since its X_L is only two-thirds that of L_3, and L_3's current is approximately 3 mA. L_2's current is 1.5 times that of the L_3 branch.
9. ≈ 6 V
10. Total inductance is 3 H. (**NOTE:** L_1 computes to be 2 H via the $L = X_L/2\pi f$ formula.)
11. L_T will RTS.
    X_{L_T} will I.
    I_T will D.
    V_{L_4} will RTS.
    I_1 will D.
12. L_T will D.
    X_{L_1} will RTS.
    I_T will I.
    V_{L_1} will I.
    f will RTS.
13. L = 0.318 H
14. L_T = 4.6 H
15. X_{L_T} = 28.89 kΩ

16. a. 0.346 mA
    b. $I_1 = 0.159$ mA; $I_2 = 0.106$ mA; $I_3 = 0.08$ mA
17. I_T would decrease.
18. I_{L_3} would decrease.
19. Voltage applied would have to decrease to one-third its original value.
20. I_T would be 0.0354 mA or 35.4 μA.

ANALYSIS QUESTIONS

1. Six times greater
2. 3/2; that is, 1.5 times greater than original value
3. Circuit current will increase, since X_L will decrease.
4. Circuit current will increase. (X_L doubled, but voltage tripled.)
5. $Q = X_L/R$ (as f decreases, X_L decreases while R remains the same); Q will decrease.
6. No.
7. Less losses (less wasted energy)
8. (c)
9. (b)

CHAPTER 15

PRACTICE PROBLEMS 1

Five pounds at an angle of 36.9 degrees from horizontal

IN-PROCESS LEARNING CHECK 1

1. In a purely resistive ac circuit, the circuit voltage and current are *in*-phase.
2. In a purely inductive ac circuit, the current through the inductance and the voltage across the inductance are *90* degrees out-of-phase. In this case, the *voltage* leads the *current*.
3. A quantity expressing both magnitude and direction is a *vector* quantity.
4. The length of the vector expresses the *magnitude* of a quantity.

PRACTICE PROBLEMS 2

$V_S = \sqrt{V_R^2 + V_L^2} = \sqrt{25 + 25} = \sqrt{50} = 7.07 \text{ V}$

PRACTICE PROBLEMS 3

1. Cos θ = adj/hyp = 5/10 = 0.5
   Angle whose cosine = 0.5 is 60 degrees.
2. Tangent of 45 degrees = 1.0
3. Sine of 30 degrees = 0.5

PRACTICE PROBLEMS 4

V applied = 150 volts

PRACTICE PROBLEMS 5

1. Diagram:

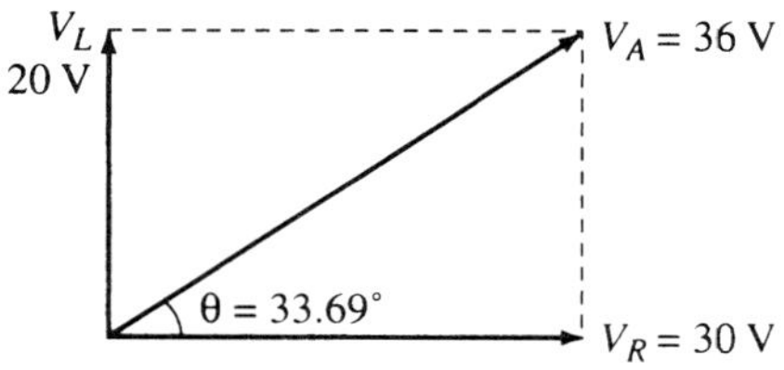

2. Angle = 33.69 degrees
   (Tan θ = 20/30 = 0.6666; arctan 0.6666 = 33.69 degrees)
3. Used the tangent function for problem in question 2.
4. Angle would have been greater (56.3 degrees).

PRACTICE PROBLEMS 6

1. $V_R = 35.35$ V
   (Cos of 45 degrees = 0.7071; 0.7071 = adj/50; adj = 35.35)
2. a. $V_R = 32.5$ V
   b. $V_L = 56.29$ V
3. a. Angle = 56.3 degrees
   b. $V_T = 43.2$ V

PRACTICE PROBLEMS 7

1. Drawing:

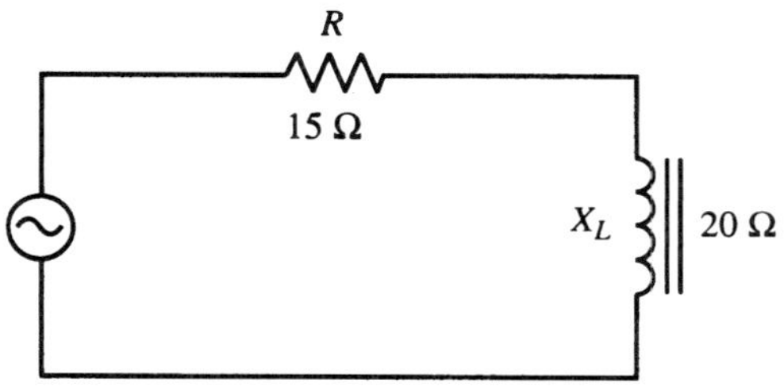

2. Drawing:

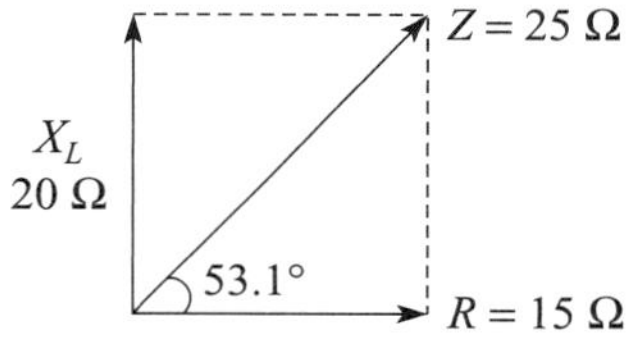

3. $Z = \sqrt{15^2 + 20^2} = \sqrt{625} = 25\ \Omega$. Yes.
4. Tan θ = X_L/R = 20/15 = 1.33
   Tan^{-1} 1.33 = an angle of 53.1 degrees
   Yes, answer agrees.

PRACTICE PROBLEMS 8

$I_T = \sqrt{I_{R_2} + I_{L_2}} = \sqrt{5^2 + 3^2} = \sqrt{34} = 5.83$ A
$Z = V_T/I_T = 300/5.83 = 51.45\ \Omega$

Phase angle = 30.9 degrees
(Tan θ = 3/5 = 0.6; $\tan^{-1}$ of 0.6 = 30.9 degrees)

PRACTICE PROBLEMS 9

$Z = 47 \times 100/(\sqrt{47^2 + 100^2}) = 4{,}700/110.5 = 42.5\ \Omega$

PRACTICE PROBLEMS 10

1. $G = 1/R = 0.0166$ S
2. $B_L = 1/X_L = 0.01$ S
3. $Y = \sqrt{G^2 + B_L^2} = 0.0194$ S
4. $Z = 1/Y = 51.55\ \Omega$
5. $I_T = V \times Y = 5.82$ A
6. $V_T = I/Y = 300$ V
7. $Y = I/V = 0.0194$ S
8. $\theta = R/X_L = 30.96°$

USING EXCEL

1. $V_A = 150$ V
2. $Z = 42.5\ \Omega$

REVIEW QUESTIONS

1. (b)
2. (a)
3. (a)
4. (b)
5. (a)
6. (b)
7. (c)
8. (a)
9. (c)
10. (b)
11. (c)
12. $V_T = \sqrt{V_R^2 + V_L^2}$
13. $Z = \sqrt{R^2 + X_L^2}$
14. $I_T = \sqrt{I_R^2 + I_L^2}$
15. $Z = R \times X_L / \sqrt{R^2 + X_L^2}$
16. (c)
17. (b)
18. (a)
19. (a)
20. (c)

PROBLEMS

1. a. 36 Ω

   b. $Z = \sqrt{R^2 + X_L^2} = \sqrt{20^2 + 30^2} = \sqrt{1{,}300}$ = approximately 36 Ω

   c. Cos $\theta = R/Z = 20/36 = 0.555$. Angle whose cosine equals 0.555 is 56.3°; thus, angle between R and $Z = 56.3°$.

2. a. $I_R = 300\text{ V}/200\ \Omega = 1.5\text{ A}$

   b. $I_L = 300\text{ V}/150\ \Omega = 2\text{ A}$

   c. $I_T = \sqrt{I_R^2 + I_L^2} = \sqrt{2.25 + 4} = \sqrt{6.26} = 2.5\text{ A}$

   d. Tan $\theta = I_L/I_R = -2/1.5 = -1.33$. Angle whose tangent equals −1.33 is −53.1°.

   e. $Z = V/I = 300\text{ V}/2.5\text{ A} = 120\ \Omega$

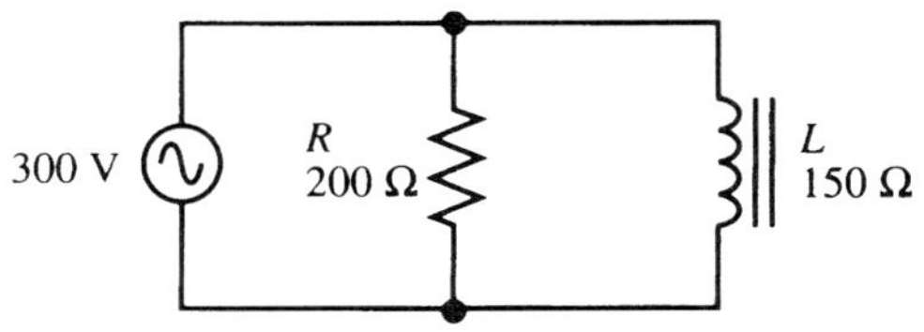

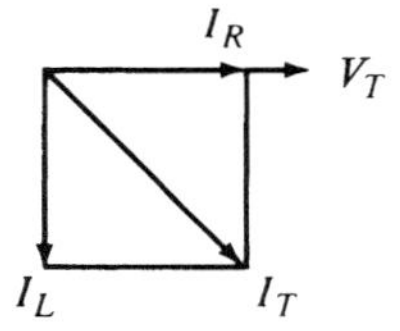

3. Diagrams for the specified series RL circuit are shown.

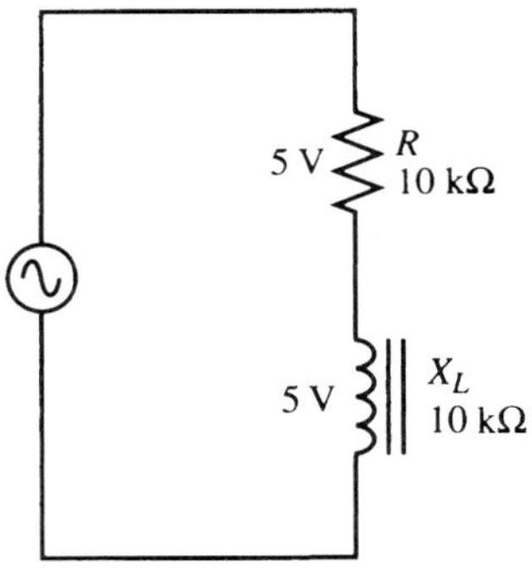

Circuit diagram

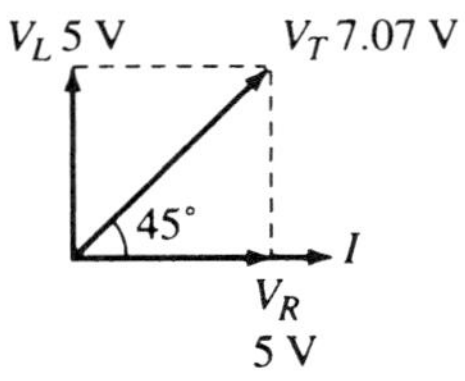

V-I vector diagram

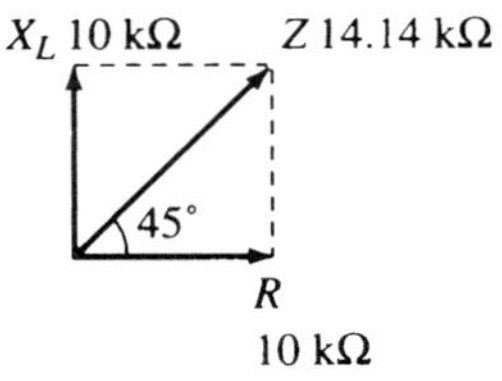

Impedance diagram

Since $\theta = 45°$, then R must $= X_L$, and V_R must $= V_L$.

$Z = \sqrt{R^2 + X_L^2}$

$Z = \sqrt{100 \times 10^6}$

$Z = 14.14 \times 10^3 = 14.14\text{ k}\Omega$

4. Diagrams for the specified parallel RL circuit are shown.

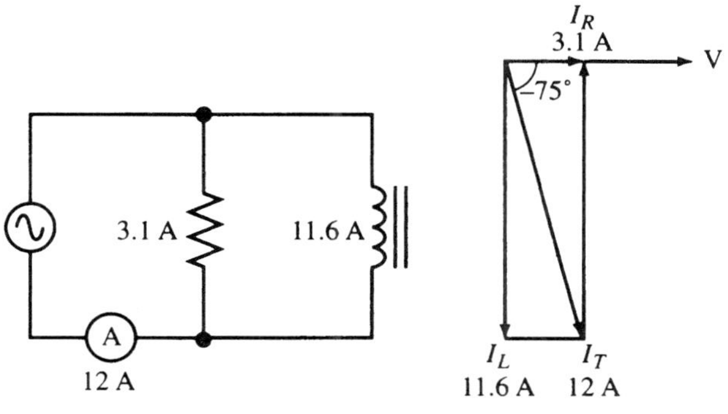

Circuit diagram

V-I vector diagram

Cos 75° × $I_T = I_R$

0.2588 × 12 = 3.1 A

Sin 75° × $I_T = I_L$

0.966 × 12 = 11.6 A

5. a. Z = vector sum of 10-kΩ R and approximate 20-kΩ X_L = 22.36 kΩ
   b. $V_R = IR$ = (67 V/22.36 kΩ) × 10 kΩ = approx 3 mA × 10 kΩ = 30 V
   c. θ = approximately 63.43°
6. a. X_L = 6.28 × 60 × 30 = 11.3 kΩ
   b. $V_L = I \times X_L$ = 10 mA × 11.3 kΩ = 113 V
   c. Cos θ = V_R/200; cos 34.4° = 0.825; 0.825 = V_R/200; V_R = 165 V
   d. Sin θ = 113 V/200 V; Sin θ = .565; θ = 34.4°
7. a. I = 5 mA
   b. Z = 40 kΩ
   c. V_L = 105.4 V
   d. X_L = 21.08 kΩ
   e. $L = X_L/2\pi f$ = 56 H
8. Diagrams and calculations for the specified series RL circuit are shown.

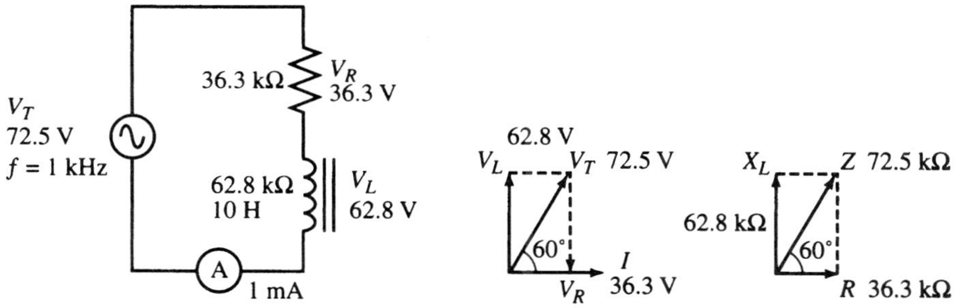

$X_L = 2\pi fL$ = 6.28 × 1 × 10^3 × 10 = 62.8 kΩ

Sin θ = X_L/Z; sin 60° = 62.8 kΩ/Z;

0.866 = 62.8 kΩ/Z; 0.866 Z = 62.8 kΩ;

Z = 62.8 kΩ/0.866 = 72.517 kΩ

Tan θ = opp/adj = X_L/R; tan 60° = 62.8 kΩ/R;

1.73 = 62.8 kΩ/R; 1.73R = 62.8 kΩ;

R = 62.8 kΩ/1.73 = 36.3 kΩ

$I = V_T/Z$; I = 72.5 V/72.5 kΩ = 1 mA

$V_R = I \times R$ = 1 mA × 36.3 kΩ = 36.3 V

$V_L = I \times X_L$ = 1 mA × 62.8 kΩ = 62.8 V

9. Diagrams and calculations for the specified parallel *RL* circuit are shown.

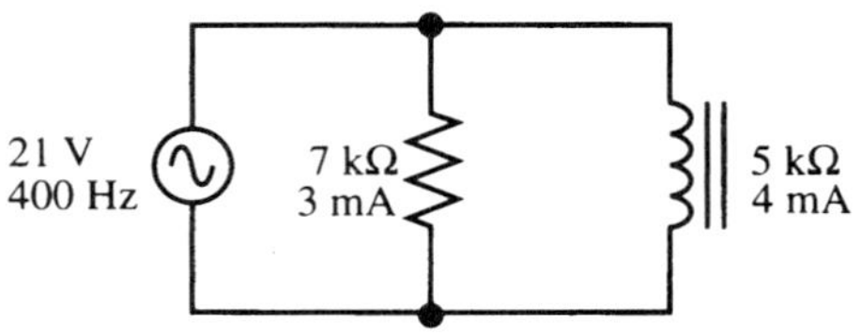

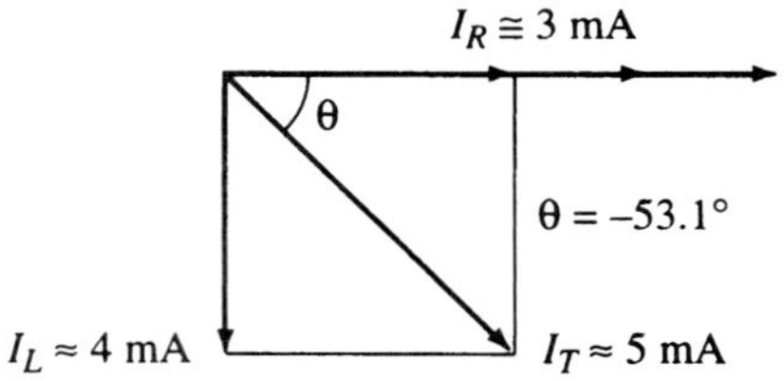

$X_L = 2\pi fL = 6.28 \times 400 \times 1.99 = 5$ kΩ (approx)

$I_R = 21$ V/7 kΩ = 3 mA

$I_L = 21$ V/5 kΩ = 4.2 mA (or 4 mA)

$I_T = \sqrt{I_R{}^2 + I_L{}^2} = \sqrt{26.6 \times 10^{-6}} = 5.16$ mA (or 5 mA)

$Z = V_T/I_T = 21$ V/5 mA = 4.2 kΩ

Cos θ = adj/hyp = 3 mA/5 mA = 0.6; θ = angle whose cos = 0.6 = –53.1° (Current lags V_T by 53.1°).

10. The circuit diagram for the specified circuit is shown.

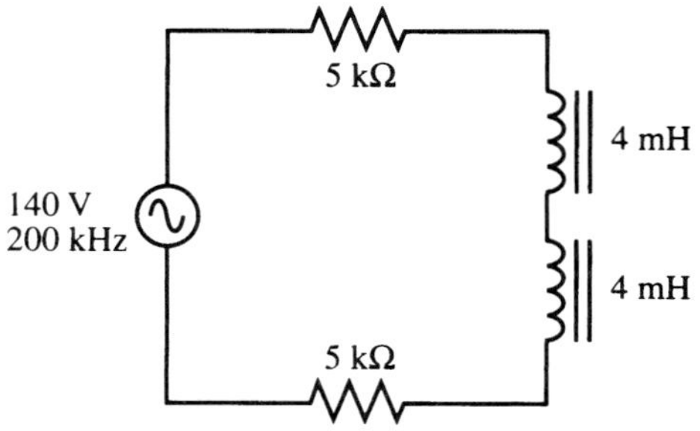

$I_T =$ approximately 140 volts/14.14 kΩ = 9.9 mA (or ≈ 10 mA)

$Z = \sqrt{R^2 + X_L{}^2}$

$X_L = 2\pi fL = 6.28 \times 200 \times 10^3 \times 4 \times 10^3 = 5.024$ kΩ (or 5 kΩ) each inductor

$R = 5$ kΩ + 5 kΩ = 10 kΩ

$Z = \sqrt{200 \times 10^6} = 14.14$ kΩ (or 14 kΩ)

$X_{L_T} \cong 10$ kΩ

ANALYSIS QUESTIONS

1. a. *R* will RTS.
   b. *L* will RTS.
   c. X_L will I.
   d. *Z* will I.
   e. θ will D.
   f. I_T will D.
   g. V_T will RTS.
2. a. *R* will RTS.
   b. *L* will I.
   c. X_L will I.
   d. *Z* will I.
   e. θ will I.
   f. I_T will D.
   g. V_T will RTS.
3. a. *R* will I.
   b. *L* will RTS.
   c. X_L will RTS.
   d. *Z* will I.
   e. θ will D.
   f. I_T will D.
   g. V_T will RTS.

4. a. R will RTS.
   b. L will I.
   c. X_L will I.
   d. Z will I.
   e. θ will D.
   f. I_T will D.
   g. V_T will RTS.
5. a. R will I.
   b. L will RTS.
   c. X_L will RTS.
   d. Z will I.
   e. θ will I.
   f. I_T will D.
   g. V_T will RTS.
6. a. R will D.
   b. L will D.
   c. X_L will D.
   d. Z will D.
   e. θ will RTS.
   f. I_T will I.
   g. V_T will RTS.
7. 57, sin
8. .8, 2nd, $\tan^{-1}$
9. 34, cos
10. .75, 2nd, $\sin^{-1}$
11. Not having to know the hypotenuse in order to calculate.
12. 1.3089 radians

MULTISIM EXERCISE

2. $V_L = 35$ V; $V_{R_1} = 35$ V; @ double f: $V_{L_1} = 45$ V; $V_{R_1} = 22$ V
3. L; R; No; $X_L = 2x$, $Z \neq 2x$; No
4. Yes; greater
5. $\approx 63°$

CHAPTER 16

IN-PROCESS LEARNING CHECK 1

1. Producing voltage via a changing magnetic field is called electromagnetic *induction*.
2. Inducing voltage in one circuit by varying current in another circuit is called *mutual inductance*.
3. The fractional amount of the total flux that links two circuits is called the *coefficient* of coupling, which is represented by the letter k. When 100% of the flux links the two circuits, the *coefficient* of coupling has a value of *1*.
4. The closer coils are, the *higher* the coupling factor produced. Compared to parallel-oriented coils, perpendicular coils have a *lower* degree of coupling.

PRACTICE PROBLEMS 1

$M = k\sqrt{L_1 \times L_2}$

$M = 0.95\sqrt{10 \times 15} = 0.95\sqrt{150} = 0.95 \times 12.25 = 11.63$ H

PRACTICE PROBLEMS 2

$L_T = L_1 + L_2 - 2M$

$L_T = 10\text{ H} + 10\text{ H} - (2 \times 5\text{ H})$

$L_T = 20\text{ H} - 10\text{ H} = 10\text{ H}$

PRACTICE PROBLEMS 3

1. $L_T = \dfrac{1}{\dfrac{1}{L_1 + M} + \dfrac{1}{L_2 + M}} = \dfrac{1}{\dfrac{1}{8+4} + \dfrac{1}{8+4}}$

   $L_T = \dfrac{1}{\dfrac{1}{12} + \dfrac{1}{12}}$ $\quad L_T = \dfrac{1}{\dfrac{1}{6}} = 1 \times \dfrac{6}{1} = 6\text{ H}$

2. $L_T = \dfrac{1}{\dfrac{1}{L_1 - M} + \dfrac{1}{L_2 - M}} = \dfrac{1}{\dfrac{1}{8-4} + \dfrac{1}{8-4}}$

   $L_T = \dfrac{1}{\dfrac{1}{4} + \dfrac{1}{4}}$ $\quad L_T = \dfrac{1}{\dfrac{1}{2}} = 1 \times \dfrac{2}{1} = 2\text{ H}$

PRACTICE PROBLEMS 4

1. $N_P / N_S = V_P / V_S$: p-s turns ratio = 100/300 or 1:3
2. Turns ratio equals voltage ratio. Since the transformer steps up the voltage six times, the secondary must have six times the turns of the primary; therefore, the p-s turns ratio is 1:6.
3. Since the p-s turns ratio is 1:5, the secondary has five times the voltage of the primary, or 5 × 50 volts = 250 volts.
4. Voltage ratio with p-s turns ratio of 4:1 is 4:1. This is a step-down transformer, since the primary voltage is four times that of the secondary.

PRACTICE PROBLEMS 5

Since the impedance is transformed in relation to the square of the turns ratio, when the secondary has twice the number of turns of the primary, the impedance at the primary looks like one-fourth the impedance across the secondary, or 2,000/4 = 500 Ω.

PRACTICE PROBLEMS 6

1. The *primary-to-secondary* impedance ratio is related to the square of the turns relationship of primary-to-secondary. Since the primary has 5 times as many turns as the secondary, the primary impedance is 5^2 times greater than that of the secondary. Therefore, the p-s impedance ratio is 25:1.
2. The impedance, looking in the primary, is $(16)^2 \times 4\ \Omega = 256 \times 4 = 1{,}024\ \Omega$.

USING EXCEL

1. $Z_P = 500\ \Omega$
2. $Z_P = 4\ \Omega$

REVIEW QUESTIONS

1. (b)
2. (b)
3. (b)
4. (c)
5. (b)
6. (c)
7. (a)
8. (b)
9. (c)
10. (c)
11. (c)
12. (a)
13. (b)
14. (c)
15. (c)
16. (b)
17. (b)
18. (d)

PROBLEMS

1. $V_S/V_P = I_P/I_S$

   240 V/120 V = I_P/100 mA

   I_P = 200 mA
2. $N_S/N_P = V_S/V_P$

   V_S/100 V = 5.5/1; V_S = 550 V
3. $N_P/N_S = V_P/V_S$; N_P/N_S = 120 V/6 V; N_P:N_s = 20:1.
4. $Z_P = Z_S\left[\frac{N_P}{N_S}\right]^2$; $Z_P = 8\,\Omega\left[\frac{5}{1}\right]^2$; $Z_P = 8\times 25 = 200\,\Omega$
5. $M = k\sqrt{L_1 + L_2} = 0.6\sqrt{2\times 2} = 0.6\times 2 = 1.2$ H
6. $L_T = L_1 + L_2 + 2M$ and $M = k\sqrt{L_1 + L_2}$

   $M = 0.3 = 0.3\sqrt{10\times 10^{-3}\times 10\times 10^{-3}}\times 10\times 10^{-3} = 3$ mH

   L_T = 10 mH + 10 mH + (2 × 3 mH) = 26 mH
7. $$L_T = \frac{1}{\frac{1}{L_1 - M} + \frac{1}{L_2 - M}}$$

   where $M = k\sqrt{L_1 L_2} = 0.5\sqrt{10\times 10} = 0.5\times 10 = 5$ H

   $$L_T = \frac{1}{\frac{1}{10-5} + \frac{1}{10-5}}$$

   $L_T = 1/0.4 = 2.5$ H
8. $V_S/V_P = N_S/N_P$

   V_S/12 = 15/1 = 12 × 15 = 180 V
9. 50/500 = 0.1 volts-per-turn
10. $I_S/I_P = N_P/N_S$

    0.1 A/I_P = 6/1

    I_P = 0.0166 A
11. 2,000 turns
12. 4 W
13. 1:4.75
14. 5.33 mA
15. 62.5 V

ANALYSIS QUESTIONS

1. A "flyback" transformer (sometimes called a horizontal transformer) is typically used in the horizontal deflection circuits of televisions. It primarily provides high voltage to the picture tube or CRT. It may also provide low filament voltage for the high voltage rectifier. The scanning circuit in which it operates uses a frequency of 15,750 Hz.
2. Laminating the core of a transformer cuts down on eddy-current losses by virtue of the fact that the thin sheets of iron (each insulated from each other) have much higher resistance to current than the core

would have if it were all one piece of iron. This means that the induced currents (eddy currents) caused by changing flux will be much lower; hence, the I^2R losses in the core will be greatly reduced.

3. Primary winding should have 3,535.5 turns.
4. Since the power delivered to the 200-Ω load is 405 mW, then the secondary current can be calculated to be 45 mA. With the p-s turns ratio of 8:1; this means that the current in the primary will be one-eighth that in the secondary. $I_P = 45$ mA/8 = 5.625 mA.
5. Number of secondary turns = 200 turns.
6. The ohmmeter only reads the dc resistance of the transformer winding, not the ac X_L value. Therefore, in real life, the current is limited to less than the 5-A value under normal conditions.

CHAPTER 17

IN-PROCESS LEARNING CHECK 1

1. A capacitor is an electrical component consisting of two conducting surfaces called *plates* that are separated by a nonconductor called the *dielectric*.
2. A capacitor is a device that stores electrical *charge* when voltage is applied.
3. Electrons *do not* travel through the capacitor dielectric.
4. During charging action, one capacitor plate collects *electrons*, making that plate *negative*. At the same time, the other plate is losing *electrons* to become *positive*.
5. Once the capacitor has charged to the voltage applied, it acts like an *open* circuit to dc. When the level of dc applied increases, the capacitor *charges* to reach the new level. When the level of dc applied decreases, the capacitor *discharges* to reach the new level.
6. When voltage is first applied to a capacitor, *maximum* charge current occurs.
7. As the capacitor becomes charged, current *decreases* through the circuit in series with the capacitor.

PRACTICE PROBLEMS 1

1. $C = Q/V = 100 \times 10^{-6}/25 = 4 \times 10^{-6} = 4\ \mu\text{F}$
2. $Q = CV = 10 \times 10^{-6} \times 250 = 2{,}500\ \mu\text{C}$
3. $V = Q/C = 50 \times 10^{-6}/2 \times 10^{-6} = 25$ V

PRACTICE PROBLEMS 2

Energy $= CV^2/2 = 10 \times 10^{-6} \times (100)^2/2 = 100{,}000 \times 10^{-6}/2 = 50{,}000\ \mu\text{J}$

PRACTICE PROBLEMS 3

1. $C = (2.25\text{ kA}/10^7) \times (n - 1) = 2.25 \times 10 \times 1/(10^7 \times 0.1)$
   $C = (22.5/10^6) \times (3) = 22.5 \times 10^{-6} \times 3 = 67.5\ \mu\text{F}$
2. $C = 8.85\text{ kA}/10^{12} = 8.85 \times 1 \times 0.2/(10^{12} \times 0.005)$
   $C = 1.77 \times 10^{-12}/0.005 = 354$ pF

PRACTICE PROBLEMS 4

1. $C_T = C_1C_2/(C_1 + C_2) = (5 \times 10^{-6} \times 20 \times 10^{-6})/25 \times 10^{-6}$
   $C_T = (100 \times 10^{-12})/(25 \times 10^{-6}) = 4 \times 10^{-6}$, or 4 μF
2. Total capacitance of four 20-μF capacitors in series is equal to one of the equal value capacitors divided by 4. Thus, 20/4 = 5 μF. ($C_T = C_1/C_N$)

3. V_{C_1} = four-fifths V_T, or 80 volts. V_{C_2} = one-fifth V_T, or 20 volts.
   (**NOTE:** $Q_T = C_T \times V_T = 4\ \mu F \times 100\ V = 400\ \mu C$)
   ($V_{C_1} = Q/C_1 = 400\ \mu C/5\ \mu F = 80\ V$)
   ($V_{C_2} = Q/C_2 = 400\ \mu C/20\ \mu F = 20\ V$)

PRACTICE PROBLEMS 5

1. C_T = sum of the individual capacitances; therefore = 60 μF.
2. $Q_{C_1} = C_1 \times V = 12\ \mu F \times 60\ V = 720\ \mu C$
   $Q_{C_1} = Q_{C_2} = Q_{C_3} = Q_{C_4} = Q_{C_5}$
   Q_T = sum of all the charges = 5 × 720 μC = 3,600 μC.

PRACTICE PROBLEMS 6

1. τ = 0.01 milliseconds or 10 microseconds
   Time allowed for charge is 20 μsec (or 2 time constants).
   Voltage after 20 μsec = 86.5 V.
2. τ = 10 microseconds
   Time allowed for charge is 15 μsec.
   V_R after 15 μsec ≅ 22 V. Charging current at that moment ≅ 2.2 mA.
3. V_C after 30 μs ≅ 95 V.
4. V_R after 25 μs ≅ 8.2 V.

USING EXCEL

1. V_R = 22 V
2. V_C = 95 V

REVIEW QUESTIONS

1. (b)
2. (c)
3. (a)
4. (d)
5. (a)
6. (c)
7. (d)
8. (c)
9. (b)
10. (a)
11. (c)
12. (d)
13. Observe polarity
14. Large capacitance per size ratio
15. Mica
16. Fixed and variable
17. Small size; very little lead length provides little stray undesired electrical parameters, such as inductance, etc.
18. Five *RC* time constants (i.e., five times the circuit's *RC* time-constant value)
19. (b)
20. LCR meter; capacitance checker, certain DMMs, impedance bridge

PROBLEMS

1. New capacitance is four times the original capacitance, since *C* is directly related to the area of the plates facing each other and inversely related to the dielectric thickness.
2. $Q = CV = 100 \times 10^{-12} \times 100$ = 10,000 pC, or 0.01 μC

3. $C = Q/V = 200\text{ pC}/50 = 4\text{ pF}$
4. $V = Q/C = 2{,}000\ \mu\text{C}/10\ \mu\text{F} = 200\text{ V}$
5. $$C_T\ (\text{pF}) = \frac{1}{\frac{1}{C_1}+\frac{1}{C_2}+\frac{1}{C_3}} = \frac{1}{1/100+1/400+1/1{,}000}$$

   $C_T = 1/(0.01 + 0.0025 + 0.001) = 1/0.0135 = 74.07\text{ pF}$

   $C_T(\text{pF}) = C_1 + C_2 + C_3 = 100 + 400 + 1{,}000 = 1{,}500\text{ pF}$
6. Find Q_T by first finding C_T, then use the $Q = CV$ formula:

   $$C_T = \frac{1}{1/10+1/20+1/50} = 1/(0.1+0.05+0.02) = 1/0.17 = 5.88\ \mu\text{F}$$

   $Q = CV = 5.88\ \mu\text{F} \times 160\text{ V} = 940.8\ \mu\text{C}$

   Find each capacitor's V value:

   $V_{C_1} = 940.8\ \mu\text{C}/10\ \mu\text{F} = 94.08\text{ V}$

   $V_{C_2} = 940.8\ \mu\text{C}/20\ \mu\text{F} = 47.04\text{ V}$

   $V_{C_3} = 940.8\ \mu\text{C}/50\ \mu\text{F} = 18.816\text{ V}$

   The results in problem 6 check since voltage distribution is inverse to capacitance distribution ($V = Q/C$). The 50-μF capacitor has a voltage equal to one-fifth that of the 10-μF capacitor; the 20-μF capacitor has a V equal to half that of the 10-μF capacitor. Sum of series voltages add to the applied voltage of 160 V. Charge on each capacitor is 940.8 μC since this is a series circuit, and the current (and charge) is the same throughout the circuit.
7. $Q_{C_1} = C_1 \times V_1 = 10\ \mu\text{F} \times 100\text{ V} = 1{,}000\ \mu\text{C}$

   $Q_{C_2} = C_2 \times V_2 = 20\ \mu\text{F} \times 100\text{ V} = 2{,}000\ \mu\text{C}$

   $Q_{C_3} = C_3 \times V_3 = 50\ \mu\text{F} \times 100\text{ V} = 5{,}000\ \mu\text{C}$

   $C_T = C_1 + C_2 + C_3 = 80\ \mu\text{F}$

   $Q_T = Q_1 + Q_2 + Q_3 = 8{,}000\ \mu\text{C}$
8. $10.9\ \mu\text{F} = C_T$

   V across 20 μF: $V_X = 100 \times (10.9/20) = 100 \times .545 = 54.5\text{ V}$

   V across 40 μF: $V_X = 100 \times (10.9/40) = 100 \times .2725 = 27.25\text{ V}$

   V across 60 μF: $V_X = 100 \times (10.9/60) = 100 \times .1816 = 18.16\text{ V}$
9. $V_R = V\text{ applied} \times \epsilon^{-t/\tau} = 200 \times 2.718^{-1.2} = 60.23\text{ V}$

   $V_C = V_A - V_R$ or approximately $200 - 60 = 140\text{ V}$
10. Total capacitance of the circuit shown is calculated using knowledge that capacitors in series add like resistors in parallel, and capacitors in parallel add like resistors in series. Thus, total capacitance of the two branches (point A to B) = 20 μF. This 20 μF is in series with C_6's 10 μF and C_1's 10 μF. Simplifying further, the two 10 μFs in series are equivalent to 5 μF. This value is in series with the 20 μF of the two branches. $20 \times 5/20 + 5 = 4\ \mu\text{F}$ total capacitance.
11. Voltage across C_5 equals $0.5Q_T/C_5$.

    $Q_T = C_T V_T = 4\ \mu\text{F} \times 80\text{ V} = 320\ \mu\text{C}$

    $1/2\ Q_T = 160\ \mu\text{C}$

    $V_{C_5} = 160\ \mu\text{C}/20\ \mu\text{F} = 8\text{ V}$
12. Q_{C_6} = same as Q_T, or 320 μC

13. 600 V
14. Voltage applied is 3,000 volts. (Since the voltages across C_1 and across C_6 are each double that across points *A* and *B,* 1,200 V + 1,200 V + 600 V = 3,000 V.) (**NOTE:** This cannot actually be done, since the voltage ratings of C_1 and C_6 were described as being 300 V each. Therefore, it would not be possible to have 1,200 volts across each without their breaking down. But, for purposes of achieving 600 volts across points *A* and *B,* it takes 3,000 volts applied, as shown.)
15. $J = 1/2\ CV^2 = 4\ \mu F \times (80\ V)^2/2 = 0.0256/2 = 0.0128$ J, or 12,800 μJ
16. Charge on first capacitor ($Q = CV$) = 15 μF × 150 V = 2,250 μC. Charge on second capacitor = 15 μF × 300 V = 4,500 μC. Total charge to be distributed on total capacitance is 6,750 μC. Since capacitors are connected in parallel, total C = 30 μF. $V = Q/C$ = 6,750 μC/30 μF = 225 volts.
17. $J = C(V^2)/2 = 30\ \mu F \times (225)^2/2 = 1.518/2 = 0.759375$ J, or 759,375 μJ, or 759.375 mJ.

ANALYSIS QUESTIONS

1. Students should describe the ranges and operational procedures for measuring capacitance with a DMM in terms of the equipment or vendor data available to them.
2. Students should describe the ranges and operational procedures for measuring capacitance with a capacitance meter, in terms of the equipment or vendor data available to them.
3. Students should describe the ranges and operational procedures for measuring capacitance with an LCR meter, in terms of the equipment or vendor data available to them.
4. Students should describe the other types of measurements they can make relative to capacitors, such as leakage current at operating voltage, etc., based upon the equipment or information to which they have access.

CHAPTER 18

IN-PROCESS LEARNING CHECK 1

1. For a capacitor to develop a potential difference between its plates, there must be a *charging* current.
2. For the potential difference between capacitor plates to decrease (once it is charged), there must be a *discharging* current.
3. When ac is applied to capacitor plates, the capacitor will alternately *charge* and *discharge.*
4. The value of instantaneous capacitor current directly relates to the value of *capacitance* and the rate of change of *voltage.*
5. The amount of charging or discharging current is maximum when the voltage rate of change is *maximum.* For sine-wave voltage, this occurs when the sine wave is near *zero* points.
6. Capacitor current *leads* the voltage across the capacitor by *90* degrees. The expression that helps you remember this relationship is “Eli the *ice* man.”

PRACTICE PROBLEMS 1

1. Yes, the X_C formula verifies the values of X_C shown in Figure 18–3.
2. $X_C = 1/(2\pi fC) = 1/(6.28 \times 1000 \times 10^{-6}) = 159\ \Omega$
3. X_C remains the same if f is doubled and C is halved simultaneously.

PRACTICE PROBLEMS 2

1. $f = 1/(2\pi CX_C) = 1/(6.28 \times 500 \times 10^{-12} \times 1{,}000) = 318$ kHz
2. $C = 1/(2\pi fX_C) = 1/(6.28 \times 2{,}000 \times 3{,}180) = 0.025\ \mu F$

PRACTICE PROBLEMS 3

$X_{C_1} = 1/(2\pi fC) = 1/(6.28 \times 1{,}000 \times 2 \times 10^{-6}) = 79.5\ \Omega$

$X_{C_2} = 1/(2\pi fC) = 1/(6.28 \times 1{,}000 \times 4 \times 10^{-6}) = 39.75\ \Omega$

$X_{C_T} = X_{C_1} + X_{C_2} + X_{C_3} + X_{C_4} = 79.5 + 39.75 + 200 + 300 = 619.25\ \Omega$

PRACTICE PROBLEMS 4

1. $X_{C_T} = 1/(2\pi fC_T) = 1/(6.28 \times 1{,}000 \times 6 \times 10^{-6}) = 26.5\ \Omega$
2. $X_{C_T} = 1/(2\pi fC_T) = 1/(6.28 \times 500 \times 25 \times 10^{-6}) = 12.7\ \Omega$
3. $X_{C_T} = 1{,}600/4 = 400\ \Omega$

PRACTICE PROBLEMS 5

1. $X_{C_1} = 1{,}000\ \Omega$; $X_{C_2} = 1{,}000\ \Omega$; $X_{C_3} = 2{,}000\ \Omega$; $X_{C_T} = 4{,}000\ \Omega$; $C_2 = 0.795\ \mu F$; $C_3 = 0.3975\ \mu F$; $V_2 = 25$ V; $I_T = 25$ mA
2. $C_1 = 0.5\ \mu F$; $X_{C_1} = 1{,}000\ \Omega$; $X_{C_2} = 1{,}000\ \Omega$; $X_{C_3} = 500\ \Omega$; $I_2 = 2$ mA; $I_T = 8$ mA

USING EXCEL

1. $X_C = 159\ \Omega$
2. $C = 0.025\ \mu F$

REVIEW QUESTIONS

1. (c)
2. (d)
3. (b)
4. (c)
5. (b)
6. (a)
7. (b)
8. (b)
9. (c)
10. (b)
11. (a)
12. (a)
13. (c)
14. (b)
15. (b)
16. (b)
17. (b)
18. (a)
19. (b)
20. $f = 1/2\pi CX_C$

PROBLEMS

1. $X_C = \dfrac{1}{6.28fC} = \dfrac{1}{7.85 \times 10^{-6}} = 127.38\ \text{k}\Omega$
2. $\dfrac{1}{6.28fX_C} = \dfrac{1}{6.28 \times 1500 \times 250} = \dfrac{1}{2{,}355{,}000} = 0.425\ \mu F$
3. X_C decreases to one-third its original value, since X_C is inverse to frequency (X_C = fiπfC).
4. Total X_C will decrease to half the original value, since X_C is inverse to C, and total C will double.
5. For the circuit shown:

   $C_T = C_1 \times C_2/C_1 + C_2 = 0.636 \times 1.59/0.636 + 1.59 = 0.454\ \mu F$

   $X_{C_T} = X_{C_1} + X_{C_2} = 25\ \text{k}\Omega + 10\ \text{k}\Omega = 35\ \text{k}\Omega$

   $V_A = 70$ V (sum of V_{C_1} and V_{C_2})

   $C_1 = 0.159/10\ \text{Hz} \times X_C$, where $X_C = V/I = 50\ \text{V}/2\ \text{mA} = 25\ \text{k}\Omega$. Thus, $C_1 = 0.159/10 \times 25\ \text{k}\Omega = 0.159/250 \times 10^3 = 0.636\ \mu F$.

$C_2 = 0.159/10\text{ Hz} \times X_C$; where $X_C = V/I = 20\text{ V}/2\text{ mA} = 10\text{ k}\Omega$. Thus, $C_2 = 0.159/10 \times 10\text{ k}\Omega = 0.159/100 \times 10^3 = 1.59\ \mu\text{F}$.

$X_{C_2} = 10\text{ k}\Omega$

6. Voltage on each capacitor is 35 volts. (**NOTE:** The current, total C, and total X_C would be changed under these circumstances.)
7. a. C_1 will RTS.
   b. C_2 will RTS.
   c. C_T will RTS.
   d. X_{C_1} will D.
   e. X_{C_2} will D.
   f. X_{C_T} will D.
   g. I will I.
   h. V_{C_1} will RTS.
   i. V_{C_2} will RTS.
   j. V_A will RTS.
8. $C_T = C_1 + C_2 = 2\ \mu\text{F} + 1\ \mu\text{F} = 3\ \mu\text{F}$
   $I_1 = V/X_{C_1} = 25\text{ V}/3.975\text{ k}\Omega = 6.29\text{ mA}$
   $I_2 = V/X_{C_2} = 25\text{ V}/7.95\text{ k}\Omega = 3.145\text{ mA}$
   $X_{C_T} = X_{C_1} \times X_{C_2}/X_{C_1} + X_{C_2} = 3{,}975 \times 7{,}950/3{,}975 + 7{,}950 = 2{,}650\ \Omega$
   $X_{C_1} = 0.159/20 \times 2 \times 10^{-6} = 3{,}975\ \Omega$
   $X_{C_2} = 0.159/20 \times 1 \times 10^{-6} = 7{,}950\ \Omega$
   $I_T = I_1 + I_2 = 6.29\text{ mA} + 3.145\text{ mA} = 9.435\text{ mA}$ (or $V_T/X_{C_T} = 25\text{ V}/2.65\text{ k}\Omega$)
9. Total current increases to 12.56 mA if $C_2 = C_1$ and all other factors remain the same.
10. a. C_1 will RTS.
    b. C_2 will RTS.
    c. C_T will RTS.
    d. X_{C_1} will D.
    e. X_{C_2} will D.
    f. X_{C_T} will D.
    g. I will I.
    h. V_{C_1} will RTS.
    i. V_{C_2} will RTS.
    j. V_A will RTS.
11. X_C of $C_1 = 1{,}000\ \Omega$
12. $X_{C_T} \cong 2{,}500\ \Omega$
13. $C_T = 0.063\ \mu\text{F}$
14. $X_{C_T} \cong 500\ \Omega$
15. increase

ANALYSIS QUESTIONS

1. A capacitor can be said to be an "open circuit" with respect to dc because the insulating dielectric material between the capacitor's plates breaks the dc current path. Only very small leakage current can pass through the insulating material.
2. A capacitor "passes" ac current via its charging and discharging action. The fact that electrons alternately gather on one plate while leaving the other, and vice versa, allows alternating current to pass through the components in series with the capacitor.
3. Capacitive voltage divider with the parameters specified is shown.

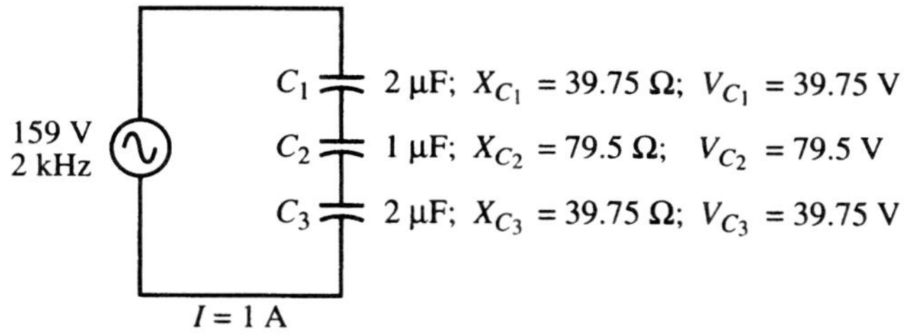

4. Circuit where six 1-μF capacitors are used and the total capacitance is 0.66 μF is shown.

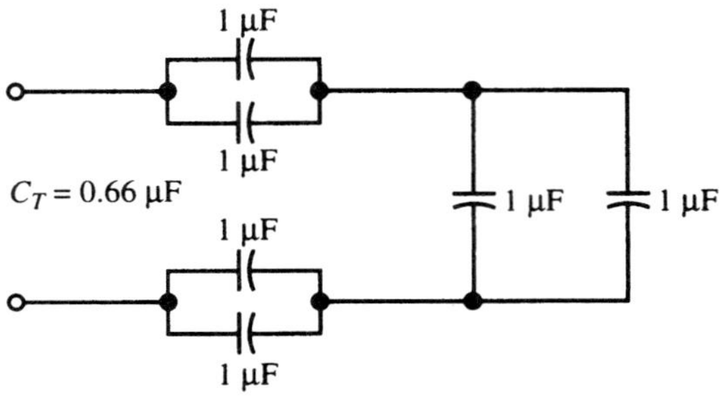

CHAPTER 19

PRACTICE PROBLEMS 1

1. $V_T = \sqrt{V_R^2 + V_C^2} = \sqrt{80^2 + 60^2} = \sqrt{10{,}000}$; $V_T = 100$ V

   Drawing:

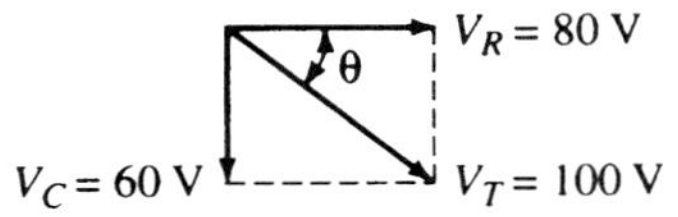

2. $Z = \sqrt{R^2 + X_C^2} = \sqrt{160^2 + 120^2} = \sqrt{40{,}000}$; Z = 200 Ω

   Drawing:

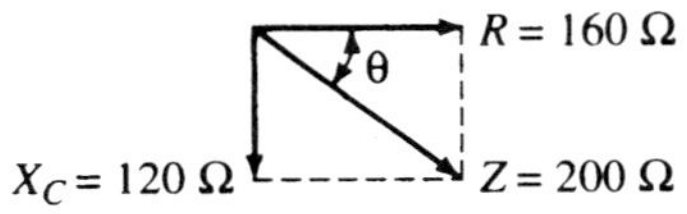

PRACTICE PROBLEMS 2

1. $V_T = 80$ V; Angle = –48.5 degrees

   Drawing:

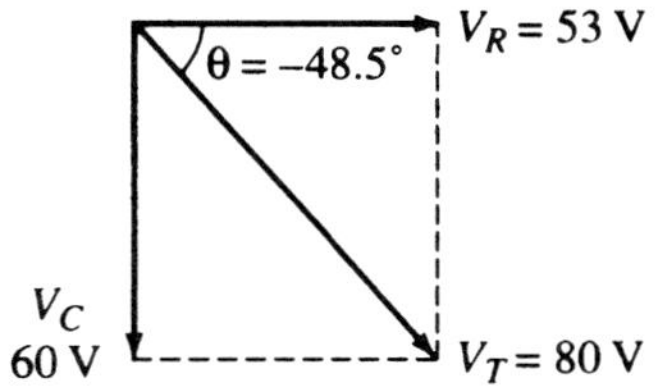

2. Angle = –48.5 degrees

   Drawing:

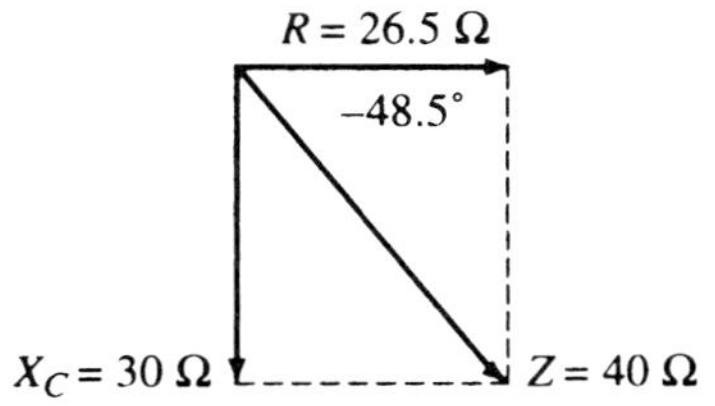

PRACTICE PROBLEMS 3

Angle of output voltage "lead" equals arctan X_C/R. With R set at 500 Ω:

$X_C = 1/(2\pi fC) = 1/(6.28 \times 400 \times 0.5 \times 10^{-6}) = 796\ \Omega$

$R = 500\ \Omega$

$X_C/R = 796/500 = 1.59$

Arctan 1.59 = angle of 57.8 degrees

Angle of lead = 57.8 degrees

PRACTICE PROBLEMS 4

Lag angle = 90 degrees minus circuit phase angle

Phase angle = arctan X_C/R

$X_C = 1/(2\pi fC) = 1/(6.28 \times 400 \times 0.5 \times 10^{-6}) = 796\ \Omega$

$R = 750\ \Omega$

$X_C/R = 796/750 = 1.06$

Arctan 1.06 = angle of 46.7 degrees

Angle of lag = 90 – 46.7 degrees = 43.3 degrees

IN-PROCESS LEARNING CHECK 1

1. In series *RC* circuits, the *current* is the reference vector, shown at 0 degrees.
2. In the *V-I* vector diagram of a series *RC* circuit, the circuit voltage is shown *lagging* the circuit current.
3. In series *RC* circuits, an impedance diagram *can* be drawn.
4. Using the Pythagorean approach, the formula to find V_T in a series *RC* circuit is $V_T = \sqrt{V_R^2 + V_C^2}$.
5. Using the Pythagorean approach, the formula to find Z in a series *RC* circuit is $Z = \sqrt{R^2 + X_C^2}$.
6. The trig function used to solve for phase angle when you do not know the hypotenuse value is the *tangent* function.

PRACTICE PROBLEMS 5

1. $I_T = 22.36$ mA
2. $\theta = 63.43$ degrees
3. $Z = 7.11$ kΩ

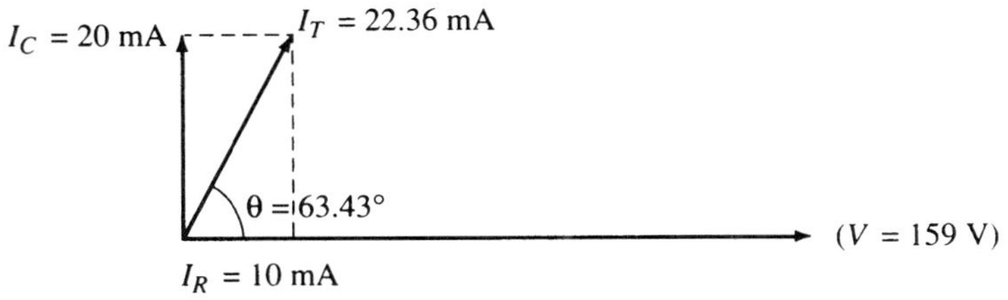

IN-PROCESS LEARNING CHECK 2

1. In parallel *RC* circuits, *voltage* is the reference vector in *V-I* vector diagrams.
2. In parallel *RC* circuits, current through the resistor(s) is plotted *out of phase* with the circuit current vector.
3. In parallel *RC* circuits, current through capacitor branch(es) is plotted at *+90 degrees* with respect to circuit applied voltage.
4. The Pythagorean formula to find circuit current in parallel *RC* circuits is $I_T = \sqrt{I_R^2 + I_C^2}$.

5. No.
6. Yes.
7. No.

PRACTICE PROBLEMS 10

1. $G = 1/R = 0.25$ S
2. $B_L = 1/X_C = 0.333$ S
3. $Y = \sqrt{G^2 + B_C^2} = 0.414$ S
4. $Z = 1/Y = 2.42\ \Omega$
5. $I_T = V \times Y = 24.84$ A
6. $V_T = I/Y = 60$ V
7. $Y = I/V = 0.414$ S
8. $\theta = R/X_L = 53.1°$

USING EXCEL

1. $V_T = 100$ V
2. $Z = 7{,}119\ \Omega$

REVIEW QUESTIONS

1. (d)
2. (b)
3. (b)
4. (a)
5. (b)
6. (a)
7. (b)
8. (b)
9. (b)
10. (c)
11. (b)
12. (a)
13. (b)
14. (c)
15. (b)
16. (c)
17. (b)
18. (a)
19. Ohm's Law and Pythagorean theorem ($I_T = V_T/Z$); and $I_T = \sqrt{I_R^2 + I_C^2}$
20. $Z = R \times X_C / \sqrt{R^2 + X_C^2}$

PROBLEMS

1. $Z = \sqrt{R^2 + X_C^2} = \sqrt{64 \times 10^6 + 225 \times 10^6} = \sqrt{189 \times 10^6} = 17\ \text{k}\Omega$
2. $V_C = I \times X_C = 2\ \text{mA} \times 15\ \text{k}\Omega = 30$ V
   $V_R = I \times R = 2\ \text{mA} \times 8\ \text{k}\Omega = 16$ V
   $V_T = I \times Z = 2\ \text{mA} \times 17\ \text{k}\Omega = 34$ V or $V_T = \sqrt{V_R^2 + V_C^2} = \sqrt{256 + 900} = \sqrt{1156} = 34$ V
3. $C = 0.159/fX_C = 0.159/1060 \times 15\ \text{k}\Omega = 0.159/15.9 \times 106 = 0.01\ \mu\text{F}$
4. Circuit voltage is lagging circuit current. Tan $\theta = V_C/V_R = 1.875$. Angle whose tan equals 1.875 is 61.92°. Voltage is lagging current by 61.92°.
5. a. R will RTS
   b. X_C will D
   c. Z will D
   d. I_T will I
   e. θ will D
   f. V_C will D
   g. V_R will I
   h. V_T will RTS
6. $I_C = V/X_C = 150\ \text{V}/30\ \text{k}\Omega = 5$ mA
   $I_T = \sqrt{I_R^2 + I_C^2} = \sqrt{100 \times 10^{-6} + 25 \times 10^{-6}} = 11.18$ mA
7. $R = V_R/I = 150\ \text{V}/10\ \text{mA} = 15\ \text{k}\Omega$
8. $Z = V_T/I_T = 150\ \text{V}/11.18\ \text{mA} = 13.41\ \text{k}\Omega$
9. $f = 0.159/CX_C = 0.159/0.02 \times 10^{-6} \times 30 \times 10^3 = 0.159/0.6 \times 10^{-3} = 265$ Hz
10. a. R will RTS
    b. X_C will D
    c. Z will D
    d. I_T will I
    e. θ will I
    f. I_C will I
    g. I_R will RTS
    h. V_T will RTS
11. $R_T = 20\ \text{k}\Omega + 18\ \text{k}\Omega + 30\ \text{k}\Omega = 68\ \text{k}\Omega$

12. $X_{C_T} = 30\ k\Omega + 22\ k\Omega = 52\ k\Omega$
13. $Z = \sqrt{R^2 + X_C^2} = \sqrt{(68 \times 10^3)^2 + (52 \times 10^3)^2} = \sqrt{7328 \times 10^6} = 85.6\ k\Omega$
14. Tan $\theta = X_C/R = \tan -52/68 = -0.764$. Angle whose tangent equals –0.764 is –37.4°.
15. $f = 0.159/CX_C = .159/.1 \times 10^{-6} \times 30 \times 10^3 = .159/.003 = 53$ Hz
16. $I_R = 320\ V/10\ k\Omega = 32$ mA
17. $I_C = 320\ V/16\ k\Omega = 20$ mA
18. $I_T = \sqrt{I_R^2 + I_C^2} = \sqrt{(32 \times 10^{-3})^2 + (20 \times 10^{-3})^2} = \sqrt{1424 \times 10^{-6}} = 37.74$ mA
19. Tan $\theta = I_C/I_R = 20/32 = 0.625$. Angle whose tangent equals 0.625 is 32°.
    $\theta = 32°$
20. $Z_T = 320\ V/37.74\ mA = 8.48\ k\Omega$
21. $R = 750\ \Omega$ $\quad V_R = 15$ V
    $X_C = 1{,}000\ \Omega$ $\quad V_C = 20$ V
    $I = 20$ mA $\quad \theta = -53.1°$
22. $R = 1.15\ k\Omega$ $\quad Z = 1\ k\Omega$
    $I_C = 2.5$ mA $\quad \theta = 30°$
    $I_R = 4.33$ mA

ANALYSIS QUESTIONS

1. Two basic circuit operation differences between series *RC* circuits and series *RL* circuits are:
   a. Circuit current leads circuit voltage in series *RC* circuits and lags circuit voltage in series *RL* circuits.
   b. As frequency of operation is increased, or *C* is increased in series *RC* circuits, the phase angle between circuit current and circuit voltage decreases. For a series *RL* circuit, as frequency increases, or *L* is increased, the circuit phase angle increases.
2. Two basic circuit operation differences between parallel *RC* circuits and parallel *RL* circuits are:
   a. Circuit current leads circuit voltage in parallel *RC* circuits and lags circuit voltage in parallel *RL* circuits.
   b. As frequency of operation is increased, or *C* is increased in parallel *RC* circuits, the phase angle between circuit current and circuit voltage increases. For a parallel *RL* circuit, as frequency increases, or *L* is increased, the circuit phase angle decreases.
3. Yes, there are similar circuit effects when comparing a series *RC* circuit and a parallel *RL* circuit. For example, as frequency of operation is increased, a series *RC* circuit will cause the phase angle to decrease. The same is true for a parallel *RL* circuit; that is, an increase in *f* will cause the phase angle to decrease.
4. Diagram:

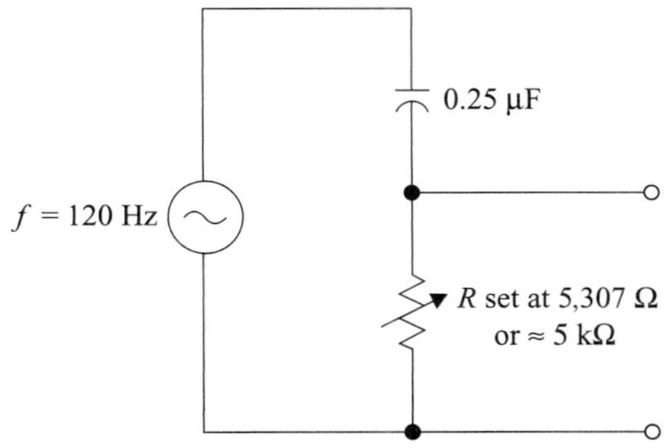

MULTISIM EXERCISE

2. $I_C \cong 20$ mA; $I_R \cong 10$ mA; $I_T \cong 22$ mA
3. Yes.
4. $I_C \cong 10$ mA; $I_R \cong 10$ mA; $I_T \cong 14.2$ mA
5. Yes; decrease

CHAPTER 20

PRACTICE PROBLEMS 1

$R = 200\ \Omega$

$X_{CT} = 500\ \Omega$

$X_L = 250\ \Omega$

$Z = \sqrt{R^2 + X^2} = \sqrt{200^2 + 250^2} = \sqrt{102{,}500} = 320.15\ \Omega$

$V_T = I \times Z = 0.25 \times 320.15 = 80.04$ V

$V_{C_2} = I \times X_{C_2} = 0.25 \times 250 = 62.5$ V

$V_L = I \times X_L = 0.25 \times 250 = 62.5$ V

$V_R = I \times R = 0.25 \times 200 = 50$ V

Tan $\theta = 250/200 = 1.25$

Arctan 1.25 = angle of 51.34 degrees

Phase angle = −51.34 degrees (V with respect to I)

IN-PROCESS LEARNING CHECK 1

1. When both capacitive and inductive reactance exist in the same circuit, the net reactance is *the difference between their reactances.*
2. When plotting reactances for a series circuit containing L and C, the X_L is plotted *upward* from the reference, and the X_C is plotted *downward* from the reference.
3. In a series circuit with both capacitive and inductive reactance, the capacitive voltage is plotted *180* degrees out-of-phase with the inductive voltage.
4. If inductive reactance is greater than capacitive reactance in a series circuit, the formula to find the circuit impedance is $Z = \sqrt{R^2 + (X_L - X_C)^2}$.
5. In a series *RLC* circuit, current is found by V/Z.
6. False.
7. If the frequency applied to a series *RLC* circuit decreases while the voltage applied remains the same, the component(s) whose voltage will increase is/are the *capacitor(s).*

PRACTICE PROBLEMS 2

$I_R = V/R = 300/200 = 1.5$ A

$I_L = V/X_L = 300/200 = 1.5$ A

$I_C = V/X_C = 300/300 = 1$ A

$I_X = I_L - I_C = 1.5 - 1 = 0.5$ A

$I_T = \sqrt{I_R^2 + I_X^2} = \sqrt{1.5^2 + 0.5^2} = \sqrt{2.5} = 1.58$ A

$Z = V/I = 300/1.58 = 189.8\ \Omega$

θ = angle whose tangent equals $I_X/I_R = 0.5/1.5 = 0.333$

Therefore, $\theta = -18.43$ degrees (I with respect to V)

IN-PROCESS LEARNING CHECK 2

1. In parallel *RLC* circuits, capacitive and inductive branch currents are *180 degrees* out-of-phase with each other.
2. The formula to find total current in a parallel *RLC* circuit when I_C is greater than I_L is:
   $I_T = \sqrt{I_R^2 + (I_C - I_L)^2}$
3. False.
4. If the frequency applied to a parallel *RLC* circuit increases while the applied voltage remains the same, the branch current(s) that will increase is/are the *capacitive* branch(es).

IN-PROCESS LEARNING CHECK 3

1. Power dissipated by a perfect capacitor over one whole cycle of input is *zero*.
2. The formula for apparent power is $S = V \times I$, and the unit expressing apparent power is *volt-amperes*.
3. True power in *RLC* circuits is dissipated by the *resistances*.
4. The formula for power factor is *p.f.* = $\cos \theta$.
5. If the power factor is one, the circuit is purely *resistive*.
6. The lower the power factor, the more *reactive* the circuit is acting.
7. In the power triangle, true power is on the *horizontal* axis, reactive power (VAR) is on the *vertical* axis, and apparent power is on the *hypotenuse* of the right triangle.

PRACTICE PROBLEMS 3

1. $20 + j30$ in rectangular form converts to $36.05 \angle 56.31$ degrees in polar form.
2. $30 - j20$ in rectangular form converts to $36.05 \angle -33.7$ degrees in polar form.

PRACTICE PROBLEMS 4

1. $66 \angle 60$ degrees converts to $33 + j57.15$ in rectangular form.
2. $75 \angle 48$ degrees converts to $50.2 + j55.7$ in rectangular form.

PRACTICE PROBLEMS 5

1. $20 + j5$
   $\underline{35 - j6}$
   $55 - j1$
2. $34.9 - j10$
   $\underline{15.4 + j12}$
   $50.3 + j2$
3. $35 - j6$
   $\underline{20 + j5}$ (change sign) $35 - j6$
   $\underline{-20 - j5}$
   $15 - j11$
4. $37.8 - j13$
   $\underline{22.9 - j7}$ (change sign) $37.8 - j13$
   $\underline{-22.9 + j7}$
   $14.9 - j6$
5. $5 \angle 20$ deg $\times$ $25 \angle -15$ deg $= 125 \angle 5$ deg
6. $24.5 \angle 31$ deg $\times$ $41 \angle 23$ deg $= 1{,}004.5 \angle 54$ deg
7. $5 \angle 20$ deg/$25 \angle -15$ deg $= 0.2 \angle 35$ deg
8. $33 \angle 34$ deg/$13 \angle 45$ deg $= 2.54 \angle -11$ deg
9. If $I = 12 \angle 25$ deg; then $I^2 = 144 \angle 50$ deg
10. If $I^2 = 49 \angle 66$ deg; then $I = 7 \angle 33$ deg

11. $\dfrac{5\angle 50\text{ deg} + (30 + j40)}{2.6\angle -15\text{ deg}}$

$= \dfrac{(3.2 + j3.8) + (30 + j40)}{2.6\angle -15\text{ deg}}$

$= \dfrac{33.2 + j43.8}{2.6\angle -15\text{ deg}}$

$= \dfrac{54.9\angle -52.8\text{ deg}}{2.6\angle -15\text{ deg}}$

$= 21.1\angle -37.8\text{ deg}$

12. $\dfrac{10\angle 53.1\text{ deg} + (25 - j25)}{4.8\angle 36\text{ deg}}$

$= \dfrac{31 - j17}{4.8\angle 36\text{ deg}}$

$= \dfrac{35\angle -28.7\text{ deg}}{4.8\angle 36\text{ deg}}$

$= 7.3\angle -64.7\text{ deg}$

USING EXCEL

1. $V_T = 80$ V
2. $jB_L = 0.00265$ s

REVIEW QUESTIONS

1. (b)
2. (c)
3. (d)
4. (c)
5. (b)
6. (d)
7. (b)
8. (a)
9. (c)
10. (c)
11. (c)
12. (c)
13. (c)
14. (c)
15. (b)
16. (c)
17. (a)
18. (c)
19. (b)
20. (b)

PROBLEMS

1. $Z = \sqrt{R^2 + X_T^2} = \sqrt{300^2 + 150^2} = \sqrt{112{,}500} = 335.4\ \Omega$

   $I = V_T/Z = 10\text{ V}/0.3354\text{ k}\Omega = 29.8\text{ mA}$

   $V_C = I \times X_C = 29.8\text{ mA} \times 250\ \Omega = 7.45\text{ V}$

   $V_L = I \times X_L = 29.8\text{ mA} \times 100\ \Omega = 2.98\text{ V}$

   $V_R = I \times R = 29.8\text{ mA} \times 300\ \Omega = 8.94\text{ V}$

   Cos θ = adj/hyp = $R/Z = 300/335.4 = 0.89$

   θ = angle whose cos = 0.89; = 26.56° (I_T will lead V_T)

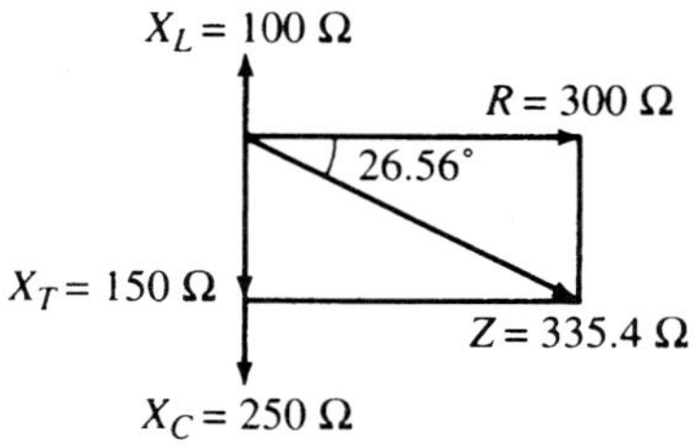

2. $Z = V_T/I_T = 10\text{ V}/68.48\text{ mA} = 146\ \Omega$

   $I_C = V_T/X_C = 10\text{ V}/250\ \Omega = 40\text{ mA}$

   $I_L = V_T/X_L = 10\text{ V}/100\ \Omega = 100\text{ mA}$

   $I_R = V_T/R = 10\text{ V}/300\ \Omega = 33.33\text{ mA}$

   $I_T = \sqrt{I_R{}^2 + I_{X_T}{}^2} = \sqrt{1{,}089 \times 10^{-6} + 3{,}600 \times 10^{-6}}$

   $= \sqrt{4{,}689 \times 10^{-6}} = 68.48\text{ mA}$

   Tan θ = opp/adj = $I_{X_T}/I_R = -60\text{ mA}/33.33\text{ mA} = -1.8$

   θ = angle whose tangent = −1.8; = approximately −61° (I_T will lag V_T)

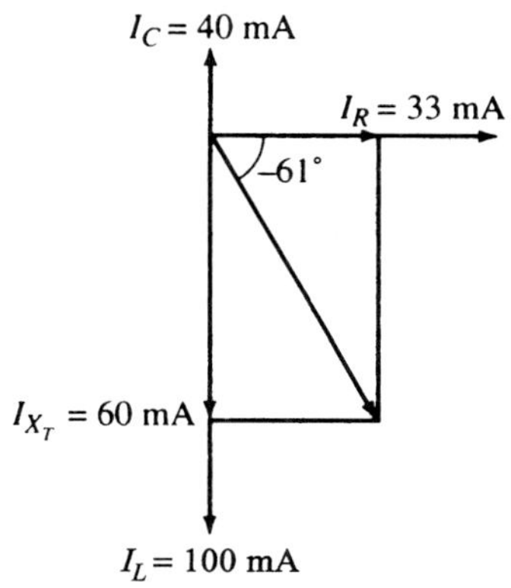

3. True power (P) = $I^2R = (0.0298)^2 \times 300 = 0.266$ W

   Apparent power (S) = $V \times I$ = 10 V × 0.0298 mA = 0.298 VA

   Reactive power (Q) = $VI \times \sin\theta$ = 10 V × 0.0298 × 0.447 = 0.133 VAR

   Power factor ($p.f.$) = true power/apparent power = 0.266/0.298 = 0.89
   (or $p.f.$ = Cos θ = cos 26.56° = 0.89)
4. Power triangle for the circuit is as shown.

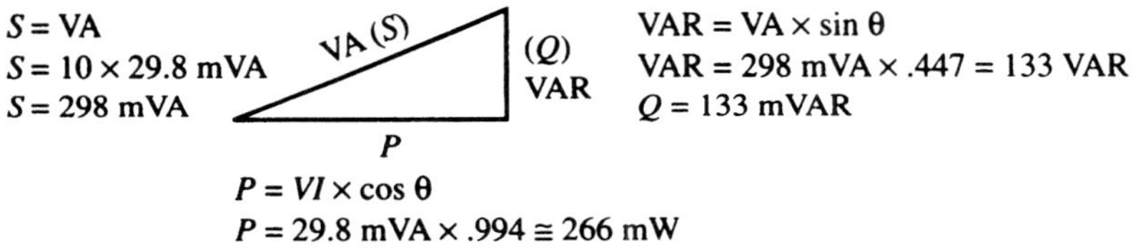

5. $Z = 20 + j40$

   If the circuit were an RC circuit, it would be stated as:

   $Z = 20 - j40$
6. $Z = 10 - j10$
7. $I_R = 20/20 = 1$ A

   $I_C = 20/20 = 1$ A

   $I_T = 1.414$ A at an angle of 45°

   $I_T = 1.414\ \angle 45°$ A
8. Converting $30 + j40$ to polar form: $50\ \angle 53.1°$

   That is, $Z = \sqrt{R^2 + X^2} = \sqrt{30^2 + 40^2} = 50$

   Tan θ = opp/adj = 40/30 = 1.33; therefore θ = 53.1°.
9. Converting $14.14\ \angle 45°$ to rectangular form: $10 + j10$

   That is, $R = Z \times \cos\theta = 14.14 \times \cos 45° = 10$, since the angle is 45°; R and X must be equal; thus $Z = R + jX_L$, $14.14 = 10 + j10$.
10. $I_T = 1.21\ \angle -14°$ mA

    True power = 11.7 mW
11. $Z = 61.43\ \angle -75.96°\ \text{k}\Omega$ (polar form)

    $Z = 14900 - j59600$ (rectangular form)
12. $V_L = 80\ \angle 90°$ volts

13. V_C will decrease.
14. True power = 166.67 W
15. $Z = 1.79 \angle{-63.4°}\ \Omega$
16. $C = 4.6\ \mu F$
17. $Z = 5.82 \angle 14°\ \Omega$
18. $G = 1/400 = 0.0025$
$-jB_L = 1/500 = 0.002$
$jB_C = 1/800 = 0.00125$
$Y = G + j(B_C - B_L)$
$Y = 0.0025 + j(0.00125 - 0.002)$
$Y = 0.0025 + j(-0.00075)$
$Y = \sqrt{0.0025^2 + (-0.00075)^2}$
$Y = 0.0026$ siemens
19. $I_R = 800\ V/400\ \Omega = 2\ A$
$I_C = 800\ V/800\ \Omega = 1\ A$
$I_L = 800\ V/500\ \Omega = 1.6\ A$
$I_{X_T} = 1.6\ A - 1\ A = 0.6\ A$
$I_T = 2^2 + 0.6^2 = 2.09\ A$
$Z = V_T/I_T = 800\ V/2.09\ A = 383.8\ \Omega$
$\theta = 16.69°$
20. $X_C = 500\ \Omega$
21. I_R will RTS
I_C will I
I_L will D
I_T will I
Z_T will D
θ will I
22. $10 - j10\ \Omega = 14.14 \angle{-45°}\ \Omega$ in polar form.
23. $1.414 \angle 45°\ A = 1 + j1\ A$ in rectangular form.
24. $1.21 \angle{-14°}\ mA = 1.17 - j0.29\ mA$ in rectangular form.
25. $1.79 \angle{-63.4°}\ \Omega = 0.8 - j1.6\ \Omega$ in rectangular form.
26. $Z_1 = 10 - j15$ (Rectangular) $= 18.02 \angle{-56.3°}$ (Polar)
$Z_2 = 15 + j10$ (Rectangular) $= 18.02 \angle 33.69°$ (Polar)

$$Z_T = \frac{Z_1 \times Z_2}{Z_1 + Z_2}$$

$$Z_T = \frac{18.02 \angle{-56.3°} \times 18.02 \angle 33.69°}{10 - j15 + 15 + j10} = \frac{325 \angle{-22.61°}}{25.5 \angle{-11.3}}$$

$Z_T = 12.75 \angle{-11.3°}\ \Omega$ (Polar)
27. $Z_T = 12.5 - j2.5\ \Omega$ (Rectangular)
28. Branch 1 current = $V/Z_1 = 50\ V \angle 0°/18.02 \angle{-56.3°} = 2.77 \angle{-56.3°}\ A$
Branch 2 current = $V/Z_2 = 50\ V \angle 0°/18.02 \angle 33.69° = 2.77 \angle{-33.7°}$
Branch 1 current (Rectangular) = $1.54 + j2.3\ A$
Branch 2 current (Rectangular) = $2.30 - j1.5\ A$
Total current (Rectangular) = $3.84 + j0.8$; therefore in polar form $I_T = 3.92 \angle 11.76°\ A$
29. I_T in rectangular form = $3.84 + j0.8\ A$

ANALYSIS QUESTIONS

1. A real number in math can be any rational or irrational number. In complex numbers related to the coordinate system, the real term in the complex number is plotted on the horizontal (x) axis of the coordinate system. Imaginary numbers are complex numbers; they involve the square roots of negative numbers. The imaginary unit, or j operator, represents the square root of −1. For our purposes, the j operator is the vertical element used in ac analysis.
2. A complex number is a mathematical term that expresses both a "real" and an "imaginary" term. Complex numbers must be added vectorially, rather than simply arithmetically. The "imaginary" term in a complex number is graphically shown on the vertical (y) axis of the coordinate system.

3. The reason simply measuring voltage with a voltmeter and current with a current meter does not provide adequate information to find the actual power dissipated by the circuit is that in ac circuits containing reactances, currents and voltages are not generally in phase with one another. Thus, taking the product of values that are not in phase will not yield the actual power being dissipated.
4. One method to make measurements in order to determine actual power dissipation in an ac circuit with reactive components is to:
   a. Measure all resistances in the circuit (including the *R*s of reactive components)
   b. Measure the circuit current
   c. Use the I^2R technique to find true power dissipation

CHAPTER 21

IN-PROCESS LEARNING CHECK 1

1. In circuits containing both inductance and capacitance, when the source frequency increases, inductive reactance *increases* and capacitive reactance *decreases*.
2. In a series *RLC* (or *LC*) circuit, when X_L equals X_C, the circuit is said to be at *resonance*.
3. In a series *RLC* (or *LC*) circuit, when X_L equals X_C, the circuit impedance is *minimum*; the circuit current is *maximum*; and the phase angle between applied voltage and current is *zero* degrees.
4. In a series *RLC* (or *LC*) circuit, when X_L equals X_C, the voltage across the inductor or the capacitor is *maximum*.

PRACTICE PROBLEMS 1

$f_r = 1/(2\pi \sqrt{LC}) = 1/(6.28 \times \sqrt{250 \times 10^{-6} \times 500 \times 10^{-12}} = 450.4\ \text{kHz}$

PRACTICE PROBLEMS 2

$L = 25{,}330/(f^2 C) \approx 10.13\ \mu\text{H}$

PRACTICE PROBLEMS 3

1. $Q = X_L/R = 1{,}000/50 = 20$
2. $V_C = I \times X_C = 0.24\ \text{A} \times 1{,}000\ \Omega = 240\ \text{V}$ (or $Q \times V_A = 20 \times 12 = 240\ \text{V}$)
   where: $I = V_T/Z = 12/50 = 0.24\ \text{A}$
3. If r_s doubles, $Q = 1{,}000/100 = 10$.
4. If r_s is made smaller in value, the skirts are steeper.

IN-PROCESS LEARNING CHECK 2

1. In a high-Q parallel resonant circuit, Z is *maximum*.
2. In a high-Q parallel resonant circuit using ideal components, X_L *does* equal X_C at resonance.
3. The power factor of an ideal parallel resonant circuit is *1*.
4. Phase angle in an ideal parallel resonant circuit is *0 degrees*.
5. In a parallel resonant circuit, the current circulating in the tank circuit is *higher* than the line current.

PRACTICE PROBLEMS 4

1. $f_r = 1/(2\pi \sqrt{LC}) = 2.517\ \text{MHz}$
2. $Q = X_L/R \approx 316$
3. $Z = Q \times X_L \approx 998\ \text{k}\Omega$
4. If r_s is $1\ \text{k}\Omega$, the f_r is lower.

PRACTICE PROBLEMS 5

1. $f_1 \approx 2{,}494$ kHz and $f_2 \approx 2{,}506$ kHz
2. $f_r = 500$ kHz and bandpass = from 495 kHz to 505 kHz
3. $f_1 = 3{,}200$ kHz − (2 kHz/2) = 3,200 kHz − 1 kHz = 3,199 kHz

USING EXCEL

1. $f = 450.4$ kHz
2. $C = 253$ pF

REVIEW QUESTIONS

1. (b)
2. (b)
3. (a)
4. (c)
5. (b)
6. (c)
7. (c)
8. (b)
9. (b)
10. (a)
11. (c)
12. (d)
13. (c)
14. (a)
15. (a)
16. (c)
17. (b), (c)
18. (c)
19. (b)
20. (b)

PROBLEMS

1. a. At resonance $Z = R$; therefore, $Z = 10\ \Omega$.
   b. $f_r = 0.159/\sqrt{LC} = 0.159/\sqrt{1 \times 10^{-6} \times 100 \times 10^{-12}} = 15.9$ MHz (or use $\frac{1}{6.28\sqrt{LC}}$)
   c. $Q = X_L/R$
   $X_L = 2\pi fL = 6.28 \times 15.9 \times 10^6 \times 1 \times 10^{-6} = 99.8\ \Omega$
   $Q = 99.8/10 = 9.98$
   d. $BW = f_r/Q = 15.9$ MHz/9.98 = 1.59 MHz
   e. Bandpass = 15.105 MHz to 16.695 MHz.
   f. Below resonance, circuit acts capacitively.
2. a. $V_C = Q \times V_A = 9.98 \times 20$; $V_C = 199.6$ V
   b. I at resonance = $V_A/R = 20\ \text{V}/10\ \Omega = 2$ A
   c. True power = $I^2R = 2^2 \times 10\ \Omega = 40$ W
   d. Net reactive power is 0 VAR.
3. a. $f_r = 0.159/\sqrt{LC} = 0.159/\sqrt{28 \times 10^{-6} \times 30 \times 10^{-12}} = 0.159/\sqrt{840 \times 10^{-18}} = 159000000/28.98$
   $= 5.486$ MHz (or use $\frac{1}{6.28\sqrt{LC}}$)
   b. $Q = X_L/R$
   $X_L = 2\pi fL = 6.28 \times 5.486 \times 10^6 \times 28 \times 10^{-6} = 964.65\ \Omega$
   $Q = 964.65/30 = 32.155$
   c. $BW = f_r/Q = 5.486$ MHz/32.155 = 0.17 MHz
   d. Bandpass = 5.316 MHz to 5.65 MHz.
   e. At resonance, $Z = Q \times X_L = 32.155 \times 964.65 = 31{,}018.6\ \Omega$.
   f. Below resonance, circuit acts inductively.
4. I line = $V/Z = 20\ \text{V}/31.018\ \text{k}\Omega = 0.6447$ mA
   $I_C = Q \times I$ line = 32.155×0.6447 mA = 20.73 mA

5. Resonant frequency is 1,060 Hz, $f_r = 0.159/\sqrt{LC}$ (or use $\frac{1}{6.28\sqrt{LC}}$).
6. Phase angle will be 45° at a frequency of 39 kHz and 41 kHz. (Start solution with $\Delta f = f_r/Q$, 45° phase angle occurs at a half power point.)
7. $L = 0.02533/f_r^2C = 45$ mH
8. $C = 0.02533/f_r^2L = 4.79$ pF
9. R value must be 3.33 Ω. (Start with $Q = f_r/\Delta f$...)
10. At resonance, voltage across $R = V$ applied or 100 V.
11. Resonant frequency is 795 kHz, $f_r = 0.159/\sqrt{LC}$ or $\frac{1}{2\pi\sqrt{LC}}$.
12. I_T will lead V_A due to higher capacitive branch current and lower inductive branch current. I_T would be capacitive.
13. Exhibited impedance equals 104 kΩ. (Solve using the formulas for resonant frequency, X_L, Q, and finally $Z = Q \times X_L$.)

ANALYSIS QUESTIONS

1. The tuned circuit in a receiver input stage is typically a parallel *LC* circuit.
2. A double-tuned transformer (as used in receivers), is a transformer that has a capacitor used in conjunction with the transformer primary, and a capacitor, used in conjunction with the transformer secondary. The *LC* circuit formed by the primary components can be adjusted to resonance at the desired signal frequency(ies). Also, the secondary components can be adjusted to be resonant at the desired signal frequency(ies) to optimally pass along the desired signals to the next stage, which is connected to the secondary side of the tuned transformer.
3. They would be considered resonant filters.
4. <u>e</u> Parallel resonant

   <u>f</u> Low-pass

   <u>d</u> Bandpass

   <u>b</u> High-pass

   <u>c</u> Bandstop

   <u>a</u> Series resonant
5. *Selectivity* is the ability to select desired signal frequencies, while rejecting undesired signal frequencies. Referring to resonant circuits, selectivity indicates the sharpness of the response curve. A narrow and more pointed response curve results in a higher degree of frequency selectiveness, or selectivity.

   Bandwidth is the frequency limit's *total difference in frequency* between upper and lower frequencies over which a specified response level is shown by a frequency-sensitive circuit, such as a resonant circuit. The cutoff level is defined at 70.7% of peak response, sometimes called the half-power points on the response curve.

   Bandpass is the band of frequencies that passes through a filter or circuit with minimal attenuation or degradation; often *designated by specific frequencies.* When specified, they are designated indicating that the lower frequency is at the lower frequency half-power point of the response curve and the upper frequency is designated as the frequency which is at the upper half-power point on the response curve. Bandpass is often expressed as *the number of hertz between these two frequency response limits.*

6. a. X_L will I.
   b. X_C will D.
   c. R will RTS.
   d. Z will I.
   e. I will D.
   f. θ will I.
   g. V_C will D.
   h. V_L will I.
   i. V_R will D.
   j. V_T will RTS.
   k. BW will RTS.
   l. Q will RTS (close to the same).
   m. I_C will D.
   n. I_L will D.
   o. I_T will D.
7. a. X_L will D.
   b. X_C will I.
   c. R will RTS.
   d. Z will I.
   e. I will D.
   f. θ will I.
   g. V_C will I.
   h. V_L will D.
   i. V_R will D.
   j. V_T will RTS.
   k. BW will RTS.
   l. Q will RTS (close to the same).
   m. I_C will D.
   n. I_L will D.
   o. I_T will D.
8. a. X_L will I.
   b. X_C will D.
   c. R will RTS.
   d. Z will D.
   e. I will I.
   f. θ will I.
   g. V_C will RTS.
   h. V_L will RTS.
   i. V_R will RTS.
   j. V_T will RTS.
   k. BW will RTS.
   l. Q will RTS.
   m. I_C will I.
   n. I_L will D.
   o. I_T will I.
9. a. X_L will D.
   b. X_C will I.
   c. R will RTS.
   d. Z will D.
   e. I will I.
   f. θ will I.
   g. V_C will RTS.
   h. V_L will RTS.
   i. V_R will RTS.
   j. V_T will RTS.
   k. BW will RTS.
   l. Q will RTS.
   m. I_C will D.
   n. I_L will I.
   o. I_T will I.
10. The *m*-derived filter is a modified version of a constant-*k* filter. Its characteristics are such that the "*m*" factor relates to the ratio of the filter cutoff frequency to the frequency of infinite attenuation. In general, *m* ranges in values between 0 and 1 (typically around 0.6).

    The constant-*k* filter is one in which the product of the impedances in the filter "arms" (typically a capacitive element and an inductive element) is constant at any frequency. The "*k*" represents whatever this "constant" factor is (e.g., $Z_1Z_2 = k^2$).

    The purpose for such filter design parameters (constant-*k* and *m*-derived) is to have filters whose input and output impedances are relatively constant over the frequency range of expected operation, or the passband—that is, to maintain their impedance match between output (load) and input (source) connected to the filter network.
11. The condition is such that $X_L = X_C$.

12. Circuit diagram of a pi-type *L-C* or *LC* filter for a power supply. (This is a low-pass type filter.)

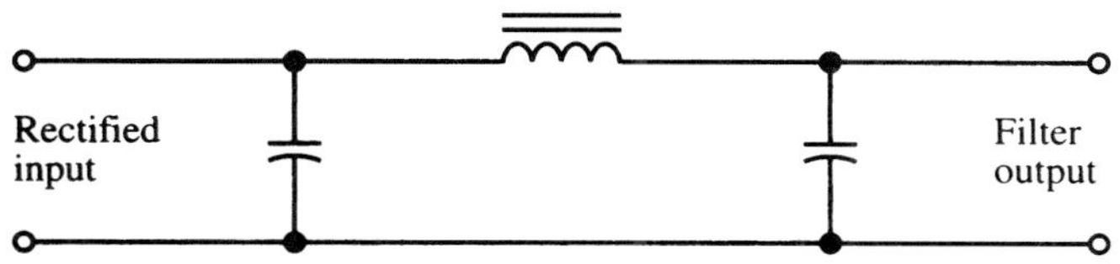

13. Circuit diagram of a bandpass filter using a series resonant and a parallel resonant circuit:

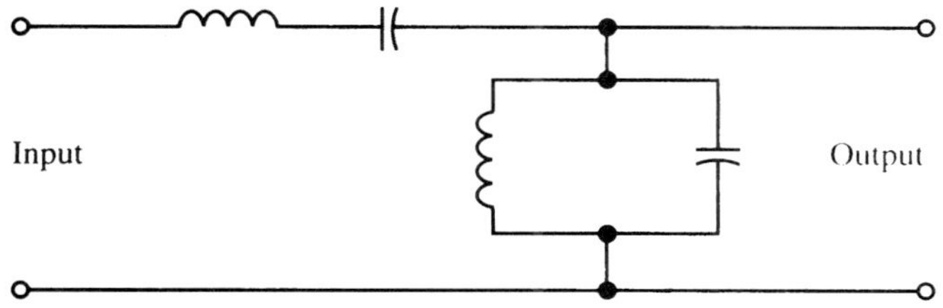

14. (d)

MULTISIM EXERCISE 1

2. $V_C = 100$ V; $V_L = 99$ V; $V_R = 10$ V
3. Yes.
4. ≈ 10
5. No. Reactances are no longer equal. ($V_C \cong 3.3$ V; $V_L \cong 13$ V)

MULTISIM EXERCISE 2

2. $V_L \cong 10$ V; $V_R \cong 31$ V; $I_L \cong 3$ mA
3. $X_L \cong 3.3$ kΩ; $Q = \dfrac{3.3\ \text{k}\Omega}{10\Omega} = 330$
4. Yes
5. $Z \cong 120$ kΩ
6. $\approx 180°$

CHAPTER 22

IN-PROCESS LEARNING CHECK 1

1. Before doping (in their pure form), semiconductor materials are sometimes called *intrinsic* semiconductors. Once they are doped with tiny amounts of an impurity atom, they are called *extrinsic* semiconductors.
2. N-type materials are formed by doping a *tetra*valent semiconductor material with a *penta*valent material. P-type materials are formed by doping a *tetra*valent semiconductor material with a *tri*valent material.
3. The doping atoms for N-type materials are called *donor* atoms because they donate extra electrons to the covalent bonds. The doping atoms for P-type materials are called *acceptor* atoms because they accept electrons that will fill the holes in the covalent bonds.
4. The majority carriers in an N-type material are *electrons*, and the minority carriers are *holes*. The majority carriers in a P-type material are *holes*, and the minority carriers are *electrons*.

USING EXCEL

$V_S = 6$ V; $V_D = 0.7$ V; $R = 1$ kΩ

Therefore, I = 0.0053 A

$V_D = 0.7$ V; $I = 0.0053$ A

Therefore, P = 0.00371 W

REVIEW QUESTIONS

1. a. A *trivalent* atom has three valence electrons.
   b. A *tetravalent* atom has four valence electrons.
   c. A *pentavalent* atom has five valence electrons.
2. a. 4
   b. 5
   c. 3
3. a. electrons
   b. holes
4. a. holes
   b. electrons
5. a. 0.7 V
   b. 0.3 V
6. (a)
7. (b)
8. (b)
9. P-N junction semiconductor showing an external dc power supply connected for forward biasing:

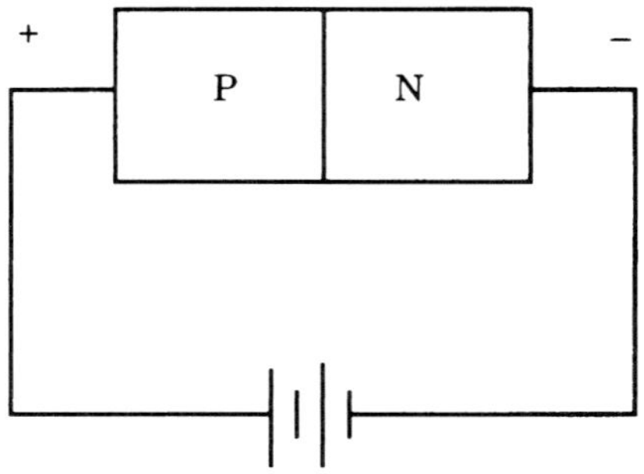

10. P-N junction semiconductor showing an external dc power supply connected for reverse biasing:

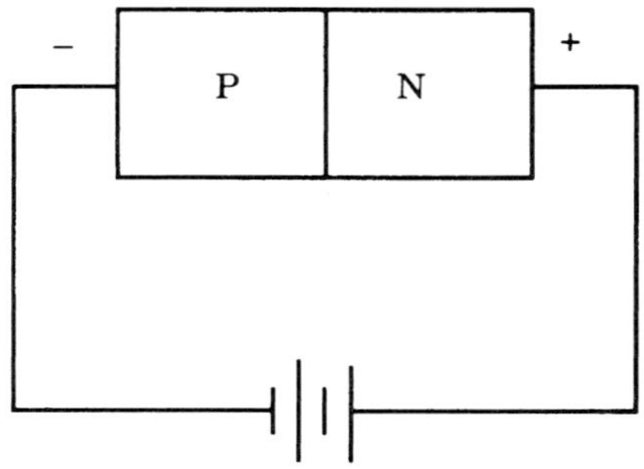

11. Forward-bias portion of the *I-V* curve for a P-N junction:

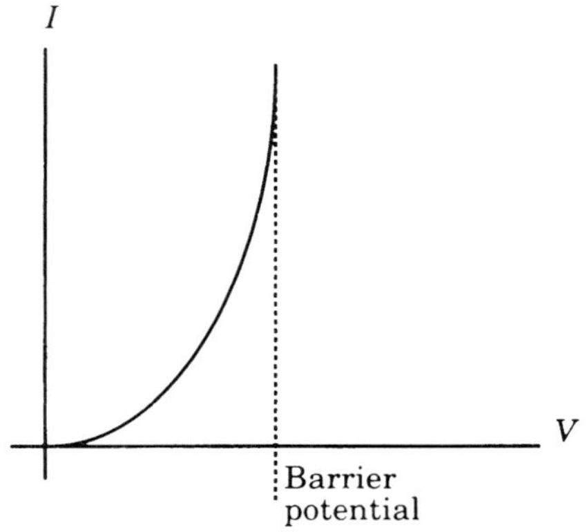

12. Reverse-bias portion of the I-V curve for a P-N junction:

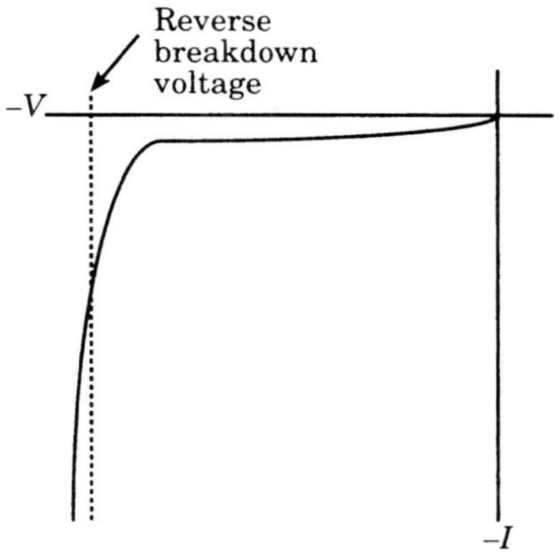

PROBLEMS

1. 32
2. 8
3. 3rd

ANALYSIS QUESTIONS

1. Valence-band electrons are bound to their atom and are not free to move from one atom to another in the material. Electrons in the conduction band are free to move. Also, conduction-band electrons possess more energy than their corresponding valence-band electrons.
2. An intrinsic semiconductor is one that has not been doped to perform as an N- or P-type material. An extrinsic semiconductor material is one that has been doped to create an N- or P-type material.
3. Increasing the amount of forward bias narrows the depletion region. Decreasing the amount of forward bias or applying a reverse bias increases the width of the depletion region.
4. The conductance of a P-N junction is greater when it is forward biased because the resistance to conduction at the depletion region is smaller.
5. In a semiconductor, holes flow from positive to negative, and electrons flow from negative to positive. Conventional current flow shows current flowing from positive to negative, and the electron-flow convention shows current flowing from negative to positive. So in a manner of speaking, hole flow is similar to conventional current flow, and electron flow is similar to the electron version of current flow.
6. Increasing the energy given to a charge carrier increases its likelihood of jumping the depletion region. It can be said that increasing the energy applied to a semiconductor makes it more conductive. In contrast, the resistance of regular conductors, like copper, increases as temperature increases.

MULTISIM EXERCISE

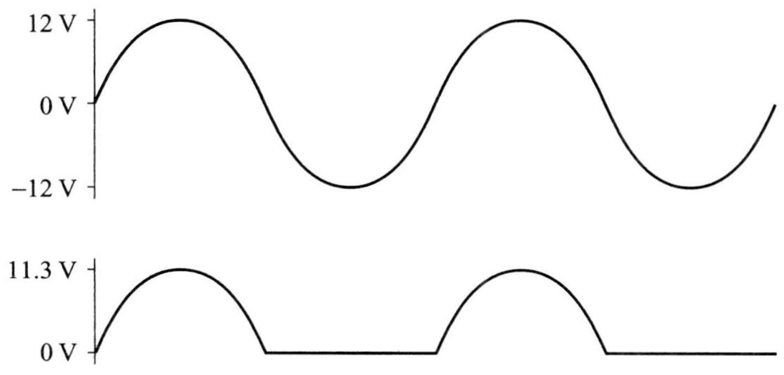

CHAPTER 23

PRACTICE PROBLEMS 1

1. −12.6 V
2. 0 V
3. 0.7 V
4. 11.9 V

PRACTICE PROBLEMS 2

1. 0.7 V
2. 75.3 V
3. 753 μA

IN-PROCESS LEARNING CHECK 1

1. To cause conduction in a diode, the diode must be *forward* biased.
2. To reverse bias a diode, connect the negative source voltage to the *P*-type material and the positive source voltage to the *N*-type material.
3. The anode of a junction diode corresponds to the *P*-type material and the cathode corresponds to the *N*-type material.
4. The forward conduction voltage drop across a silicon diode is approximately *0.7* volt.
5. When a diode is reverse biased in a circuit, it acts like an *open* switch.
6. When the diode is *reverse* biased, there will be no current flow through the diode.
7. When a diode is connected in series with a resistor, the voltage across the resistor is very nearly equal to the dc source voltage when the diode is *forward* biased.
8. The four most general diode ratings are *forward voltage drop*, *average forward current*, *peak reverse voltage*, and *reverse breakdown voltage*.

IN-PROCESS LEARNING CHECK 2

1. Rectifier diodes are used where it is necessary to change *ac* power into *dc* power.
2. Where additional cooling is necessary, a rectifier diode can be connected to a *heat sink* in order to dissipate heat more efficiently.
3. When an ac waveform is applied to a rectifier diode, the diode's *reverse breakdown voltage* rating must be greater than the peak voltage level.
4. The main current specification for rectifier diodes is *average forward current.*
5. To test the forward conduction of a diode with an ohmmeter, connect the *negative* lead of the meter to the cathode and the *positive* lead to the anode.
6. The forward resistance of a good diode should be much *less* than its reverse resistance.
7. Switching diodes have a *reverse recovery time* rating that is hundreds of times less than most rectifier diodes.
8. The type of diode circuit that removes the peaks from an input waveform is called a *clipper* circuit.
9. The type of diode circuit that changes the baseline level of an input waveform is called a *clamp* circuit.

PRACTICE PROBLEMS 3

1. 6 V
2. 12 V
3. 5 V, 14 V
4. 5 V, 6 V
5. 5 mA, 6 mA
6. 5 mA, 14 mA
7. 0 mA, 8 mA
8. 0 W, 48 mW

REVIEW QUESTIONS

1.

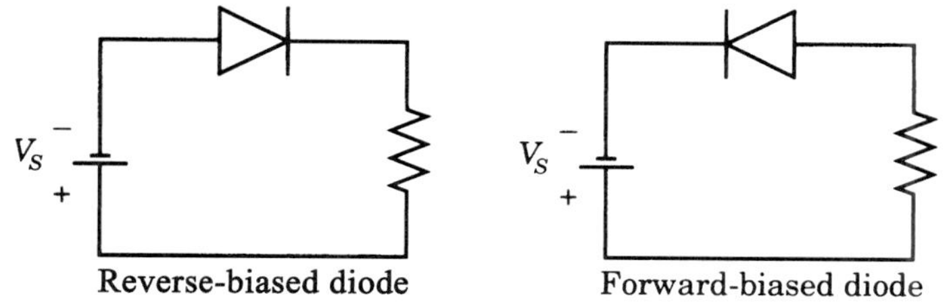

2.

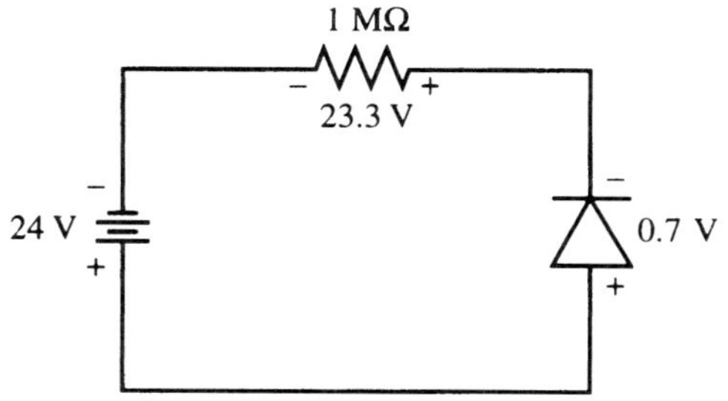

3.

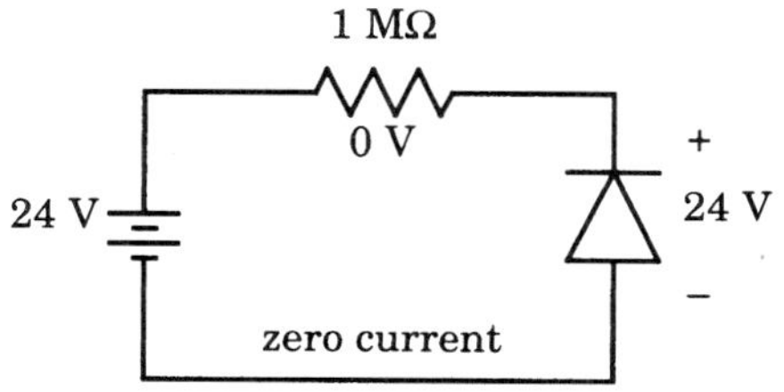

4. Maximum reverse voltage
$= 1.41 \times 28 \text{ V} = 39.5 \text{ V}$

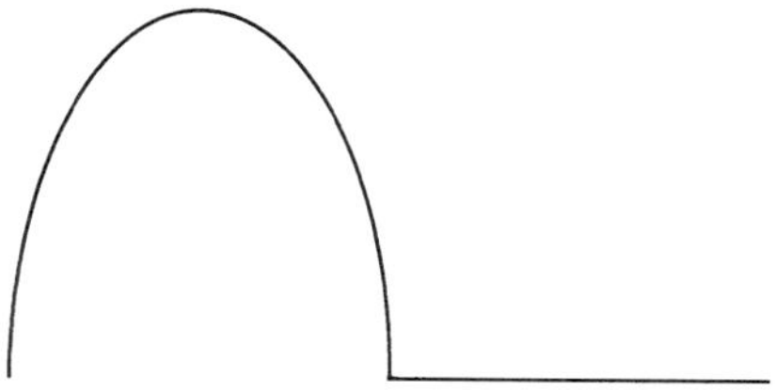

5.

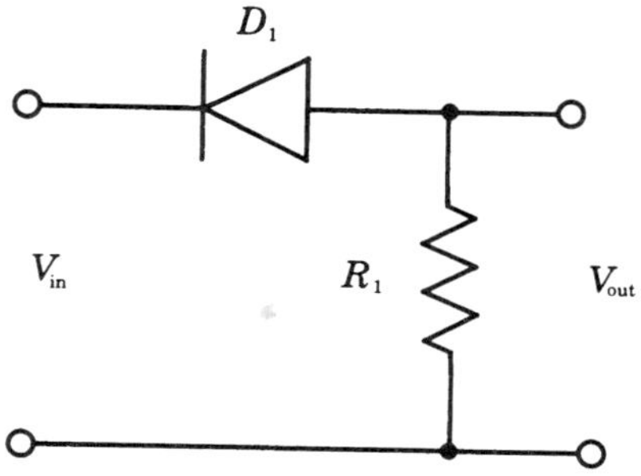

6.

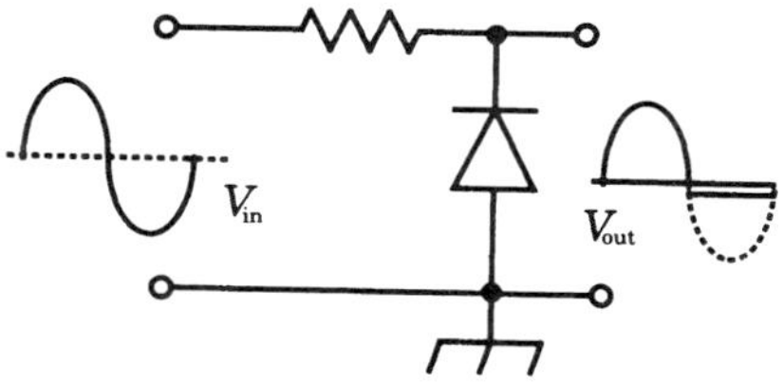

7. A diode clamp circuit shifts the baseline of an input waveform from zero volts to a different level. Whether the baseline is shifted positive or negative depends on the direction the diode is connected into the circuit.
8. Biased clippers and clamper circuits both shift the baseline level. The clipper, however, clips off a portion of the input signal, while a clamper leaves the signal unchanged.
9. a. The reverse breakdown voltage is the maximum amount of reverse bias a diode can withstand indefinitely.
   b. The forward conduction voltage is the voltage drop across a diode when it is carrying forward-bias current. This value is fairly constant at 0.3 V for germanium diodes and 0.7 V for silicon diodes.
   c. The average forward current specification is the maximum amount of forward current a diode can carry for an indefinite time.
   d. The reverse recovery time is the time that is required for changing the condition of a diode from reverse bias to forward bias.
10. The source voltage must be equal to or greater than the value of V_Z.
11. The voltage across the zener diode will be unchanged. The voltage across the resistor will increase. Increasing the applied voltage will cause the current through the circuit to increase.
12. The negative-resistance region is the peculiar feature of the I-V curve for tunnel diodes.

13.

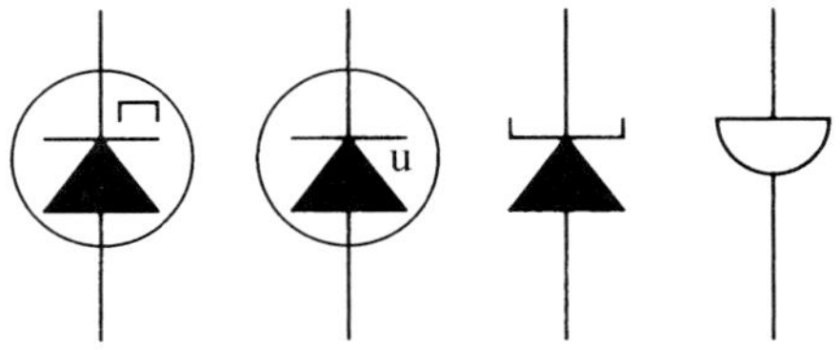

14. Oscillator element and electronic switch.

15. As the reverse-bias voltage increases, the depletion region becomes wider. The increased width of the depletion region causes a decrease in the capacitance of the varactor diode. With a decrease of the reverse-bias voltage, the depletion region becomes narrower, resulting in an increase of the varactor diode capacitance.

16. Schottky diodes use a hot-carrier junction that allows them to operate much faster than conventional diodes.

PROBLEMS

1. a. 5.3 mA
   b. 5.3 V
   c. 5.3 mA
2. a. 22.6 Ω
   b. 5.65 W
3. 100.7 V
4. a. 12 V
   b. 8 V
   c. 80 mA
   d. 960 mW
5. a. 3 V
   b. 25 mA
   c. 75 mW
   d. 225 mW
6. a. 3 Ω
   b. 3 V
   c. 3 W
   d. 9 W
7. 25 mW
8. Change in $V_R = 2$ V; change in $V_Z = 0$ V

ANALYSIS QUESTIONS

1. It is unlikely that any two silicon diodes will have exactly the same forward conduction voltage. This means that the diode with the lower forward conduction voltage will begin conducting first, and the voltage across the two diodes is clamped at that voltage. The result is that there is never enough forward voltage to allow the second diode to begin conduction, and the first diode carries the full amount of current. Placing small-valued resistors in series with both resistors fixes the problem by allowing the voltages across them to rise above the conduction voltage for both of them.

2. Answers to this question will vary, depending largely upon the type and quality of the catalog the student uses.

3. The highest V_Z in a common set of parts catalogs is 51 V, and the lowest is 3.3 V. The highest power rating is 1.0 W, and the lowest is 350 mW.

4.

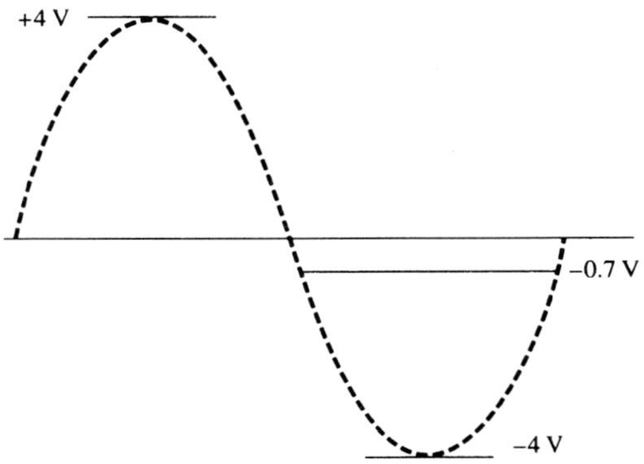

5. Tunnel diodes are sometimes called Esaki diodes. Applications of tunnel diodes are in UHF oscillators and amplifiers, and in very high-speed switching circuits. Typical operating frequencies are between 2 MHz and 110 GHz. Switching times are better than 1 nanosecond.
6. Responses should include at least PRV and recovery time.
7. Discrete devices contain a single operating element, such as a single transistor. An integrated circuit contains two or more operating elements.

CHAPTER 24

PRACTICE PROBLEMS 1

$V_{dc} = 0.45 \times V_{rms} = 0.45 \times 200\text{ V} = 90\text{ V}$

PRACTICE PROBLEMS 2

$V_{dc} = 0.45 \times V_{rms} = 0.45 \times 440\text{ V} = 198\text{ V}$

$\text{PIV} = 1.414 \times V_{rms} = 1.414 \times 440\text{ V} = 622\text{ V}$

Ripple frequency = same as ac input = 60 pps (pulse per second)

Polarity of output = positive with respect to ground reference

PRACTICE PROBLEMS 3

$V_{dc} = 0.9 \times V_{rms}$ of half the secondary $= 0.9 \times 150\text{ V} = 135\text{ V}$

Ripple frequency = 2 × ac input frequency = 2 × 400 = 800 pps

PIV = $1.414 \times V_{rms}$ full secondary $= 1.414 \times 300\text{ V} = 424\text{ V}$

PRACTICE PROBLEMS 4

1. $V_{dc} = 0.9 \times V_{rms}$ input $= 0.9 \times 120\text{ V} = 108\text{ V}$
2. $I_{dc} = V_{dc}/R_L = 108\text{ V}/50\ \Omega = 2.16\text{ A}$
3. PIV (Bridge) = $1.414 \times V_{rms}$ input $= 1.414 \times 120 = 169.7\text{ V}$
4. Ripple frequency (Bridge) = 2 × ac input frequency = 2 × 60 Hz = 120 pps

IN-PROCESS LEARNING CHECK 1

1. The simplest rectifier circuit is the *half-wave* circuit.
2. The ripple frequency of a bridge rectifier circuit is *two times* the ripple frequency of a half-wave rectifier.
3. The circuit having the highest output voltage for a given transformer secondary voltage is the *bridge* rectifier.
4. For a given full transformer secondary voltage, the full-wave rectifier circuit unfiltered dc output voltage is *equal to* the dc output of a half-wave rectifier that is using the same transformer.
5. To find the average dc output (unfiltered) of a center-tapped, full-wave rectifier, multiply the full secondary rms voltage by *0.45*.

IN-PROCESS LEARNING CHECK 2

1. In a C-type filter, the output of the rectifier circuit is connected in *parallel* with a filter *capacitor* and the load resistance.
2. In an L-type filter, the output of the rectifier circuit is connected in *series* with a filter *choke (inductor)* and the load resistance.
3. In an L-type filter, energy is stored in the form of a *magnetic* field.

4. The five main considerations in the selection and quality of a power supply filter network are: *output ripple*, *regulation*, *rectifier peak current limits*, *load current*, and *output voltage*.
5. The two main causes of swings in the output voltage of a power supply are changes in the *input* voltage level and changes in the *output* current demand.
6. The three terminals on an integrated-circuit voltage regulator are the unregulated *dc input*, regulated *dc output*, and *common ground*.
7. The *lower* the percent of ripple, the higher the quality of the power supply.
8. Full-load output voltage is taken when the current from the power supply is *maximum*.
9. The *lower* the percent of regulation, the higher the quality of the power supply.

USING EXCEL

$V_{pri} = 110\ V_{rms}$

$V_{sec} = 4 \times V_{pri} = 440\ V_{rms}$

$V_{sec(pk)} = 1.414 \times V_{sec(rms)} = 622.16\ V_{pk}$

Therefore, $V_{dc} = 0.318 \times V_{pk} = 197.85$ V

$V_{NL} = 26$ V

$V_{FL} = 23.5$ V

Therefore, % regulation = 9.62%

REVIEW QUESTIONS

1.

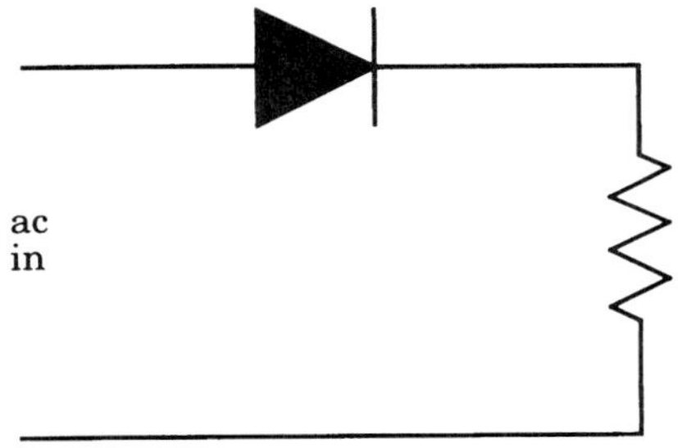

2.

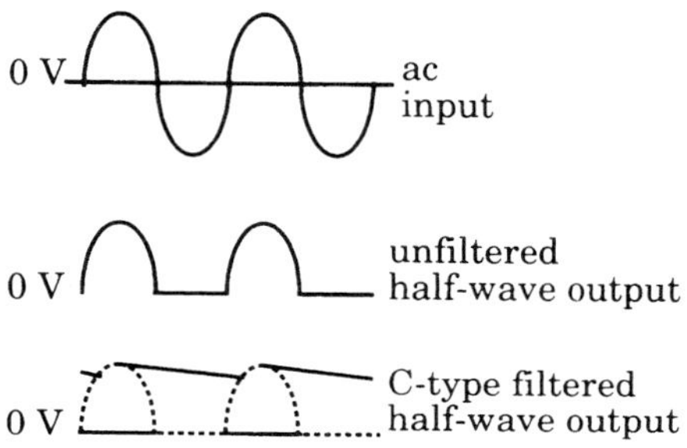

3. When you turn around a rectifier diode, the polarity of the output from a half-wave rectifier is reversed. Switching the connections to the ac source has no effect on the operation of the rectifier however.
4. 60 pps

5.

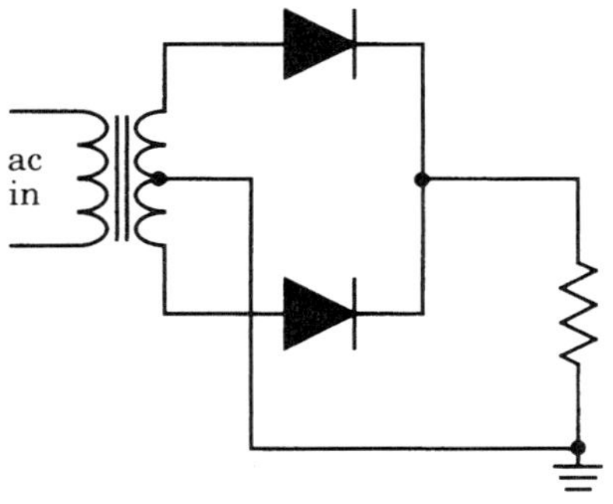

6.

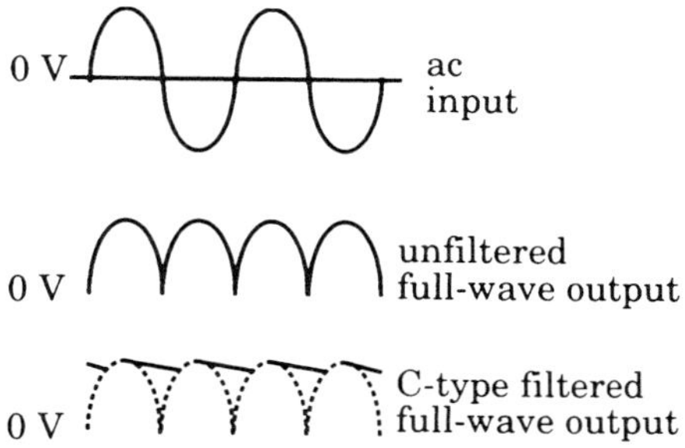

7. 120 pps

8.

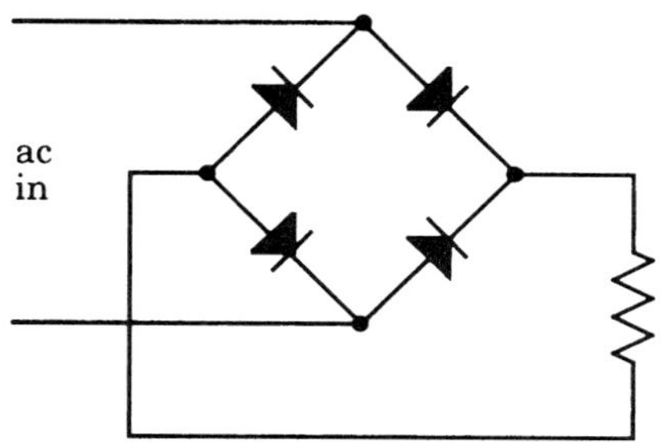

9.

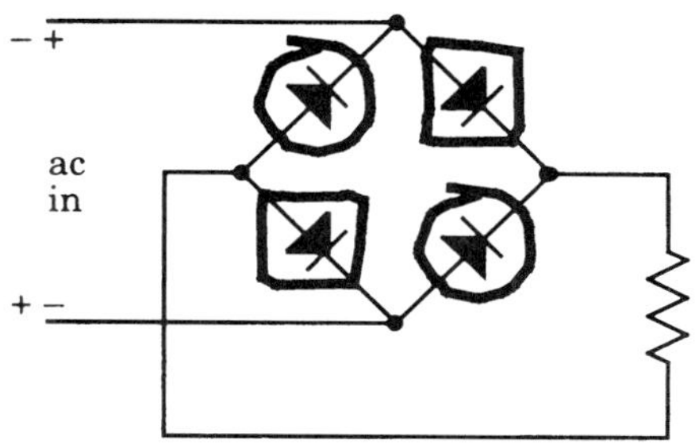

10.

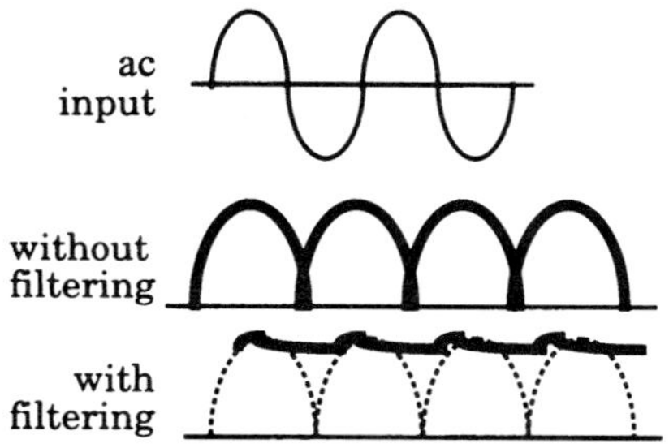

12. The dc output is greater.

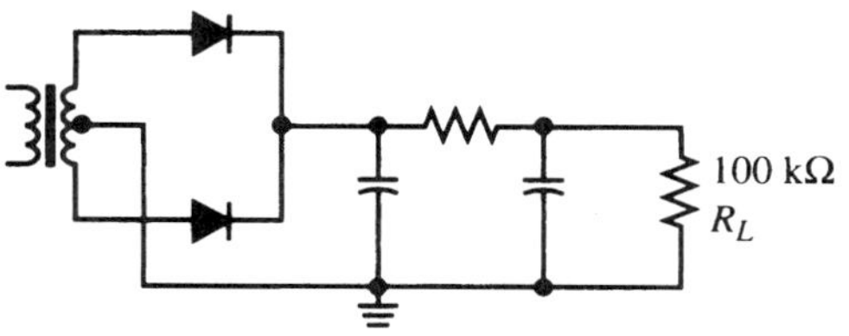

11. 800 pps
13. Excessive ripple is most likely caused by an open capacitor or shorted inductor in the power supply filter network.
14. A voltage regulator maintains a constant output voltage level in spite of (a) changes in the input voltage level and (b) changes in the amount of output load.
15.

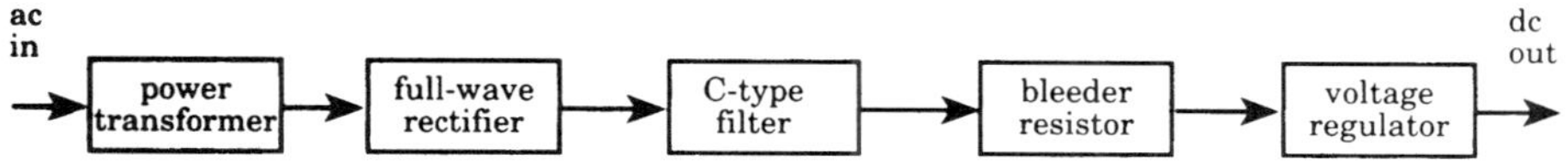

16. *Percent of ripple* is a measure of the amount of ripple on the output voltage of a power supply. *Percent of regulation* is a measure of how well the power supply maintains its rated output voltage when there are changes in input voltage and output load.
17. A no-load condition occurs when no current is drawn from the power supply. A full-load condition occurs when the maximum allowable amount of current is drawn from the power supply.
18. The output voltage will be three times the peak value of the input: 53.5 V.
19. The RMS input voltage is 212 V.
20. Worse

PROBLEMS

1. a. 54 Vdc
   b. 169.7 V (unfiltered)
   c. 60 Hz
2. The required PIV is 636 V, but the actual selection should be about twice that value (approximately 1,269 V).
3. 1.7 A
4. Transformer secondary is 400 Vrms; dc output is 180 V.
5. 13.9:1 step-down
6. 108 Vdc
7. 10%

8. 600 mV of ripple
9. 10.6% regulation
10. 5.19 V

ANALYSIS QUESTIONS

1.

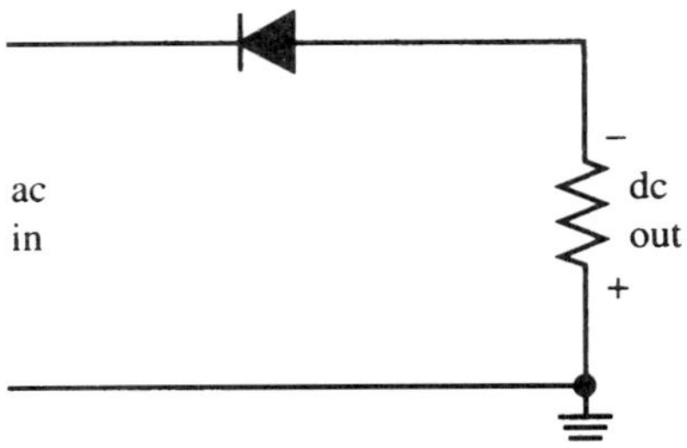

2.

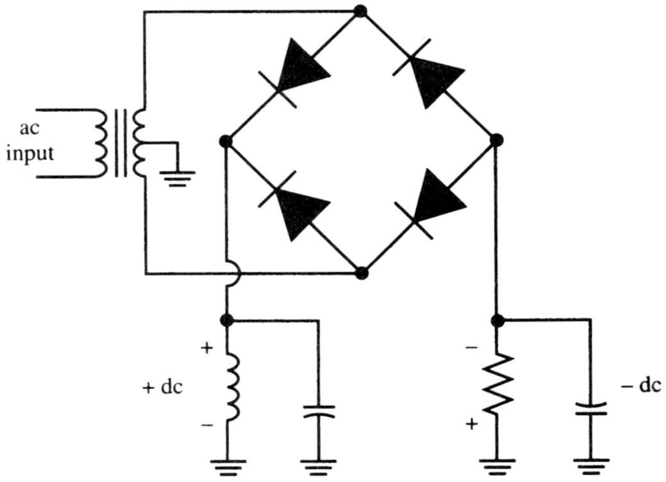

3. A larger filter capacitor provides a longer time constant. There will be less ripple because the capacitor can "hold up" the output voltage level better. This means better regulation because the capacitor responds more slowly to changes in input voltage and output load.
4. Critical inductance is the value of inductance required for the first inductor of a filter network in order to keep the output voltage from going above the average value of the rectified wave at the input of the inductor when the load current is minimum.
5. Optimum inductance is that value of the first inductor where peak rectifier current is kept from exceeding the dc load current by more than 10% at maximum load levels.
6. The voltage rating of a filter capacitor must exceed the peak level voltage from the rectifier because the voltage to which the capacitor is charged from a previous alternation is additive with the voltage seen from the source, through the rectifier, during the present alternation. This means the capacitor must withstand at least the sum of the two voltages.
7. A shorted diode can cause the source of ac for the rectifier to be shorted and may also cause ac to be applied to electrolytic filter capacitors, which may eventually destroy them should the power supply fuse not blow in time.
8. A leaky filter capacitor will cause lower-than-normal dc output, and higher-than-normal ripple in the output of a power supply. This also deteriorates the ability of the power supply to maintain constant dc output under varying load conditions, thus degrading the power supply's regulation.

MULTISIM EXERCISE

CHAPTER 25

PRACTICE PROBLEMS 1

1. $I_E = I_C + I_B = 10\text{ mA} + 100\ \mu\text{A} = 10.1\text{ mA}$
2. $\beta_{dc} = I_C / I_B = 2\text{ mA}/15\ \mu\text{A} = 133$ for both NPN and PNP transistors
3. $I_C = \beta_{dc} \times I_B = 100 \times 10\ \mu\text{A} = 1\text{ mA}$
   $I_E = I_C + I_B = 1\text{ mA} + 10\ \mu\text{A} = 1.01\text{ mA}$
   $\alpha = I_C / I_E = 1\text{ mA}/1.01\text{ mA} = 0.99$

IN-PROCESS LEARNING CHECK 1

1. The symbol for an NPN transistor shows the emitter element pointing *away from* the base element.
2. The symbol for a PNP transistor shows the emitter element pointing *in toward* the base element.
3. For BJTs:
   V_{BE} is the *base-emitter* voltage.
   V_{CB} is the *collector-base* voltage.
   V_{CE} is the *collector-emitter* voltage.
4. For a typical BJT circuit:
   I_C stands for the *collector* current.
   I_B is the *base* current.
   I_E is the *emitter* current.

IN-PROCESS LEARNING CHECK 2

1. Under normal operating conditions, the base-emitter junction of a BJT must be forward biased, while the collector-base junction is reverse biased.
2. Current must be flowing through the *base-emitter* junction of a BJT before current can flow between the emitter and collector.
3. Stated in words, the dc beta of a BJT is the ratio of *collector current* divided by *base current.*
4. Stated in words, the alpha of a BJT is the ratio of *collector current* divided by *emitter current.*
5. The α of a BJT is *less* than 1, while the β_{dc} is *greater* than 1.

IN-PROCESS LEARNING CHECK 3

1. BJTs use a small amount of base *current* to control a larger amount of collector *current.* It can be said that BJTs are *current* controllers.
2. When base current in a BJT is zero, the collector current is *zero.*
3. BJTs are basically used as *switches* and *amplifiers.*
4. When a BJT is being used as a switch (as in Figure 26–6), V_{CE} is maximum when I_B is *minimum*, and V_{CE} is minimum when I_B is *maximum.* In the same circuit, I_C is maximum when I_B is *maximum*, and I_C is minimum when I_B is *minimum.*
5. When a BJT is being used as a voltage amplifier (as in Figure 26–7), an increase in base current causes a *decrease* in V_{CE}.
6. On a family of collector-characteristic curves, the horizontal axis represents the *collector-emitter voltage* and the vertical axis represents the *collector current.* Each curve in the family represents a different level of *base current.*

USING EXCEL

$I_C = 0.010\text{ A}$

$I_B = 0.0001$ A

Therefore, $I_E = 0.0101$ A

$I_C = 0.002$ A

$I_B = 0.000015$ A

Therefore, $\beta_{dc} = 133.3$ NPN; $\beta_{dc} = 133.3$ PNP

REVIEW QUESTIONS

1. Two, two
2. P, N
3. N, P
4 Positive to base, negative (or less positive) to emitter; positive to collector
5. Negative to base, positive (or less negative) to emitter; negative to collector
6. 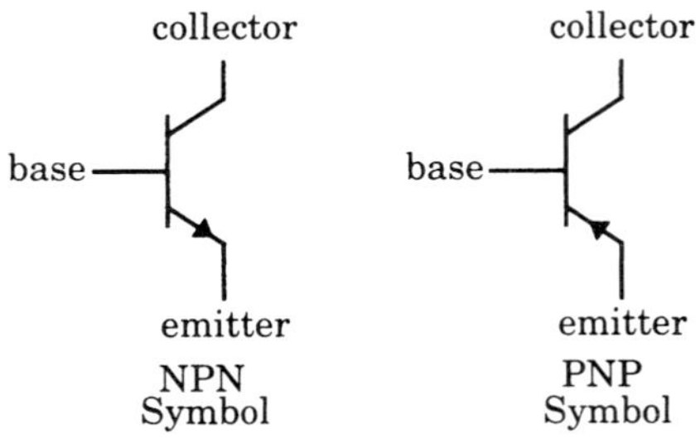

7. Approximately 96–99% of a BJT's emitter current appears in the collector circuit.
8. Forward biasing allows current flow to occur through a P-N junction. This is accomplished by applying a positive potential to the P-type material and a negative (or less positive) potential to the corresponding N-type material.
9. A reverse-biased P-N junction is one in which the P-type material is connected to the negative side of the source and the N-type material is connected to the positive side of the source.
10. Forward
11. Reverse
12. Low
13. False
14. A basing diagram is an outline sketch of a device, showing electrical elements of the device to which each "pin" is connected and the physical layout, or configuration, of the connection points relative to the packaging of the device.
15. Collector current (I_C) increases as base current (I_B) increases.
16. When the base current for a BJT switch is maximum, the transistor is in its saturation state. This means the transistor is conducting at its maximum current level or, in other words, it is operating in its lowest-resistance state. The voltage between emitter and collector is thus at a minimum level.
17. Increasing the forward-biasing current (I_B) in a typical BJT voltage amplifier causes the collector-emitter voltage drop (V_{CE}) to decrease.
18. a. The maximum amount of voltage that can be applied between the collector and emitter without causing a switched-off transistor to break down and conduct current.
    b. The maximum amount of reverse-bias voltage that can be applied across the collector-base junction.
    c. The maximum amount of reverse-bias voltage that can be applied across the emitter-base junction.
19. a. $I_{C_{(max)}}$ = maximum collector current; the greatest amount of current the transistor can handle for an indefinitely long period of time.

b. $P_{D(\max)}$ = maximum power dissipation; the greatest amount of power the transistor can dissipate without the help of a heat sink.

20. Dc beta (β_{dc}) is determined by taking the ratio of an I_C to the corresponding amount of I_B. It is a static value. Ac beta (β_{ac}) is also a ratio of collector and base currents, but it is the ratio of the difference between two I_C values divided by the corresponding difference between two I_B values. It is a dynamic value.

PROBLEMS

1. a. $I_E = 40.5$ mA
   b. $\beta_{dc} = 80$
2. a. $I_E = 49.9$ mA
   b. $\beta_{dc} = 417$
3. $\beta_{dc} = 500$
4. a. $I_B = 8$ mA
   b. $I_E = 808$ mA
5. $I_C = 1.6$ mA
6. $\alpha = 0.975$
7. $\alpha = 0.99$
8. a. $I_E = 420$ mA
   b. $I_B = 20$ mA
   c. $\beta_{dc} = 20$
9. a. $I_C = 38.4$ mA
   b. $I_B = 1.6$ mA
   c. $\beta_{dc} = 24$
10. $I_C = 8$ mA; $I_E = 8.04$ mA; $\alpha = 0.995$
11. $\beta_{ac} = 100$
12. $\Delta I_C = 4$ mA
13. $P_d = 480$ mW
14. $I_C = 500$ mA (The 500-mA current may exceed the BJT's maximum collector-current rating.)
15. $\beta_{ac} = (10.5\text{ mA} - 1.5\text{ mA})/(120\ \mu\text{A} - 40\ \mu\text{A}) = 112.5$

ANALYSIS QUESTIONS

1. Outline drawings for TO-3, TO-5, TO-92, TO-220, and SOT-89 transistor packages are shown.

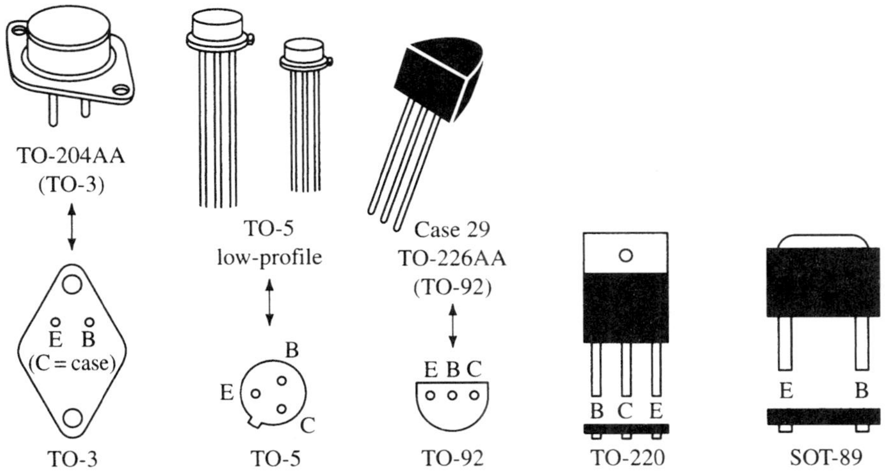

2. Listings will vary.
   Examples of NPN transistors: 2N697, 2N2222, 2N3904, 2N4238, 2N4401
   Examples of PNP transistors: 2N2904, 2N4030, 2N5086, 2N4402, 2N2907
3. The purpose of a heat sink is to draw heat away from a semiconductor device and dissipate it into the air by means of heat convection. Heat sinks are made from extruded aluminum stock and usually have rows of fins that greatly increase the effective area exposed to surrounding air. The semiconductor device to

be cooled is securely bolted to the heat sink, sometimes with a thin layer of mica serving to electrically isolate the case of the semiconductor from the heat sink. Heat sinks are sometimes anodized with a flat black color to enhance heat dissipation according to the principle of black-body radiation.

4. Responses will vary. The objective is to familiarize the student with locating, reading, and interpreting BJT data sheets.
5. Formula 26–1 shows that I_E is greater than I_C (unless all terms are zero): $I_E = I_C + I_B$. And since I_E is always greater than I_C, the quotient I_C/I_E has to be less than 1. Therefore $\alpha = I_C/I_E$ (Formula 26–5) is always less than 1.
6. When the voltage at IN is 5 V, the emitter-base voltage is 0 V, the PNP transistor is in cutoff, the collector current is 0 mA, and the LED (D_1) is off. When the voltage at IN is 0 V, the emitter-base voltage is 0.7 V, the PNP transistor is in saturation, the emitter-collector voltage is less than 0.3 V, the collector voltage is greater than 4.7 V, and the LED is on.

CHAPTER 26

IN-PROCESS LEARNING CHECK 1

1. The input signal to an amplifier is 4 mV_{PP} and the output signal is 1 V_{PP}. What is the voltage gain of the amplifier? *250*
2. A signal causes a common-emitter base current to swing by 0.03 mA and the collector current varies 1.5 mA peak-to-peak. What is the current gain of the amplifier? *50*
3. A common-emitter amplifier has a voltage gain of 60 and a current gain of 30. What is the power gain? *1800*

IN-PROCESS LEARNING CHECK 2

1. The BJT amplifier circuit having the base terminal common to both the input and output circuits is the *common-base (CB)* amplifier.
2. The BJT amplifier circuit providing both voltage and current gain is the *common-emitter (CE)* amplifier.
3. The BJT amplifier circuit having 180° phase difference between input and output signals is the *common-emitter (CE)* amplifier.
4. The BJT amplifier circuit having a voltage gain of less than one is the *common-collector (CC)* amplifier.
5. The BJT amplifier circuit having an input to the base and output from the emitter is called a common-*collector* amplifier.
6. In an amplifier that uses more than one stage, the overall gain is found by *multiplying* the gains of the individual stages.
7. The voltage-follower amplifier is another name for a common-*collector* amplifier.
8. The Darlington amplifier is made up of two common-*collector* amplifiers.

PRACTICE PROBLEMS 1

Cutoff point is 5 V (from Formula 26–25).

Saturation current is (from Formula 26–26): $I_{C(sat)} = V_{CC}/(R_C + R_E) = 5\text{ V}/517\ \Omega = 9.67\text{ mA}$

IN-PROCESS LEARNING CHECK 3

1. Class A amplifiers conduct during *360* degrees of the input cycle; Class B amplifiers conduct for *180* degrees of the input cycle; and Class C amplifiers conduct for approximately *120* degrees of the input cycle.
2. The dc load line shows all operating points of a given amplifier from *cutoff* to *saturation*.

3. What is the amount of saturation current for a Class A CE amplifier when the collector resistance is 1 kΩ, the emitter resistance is zero (this resistor not used), and the supply voltage is 12 V? *12 mA*. What is the value of the cutoff voltage for this circuit? *12 V*.
4. The *Q* point for a Class A amplifier is located at the *halfway point* of the load line.
5. The Class *C* amplifier has the *Q* point located below the cutoff.
6. Collector current flows in a Class *A* amplifier, even when there is no signal applied to the input.

PRACTICE PROBLEMS 2

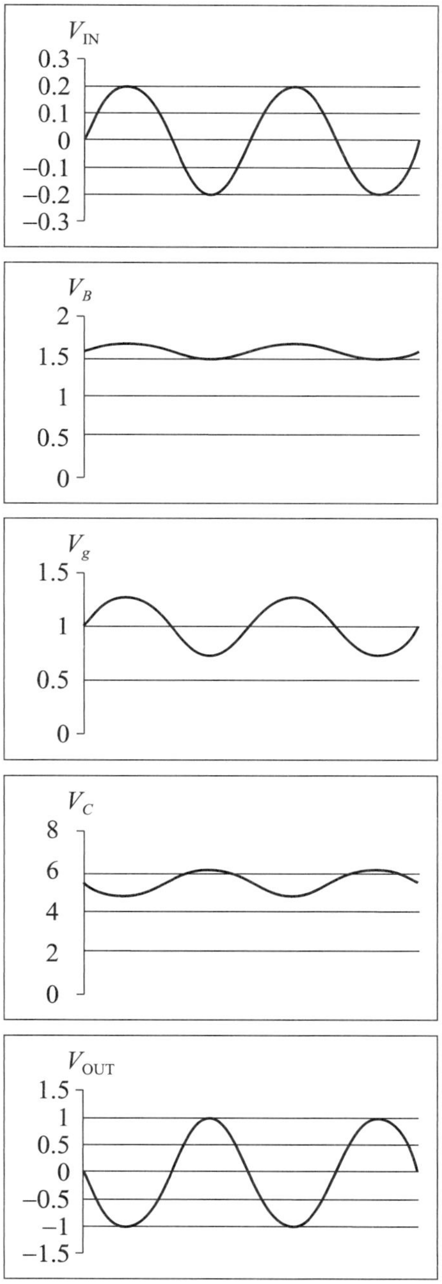

Divider current: $I = V_{CC}/(R_1 + R_2) = 10\ \text{V}/12\ \text{k}\Omega = 0.83\ \text{mA}$

Base voltage: $V_B = I \times R_2 = 0.83\ \text{mA} \times 2\ \text{k}\Omega = 1.67\ \text{V}$

Emitter voltage: $V_E = V_B - 0.7\ \text{V} = 0.97\ \text{V}$

Emitter current: $I_E = V_E/R_E = 0.97\ \text{V}/1\ \text{k}\Omega = 0.97\ \text{mA}$

Collector voltage: $V_C = V_{CC} - (I_C \times R_C) = 10\ \text{V} - 4.6\ \text{V} = 5.4\ \text{V}$

Transistor voltage: $V_{CE} = V_{CC} - (V_{RC} + V_{RE}) = 10\ \text{V} - 5.57\ \text{V} = 4.43\ \text{V}$

Dynamic emitter resistance: $r'_e = 25.86\ \Omega$

Input impedance: $Z_{in} = 1.64\ \text{k}\Omega$

Output impedance: $Z_{out} = 4.7\ \text{k}\Omega$

Voltage gain: $A_v = -4.58$

Current gain: $A_i = 1.58$

Power gain: $A_p = 7.25$

Output voltage: $V_{out} = -0.916\ \text{V}_{PP}$

(The negative sign indicates a 180° phase shift.)

USING EXCEL

$V_{CC} = 5\ \text{V}$

$R_C = 470\ \Omega$

$R_E = 47\ \Omega$

Therefore, $I_{C(snt)} = 0.00967\ \text{A}$

$V_{in} = 0.05\ \text{V}$

$V_{out} = 7.1\ \text{V}$

Therefore, $A_V = 142$

REVIEW QUESTIONS

1. Emitter
2. a. Input is between base and emitter; output is between collector and emitter.
   b. Input is between emitter and base; output is between collector and base.
   c. Input is between base and collector; output is between emitter and collector.
3. a. Common-emitter: input *Z* = low, output *Z* = low to medium
   b. Common-base: input *Z* = low, output *Z* = fairly high
   c. Common-collector: input *Z* = high, output *Z* = low

4. Common-emitter
5. Common-emitter
6. Darlington pair
7. A common-collector amplifier stage followed by a common-emitter stage provides a high input impedance followed by high voltage gain.
8. Common-base
9. Small-signal transistor amplifier stages may be used in the first stages of radio and TV receivers; large-signal transistor amplifier stages may be used in audio power amplifier stages for driving loudspeakers.
10. Class A
11. Class-A amplifiers have the least distortion.
12. Class C
13. The Q point is usually set at the mid-point of the load line for Class A operation.
14. At cutoff
15. For Class A operation of a common-emitter circuit, the quiescent value of V_{CE} is about one-half the value of V_{CC}, or 10 volts.
16. a. Collector current equals emitter current.
    b. Base current is much less than the voltage-divider current.
    c. Voltages across the divider resistors add up to the source voltage.
    d. Sum of emitter-resistor voltage, emitter-to-collector voltage, and collector-resistor voltage is equal to the source voltage.
17. a. Divider current: 1.49 mA
    b. Base voltage: 1.49 V
    c. Emitter voltage: 0.79 V
    d. Collector voltage: 7.1 V
    e. Collector-to-emitter voltage: 6.31 V
18. Common-emitter
19. A common-collector BJT amplifier circuit is needed to match a high impedance to a low impedance.
20. Class B

PROBLEMS

1. a. $V_B \approx 5.41$ V
   b. $I_E = 4.71$ mA
2. a. $V_B = 6.59$ V
   b. $I_E = 5.89$ mA
3. a. $V_B \approx 4.5$ V
   b. $I_E = 31.7$ mA
   c. $V_{CE} = 1.4$ V
   d. $V_{BE} = 0.7$ V
4. a. $I_{C(\text{sat})} = 5.45$ mA
   b. $V_{CE(\text{max})} = 6$ V
5. a. 37.5 mA
   b. 9 V
6. a. $I_{R_2} = 0$
   b. $V_{RE} = 0$
   c. $V_{CE} = 10$ V
   d. $V_{RC} = 0$
   e. $I_E = 0$
   f. Class C
   g. $V_{\text{cutoff}} = 10$ V
7. a. $I_{R_1} = 180$ μA
   b. $V_{R_1} = 18$ V
8. $A_v = 18.3$
9. $A_v = 142$
10. $A_p = 1.98$

ANALYSIS QUESTIONS

1. The steeper the slope of the load line, the less change in collector-emitter voltage for a given change in base current. The diagram shows that a steeper curve is created by lowering the collector resistance. You can conclude that the voltage gain of an amplifier changes in proportion to changes in the collector resistance.
2. Approximately 1.7 to 1.8 mA
3. Cutoff voltage = 15 V

   Saturation current = 11.7 mA
4. I_E is about 6 mA; $V_E = I_E \times R_E = 6\text{ mA} \times 220\ \Omega = 1.32\text{ V}$
5. The ratio of divider resistances is 6.5:1, with the larger value connected to V_{CC}.

MULTISIM EXERCISE

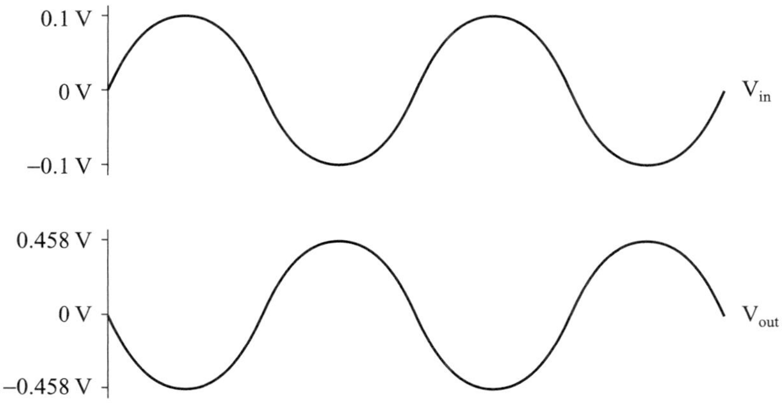

CHAPTER 27

IN-PROCESS LEARNING CHECK 1

1. *JFET* is the abbreviation for *Junction Field-Effect Transistor.*
2. The three terminals on a JFET are called *source*, *gate*, and *drain*.
3. The arrow in the symbol for an N-channel JFET points *in toward* the source-drain connections, while the arrow for a P-channel JFET points *away from* source-drain connections.
4. In an N-channel JFET, the drain-supply voltage must be connected so that the positive polarity is applied to the *drain* terminal of the JFET and the negative polarity is applied to the *source* terminal.
5. In a P-channel JFET, the drain-supply voltage must be connected so that the positive polarity is applied to the *source* terminal of the JFET and the negative polarity is applied to the *drain* terminal.
6. The charge carriers in the channel material of a JFET always flow from the *source* terminal to the *drain* terminal.
7. In an N-channel JFET, the gate-supply voltage must be connected so that the positive polarity is applied to the *source* terminal of the JFET and the negative polarity is applied to the *gate* terminal.
8. In a P-channel JFET, the source-supply voltage must be connected so that the positive polarity is applied to the *gate* terminal of the JFET and the negative polarity is applied to the *source* terminal.
9. V_{DS} stands for *voltage between the drain and source terminals.*

   V_{GS} stands for *voltage between the gate and source terminals.*

   I_D stands for *drain current.*

   g_m stands for *transconductance.*

10. The greatest amount of current flows through the channel of a JFET when V_{GS} is at its *zero* level, while the least amount of drain current flows when V_{GS} is at its *maximum* (or $V_{GS(\text{off})}$) level.

IN-PROCESS LEARNING CHECK 2

1. When a polarity opposite the drain polarity is applied to the gate of a JFET through a large-value resistor, the bias method is called *gate bias*.
2. When the gate and source of JFET are both connected to ground through resistors, the bias method being used is called *self bias*.
3. When the gate of a JFET is connected to a voltage divider and the source is connected through a resistor to ground, the bias method being used is called *voltage-divider* bias.
4. The common-*gate* FET amplifier has the input signal applied to the source terminal and the signal taken from the drain.
5. The common-*source* FET amplifier has the signal applied to the gate terminal and the output signal taken from the drain.
6. The common-*drain* FET amplifier has the signal applied to the gate terminal and output taken from the source terminal.

IN-PROCESS LEARNING CHECK 3

1. The term *MOS* stands for *metal-oxide semiconductor*.
2. The gate terminal in a D-MOSFET is separated from the channel material by a thin layer of *silicon dioxide* (or *metal-oxide insulation*).
3. In a MOSFET, the direction of flow of charge carriers is always from the *source* terminal to the *drain* terminal.
4. An outward-pointing arrow on a D-MOSFET symbol indicates an *N*-type substrate and a *P*-type channel. An inward-pointing arrow on a D-MOSFET symbol indicates a *P*-type substrate and an *N*-type channel.
5. The proper polarity for V_{DD} of an N-type D-MOSFET is *negative* to the source and *positive* to the drain. For a P-type D-MOSFET, the proper polarity for V_{DD} is *positive* to the source and *negative* to the drain.
6. When operating a D-MOSFET in the depletion mode, the polarity for V_{GG} is *negative* to the gate terminal. And when V_{GS} is at its 0-V level, I_D is at its *maximum* level.
7. The *E-MOSFET* is the only FET that is nonconducting when the gate voltage is zero.
8. The proper polarity for V_{DD} of an N-type E-MOSFET is *negative* to the source and *positive* to the drain. The proper polarity for V_{GS} is *positive* to the gate terminal.

REVIEW QUESTIONS

1. V_{DS} = voltage measured between the drain and source terminals

   V_{GS} = voltage measured between the gate and source terminals

   I_D = current at the drain terminal (usually the same found at the source terminal)

   I_{DSS} = drain current that flows when no bias voltage is applied to the gate

   $V_{GS_{(\text{off})}}$ = the amount of gate-source voltage required to turn off drain current

2. 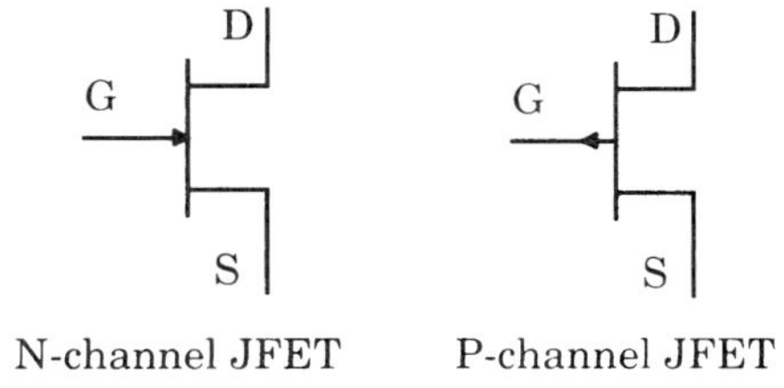

3. N-channel: V_{DD} is positive and V_{GG} is negative
   P-channel: V_{DD} is negative and V_{GG} is positive

4.

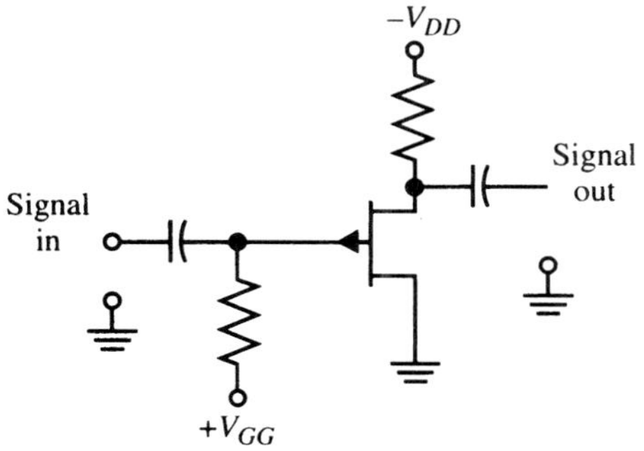

5. A JFET uses a voltage (V_{GS}) to control a current (I_D), whereas a BJT uses a current (I_B) to control a current (I_C). Also, a JFET is fully conducting when there is no bias applied to the gate, whereas a BJT is switched off when there is no bias applied to the base.
6. $V_{GS} = -4$ V

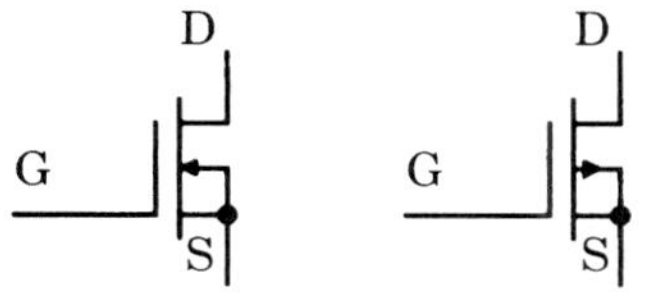

N-channel D-MOSFET   P-channel D-MOSFET

7. Transconductance is equal to a change in drain current divided by the corresponding change in gate-source voltage.
8. Drain characteristic curves show how V_{DS} and V_{GS} control drain current. The transconductance curves show how V_{GS} affects drain current.
9. A D-MOSFET makes an ideal Class A amplifier because the required gate-source bias voltage is zero. This greatly simplifies the required bias circuitry.
10.

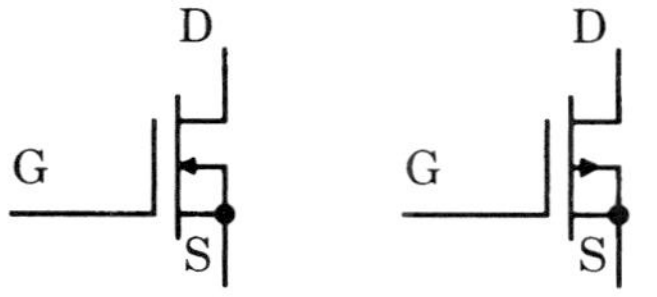

N-channel D-MOSFET   P-channel D-MOSFET

11. a. N-channel depletion mode: V_{DD} = positive and V_{GS} = negative
    b. P-channel depletion mode: V_{DD} = negative and V_{GS} = positive
    c. N-channel enhancement mode: V_{DD} = positive and V_{GS} = positive
    d. P-channel enhancement mode: V_{DD} = negative and V_{GS} = negative

12. 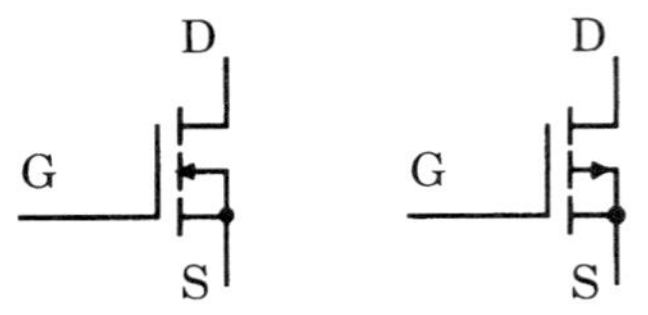

N-channel E-MOSFET | P-channel E-MOSFET

13. In the depletion mode, increasing the amount of gate-source voltage depletes the channel of charge carriers, thus tending to decrease the amount of drain current. In an enhancement mode, increasing the amount of gate-source voltage enhances the density of charge carriers in the channel, thus tending to increase the amount of drain current.
14. For a Class A, N-channel E-MOSFET amplifier, V_{DD} and V_G are positive.
    For a Class A, P-channel E-MOSFET amplifier, V_{DD} and V_G are negative.
15. An E-MOSFET makes an ideal electronic switching circuit (Class B amplifier) because it is off when there is no gate-source voltage applied, and it is turned on when the gate-source voltage is applied.
16. Threshold gate-source voltage is the minimum amount of V_{GS} required for channel conduction.
17. a. N-channel JFET, $V_{GS(\text{off})}$ is negative; P-channel JFET, $V_{GS(\text{off})}$ is positive.
    b. N-channel D-MOSFET, $V_{GS(\text{off})}$ is negative; P-channel D-MOSFET, $V_{GS(\text{off})}$ is positive.
    c. N-channel E-MOSFET, $V_{GS(\text{off})}$ is zero V; P-channel E-MOSFET, $V_{GS(\text{off})}$ is zero V.
18. JFETs must be operated in the depletion mode, and D-MOSFETs can be operated in the depletion mode. E-MOSFETS must be operated in the enhancement mode, and D-MOSFETs can be operated in the enhancement mode.
19. The zener diodes that are internally connected between the gate and source terminals of some MOSFET devices protect the device from damage that could otherwise be caused by an external build-up of static voltage.
20. MOSFETs are usually packaged and shipped in conductive foam in order to short-circuit any potential build-up of static electrical charges that could destroy the insulation layers.

PROBLEMS

1. a. $V_G = -4$ V
   b. $V_{RD} = 8$ V
   c. $V_{DS} = 4$ V
   d. $I_{RG} = 0$
2. a. $I_D = 2.72$ mA
   b. $V_{RD} = 6$ V
   c. $V_G = -5$ V
   d. $I_{RG} = 0$
3. a. $g_m = 0.5$ mS (or 500 μS)
   b. $\Delta V_{DS} = 2$ V
4. a. $V_{RS} = -5$ V
   b. $V_{RD} = 6$ V
   c. $V_{DS} = 1$ V
   d. $V_{RG} = 0$
   e. $V_{GS} = -5$ V
   f. $I_{RG} = 0$
5. a. $I_D = 1.67$ mA
   b. $V_{DS} = 6.33$ V
   c. $V_G = 0$
   d. $V_{GS} = -2$ V
   e. $V_{RG} = 0$
   f. $I_{RG} = 0$
6. a. $I_{R_2} = 164$ μA
   b. $V_{R_2} = 1.64$ V
   c. $V_{R_1} = 16.4$ V
   d. $V_{RS} = 3.3$ V
   e. $V_{GS} = -1.66$ V
   f. $I_D = 10$ mA
   g. $V_{DS} = 4.7$ V

7. a. $I_{R_2} = 90.2\ \mu A$
   b. $V_{R_2} = 2.98$ V
   c. $V_{R_1} = 9.02$ V
   d. $V_{RS} = 4$ V
   e. $V_{GS} = -1.02$ V
   f. $I_D = 8.51$ mA
   g. $V_{DS} = 2.21$ V

ANALYSIS QUESTIONS

1. Attempting to operate a JFET in an enhancement mode would forward bias the gate-source junction. This means the device could not operate as a FET, and there would also be a risk of overheating the gate junction due to excessive forward current flow.
2. A dual-gate MOSFET has lower interelectrode capacitance, thereby allowing higher-frequency operation than a conventional, single-gate MOSFET.
3. A D-MOSFET has an uninterrupted channel of N- or P-type material which makes it possible to operate it in a depletion mode. It can also be operated in an enhancement mode because of a region of the opposite type of material that can become part of the channel when the gate-source voltage contributes charge carriers to the opposite-type material. An E-MOSFET does not have an uninterrupted channel of a type of semiconductor material, so it is not possible to operate it in a depletion mode. An E-MOSFET can be operated in the enhancement mode because it has an opposite-type material that can take on charge carriers that form a conductive channel, or inversion layer.
4. The source and drain terminals of a VMOSFET (vertical MOSFET or VFET) are physically larger than those of an ordinary MOSFET, thus the capability of carrying larger amounts of current. Also, the source and drain terminals are arranged so that the channel forms at a right angle (vertically), as opposed to the JFET and MOSFET arrangement that favors the channel forming in the same narrow plane as the source and drain. Finally, the gate electrode has a "V" shape that allows it to influence a larger area of the channel region. VMOSFETS are enhancement-mode devices. Other types of power MOSFETS are TMOS and HEXFET.

CHAPTER 28

IN-PROCESS LEARNING CHECK 1

1. An op-amp is a high gain *direct*-coupled amplifier.
2. The input stage of an op-amp uses a *differential* amplifier configuration.
3. The name operational amplifier derives from their early use to perform mathematical *operations*.
4. The open-loop gain of an op-amp is much *higher* than the closed-loop gain.
5. To achieve an output that is the inversion of the input, the input signal is fed to the *negative* input of the op-amp.
6. Input signals to the op-amp can be fed to the *inverting* input, the *noninverting* input, or to both inputs.

PRACTICE PROBLEMS 1

1. $A_v = -(R_f / R_i) = -(1\ M\Omega/100\ k\Omega) = -10$
2. $R_f = -A_v \times R_i = -(-100 \times 2\ k\Omega) = 200\ k\Omega$
3. $R_i = -R_f / A_v = -(270\ k\Omega/-20) = 13.5\ k\Omega$

PRACTICE PROBLEMS 2

1. A_v (noninverting) = $(R_f / R_i) + 1 = (20{,}000\ \Omega/2{,}000\ \Omega) + 1 = 11$
2. $R_f = R_i(A_v - 1) = (10\ k\Omega)(100 - 1) = 990\ k\Omega$
3. $R_i = R_f / (A_v - 1) = 270\ k\Omega/19 = 14.2\ k\Omega$

IN-PROCESS LEARNING CHECK 2

1. For an op-amp inverting amplifier the input signal is applied to the *inverting* input, and the feedback resistor is connected from the output to the *inverting* input.
2. The voltage gain of an inverting amplifier is sometimes expressed as a negative value in order to indicate *inversion* of the signal.
3. In the schematic for an op-amp inverting amplifier, resistor R_f is located between the *output* of the circuit and *inverting* input. Resistor R_i is connected to the *inverting* terminal.
4. For an op-amp noninverting amplifier the input signal is applied to the *noninverting* input, and the feedback resistor is connected from the output to the *inverting* input.
5. The special case of a noninverting op-amp circuit that has a voltage gain of 1 is called a *voltage-follower* circuit.
6. An op-amp circuit that has one signal input and one grounded input is called a *single*-ended circuit.
7. An op-amp circuit that has a different signal source connected to its inverting and noninverting inputs is called a *double*-ended circuit.
8. A comparator circuit is an example of an op-amp operating in the *double*-ended input mode.

USING EXCEL

$R_i = 100{,}000\ \Omega$

$R_f = 1{,}000{,}000\ \Omega$

Therefore, $A_v = -10$

$R_i = 2{,}000\ \Omega$

$R_f = 20{,}000\ \Omega$

Therefore, $A_v = 11$

REVIEW QUESTIONS

1. The amplifier circuit typically used as the input stage for an operational amplifier is a differential amplifier.
2. A linear op-amp circuit does not change the quality of the input signal, whereas a nonlinear op-amp circuit significantly changes the quality of the input signal. Examples of linear circuits are inverting and noninverting amplifiers, voltage followers, summing and differential amplifiers, differentiators and integrators, sine-wave oscillators, and active filters. Examples of nonlinear are comparators, multivibrators, and hysteresis oscillators.
3. Infinite voltage gain

   Infinite bandwidth

   Infinite input impedance

   Zero output impedance
4. Open-loop gain is the voltage gain of an op-amp that has an infinite amount of feedback resistance—an open feedback path. It is the maximum amount of voltage gain the op-amp circuit can have.
5. A comparator circuit is an example of an op-amp circuit that is operated in the open-loop gain mode.
6. The gain is equal to the ratio, $-R_f/R_i$. So when the ratios are 15 and 10, the gains are –15 and –10.
7. Decrease

8. The output voltage waveform is equal and 180 degrees out of phase.
9. 16, 11
10. Decrease
11. The output voltage waveform is twice the input voltage and has the same phase.
12. Op-amp voltage followers always have zero ohms of feedback resistance.
13. The output signal has a phase that is opposite the input.
14. The output level is small for an op-amp in common-mode operation.
15. The feedback component is a capacitor, and the input component is a resistor.
16. An op-amp differentiator uses a resistor as the feedback component and a capacitor as the input component.
17. Integrator
18. A differentiator op-amp circuit should be selected where it is required to change a rectangular waveform into brief pulses.
19. $V_{out} = V_1 + V_2 + V_3$ or $V_{out} = -(V_1 + V_2 + V_3)$
20. When V_{out} attempts to increase in voltage, the input of the inverting op-amp goes less positive. A less positive level at the base of the BJT reduces its conductance, thereby lowering the amount of V_{in} voltage that reaches the V_{out} terminal.

PROBLEMS

1. $A_v = -208$
2. $A_v = -1$
3. $R_i = 8.33\ \text{k}\Omega$
4. $A_v = 209$
5. $A_v = 2$
6. $R_i = 8.04\ \text{k}\Omega$
7. a. $V_{out} = +5$ V
   b. $V_{out} = +1$ V
   c. $V_{out} = -1$ V
   d. $V_{out} = -5$ V
8. a. $V_{out} = -5$ V
   b. $V_{out} = -1$ V
   c. $V_{out} = +1$ V
   d. $V_{out} = +5$ V
9. a. $V_{out} = -8$ V
   b. $V_{out} = 0$ V
   c. $V_{out} = 0$ V
   d. $V_{out} = +8$ V
10. UTP = +9.54 V;
    LTP = 0 V; Hysteresis = 9.54 V

ANALYSIS QUESTIONS

1. From Formula 28–2: A_v (noninverting) = $(R_f/R_i) + 1$
   For a voltage follower, $R_f = 0$. Substituting this into the formula, we get: A_v (noninverting) = $(0/R_i) + 1 = 0 + 1 = 1$
   Thus the voltage gain of the circuit is 1, or unity.
2. Input resistance: 2 MΩ
   Output resistance: 75 Ω
   Supply voltage: $+V = +22\ \text{V}_{max}$; $-V = -22\ \text{V}_{max}$
   Large signal voltage gain: 200 V/mV (or open-loop gain = 200,000)
3. Voltage gain = V_{out}/V_{in} = 2 V/10 mV = 200
   $R_f = A_v \times R_i = 200 \times 2.2\ \text{k}\Omega = 440\ \text{k}\Omega$

MULTISIM EXERCISE

$V_{out} = 6.5$ V

CHAPTER 29

PRACTICE PROBLEMS 1

1. JFET Q_1 is the active amplifying device.
2. L_{1_A}, L_{1_B}, and C_1 make up the frequency-determining part of the circuit.
3. Capacitor C_3 is part of the *feedback* portion of this oscillator circuit.
4. From Formula 29–3:

   $L_t = L_{1_A} + L_{1_B} = 0.1 \text{ mH} + 0.1 \text{ mH} = 0.2 \text{ mH}$
5. From Formula 30–2:

   $f_r = 1/2\pi\sqrt{L_tC} = 1/2\pi\sqrt{(0.2 \text{ mH})(1{,}000 \text{ pF})}$

   $f_r = 356 \text{ kHz}$
6. Decreasing the value of C_1 increases the oscillating frequency. Increasing the value of L_{1_A} decreases the oscillating frequency.

PRACTICE PROBLEMS 2

1. The active amplifying device is JFET Q_1.
2. C_1, C_2, and L_p, the inductance of the primary of transformer T_1, make up the frequency-determining part of the circuit.
3. From Formula 30–5:

   $C_t = C_1C_2/(C_1 + C_2) = 50 \text{ pF}$
4. From Formula 30–4:

   $f_r = 1/2\pi\sqrt{LC_t}$

   $f_r = 1/2\pi\sqrt{(750\ \mu\text{H})(50 \text{ pF})} = 822 \text{ kHz}$
5. When the total capacitance is decreased, the frequency of oscillation increases. When the value of inductance increases the frequency of oscillation decreases.

IN-PROCESS LEARNING CHECK 1

1. The *LC* oscillator that uses a tapped coil or a set of two coils in the tank circuit is called the *Hartley* oscillator.
2. The *LC* oscillator that uses a set of two capacitors and a single inductor is called the *Colpitts* oscillator.
3. The *LC* oscillator that has three capacitors in the tank circuit is called the *Clapp* oscillator.
4. The output of a *Colpitts LC* oscillator is often coupled to the next stage by means of a transformer whose primary winding is part of the tank circuit.
5. When $C_1 = 100$ pF: $C_t = 83.3$ pF, $f_r = 1.74$ MHz

   When $C_1 = 850$ pF: $C_t = 315$ pF, $f_r = 897$ kHz or 0.897 MHz

PRACTICE PROBLEMS 3

1. From Formula 30–8:

   $f_r = 1/2\pi RC$

   $f_r = 1/2\pi(4.7 \text{ k}\Omega)(0.002\ \mu\text{F}) = 16.9 \text{ kHz}$
2. Doubling the values of the capacitors reduces the operating frequency by one-half. Cutting the resistor values in half causes the operating frequency to double.
3. Rearranging Formula 30–8 to solve for the value of R:

   $R = 1/2\pi f_rC$

   $R = 1/2\pi(1 \text{ kHz})(0.1\ \mu\text{F}) = 1.6 \text{ k}\Omega$

IN-PROCESS LEARNING CHECK 2

1. Whereas LC oscillators operate according to resonance, RC oscillators operate according to *phase shift*.
2. The four basic elements and conditions for starting and sustaining oscillation in RC sine-wave oscillators are *power source*, *frequency-determining elements*, *amplifier*, and *positive feedback*.
3. The sine-wave oscillator that uses a three-stage RC network for achieving a 180° phase shift at the frequency of oscillation is the *phase-shift* oscillator.
4. The sine-wave oscillator that uses a lead-lag RC network to produce 0° phase shift at the frequency of oscillation is the *Wien-bridge* oscillator.
5. The amplifier element of a phase-shift oscillator must produce a phase shift of *180°*, while the amplifier element of a Wien-bridge oscillator must produce a phase shift of *zero*°.

USING EXCEL

$L_{1_A} = 90\ \mu H$

$L_{1_B} = 40\ \mu H$

$C = 500\ \mu F$

Therefore, $f = 625$ Hz

$C_1 = 90$ pF

$C_2 = 40$ pF

$L = 5\ \mu H$

Therefore, $f = 13.5$ MHz

REVIEW QUESTIONS

1. A power source, a device or components that determine the frequency of oscillation, amplification, and positive feedback.
2. $$f_r = \frac{1}{2\pi\sqrt{LC}}$$

   where: f_r = resonant frequency of the tank circuit
   L = total inductance in the tank circuit
   C = total capacitance in the tank circuit
3. The Hartley oscillator uses a tapped coil. The Colpitts oscillator has two capacitors with a ground connection between them. The Clapp oscillator uses three capacitors and a single inductor in the tank circuit.
4. 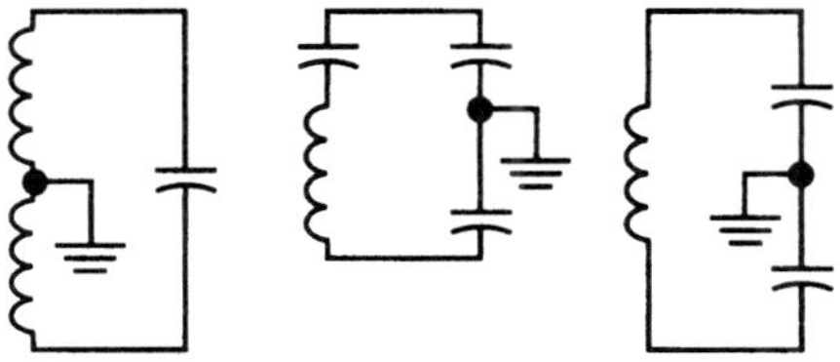

   (a) (b) (c)
5. $L_t = 20$ mH; $f_r = 3.56$ kHz
6. The student assigns values and solves the following two equations:

$$C_T = \frac{1}{\frac{1}{C_1} + \frac{1}{C_2} + \frac{1}{C_3}}$$

$$f_r = \frac{1}{2\pi\sqrt{LC}}$$

7. $C_T = 0.05\ \mu F$; $f_r = 7.12$ kHz
8. Decreasing, decreasing
9. The phase-shift oscillator uses a network of three identical *RC* elements. The Wien-bridge oscillator uses a lead-lag network.
10. The crystal
11. 180 degrees, 180 degrees
12. 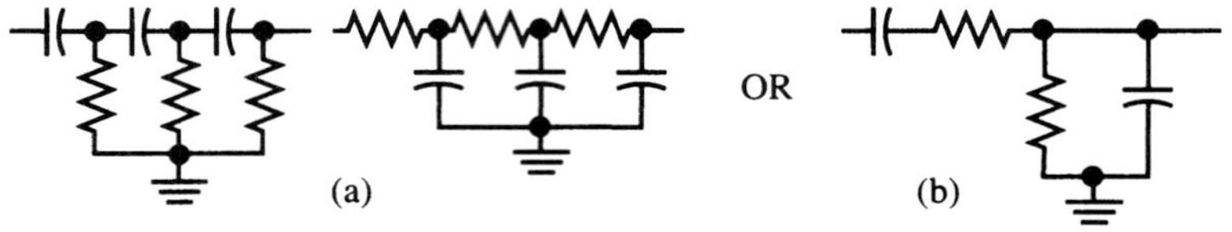

13. $f_r = 159$ Hz
14. One-shot multivibrator
15. Decrease, decrease
16. Free-running multivibrator
17. $T = 18\ \mu s$, $f = 55.6$ kHz
18. *RC* time constant for the V_{CC} interval includes both timing resistors. The *RC* time constant for the 0-V interval includes only one of the two timing resistors.
19. $f = 483$ Hz
20. Decrease, decrease

PROBLEMS

1. 160 kHz
2. 12.2 pF
3. 625 Hz
4. 15.9 kHz
5. 13.5 MHz
6. 5.87 MHz; 16.1 MHz
7. 7.33 nF
8. 175 pF
9. 65 Hz
10. 1.59 pF

ANALYSIS QUESTIONS

1. 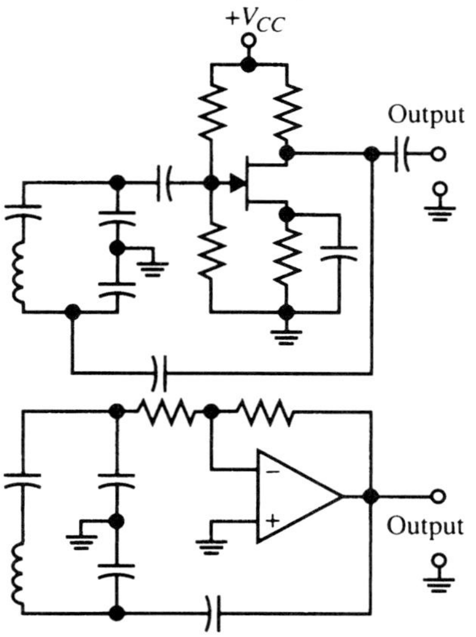

2. The Armstrong oscillator uses a transformer to accomplish positive feedback between the output and input of the amplifier. The frequency is determined by the values in a tank circuit comprised of a capacitor and the inductance of the transformer's primary winding.
3. The total phase shift from the output to the input of the amplifier element in an oscillator must be 0° (or 360°) at the operating frequency in order to provide positive feedback. In a phase-shift oscillator, the *RC* network shifts the waveform by 180°, so the amplifier element must also shift the signal by 180° in order to satisfy the requirement for positive feedback. The *RC* network in a Wien-bridge oscillator, on the other hand, does not shift the phase of the waveform at the oscillator frequency, so the amplifier must not shift the waveform either, in order to provide a total of 0° for positive feedback.
4.

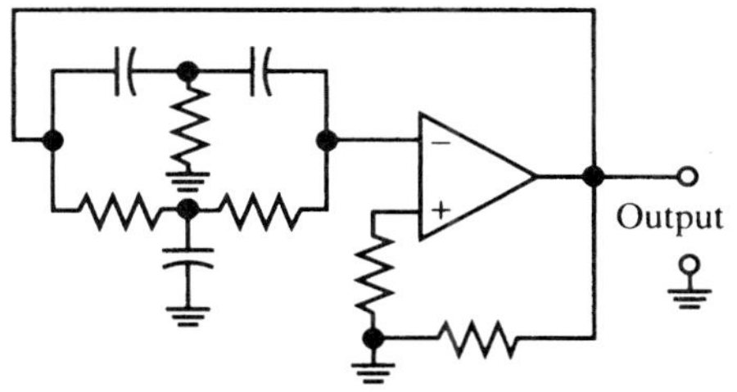

5. The feedback path through the three-section *RC* network of a phase-shift oscillator reduces the signal level a great deal by the time it reaches the input of the amplifier element. None of the other oscillators described in this chapter reduce the signal so much. The amplifier must have a high gain in order to overcome the unusual amount of signal loss through the *RC* network.
6. An *interruptible* monostable multivibrator can be retriggered during its timing interval in order to restart the timing interval. By contrast, a *noninterruptible* monostable multivibrator must complete its timing interval before a trigger can start the next timing interval.
7. A *bistable multivibrator*, or flip-flop, is a type of multivibrator that is stable in both of its two states—on and off. It changes state each time it is triggered.
8. The charge time for the timing capacitor is always longer than its discharge time because the charge-current path includes two resistors in series, and the discharge path includes only one of them. This makes the charging time constant longer than the discharging time constant.
9. The operating frequency of a 555 astable multivibrator decreases as the value of the timing capacitor increases. The operating frequency increases as the value of the timing capacitor decreases.
10. The frequency of operation of a 555 astable multivibrator depends upon the charge and discharge time constants, and time constants are independent of supply voltage.

CHAPTER 30

IN-PROCESS LEARNING CHECK 1

1. A thyristor is a *four*-layer device that has just *two* operating states.
2. An SCR is gated on by applying a *forward*-biasing voltage to the *gate-cathode* P-N junction. A *LASCR* (or *light-activated SCR*) can also be gated on by shining a light onto its lens.
3. An SCR will conduct when it is gated and a voltage is applied between the cathode and anode such that the anode polarity is *positive* and the cathode is *negative*.
4. The SCR rating that specifies the maximum forward voltage an SCR can handle is called the *peak forward blocking voltage*.
5. The SCR rating that specifies the maximum reverse voltage an SCR can handle is called the *peak reverse voltage*.
6. The smallest amount of cathode-anode forward current that can sustain conduction of an SCR is called the *holding current*.
7. The *GCS* (or *gate-controlled switch*) is a type of SCR that can be off as well as gated on.

8. The *silicon-controlled switch* is a type of SCR that has two gate terminals.
9. In order to control both cycles of ac power, two SCRs must be connected in reverse with respect to one another in an arrangement that is known as *inverse parallel* or *back-to-back*.
10. When a pair of SCRs are gated halfway through their respective half-cycles of the ac power waveform, the circuit is said to be firing at *90°*.

IN-PROCESS LEARNING CHECK 2

1. Thyristors that can conduct in two directions (such as diacs and triacs) are said to be *bidirectional* devices.
2. A diac has *no* gate terminals, while a triac has *one* gate terminal.
3. The only way to get a diac to conduct is by exceeding its *forward* blocking and reverse *breakdown* ratings.
4. A triac operates like a pair of *SCRs* connected back to back.
5. The most common commercial and industrial applications of *diacs* are as triggering devices for triacs.
6. In order to apply full ac power to a load, a triac should be fired at the *zero*° point on each half-cycle.

REVIEW QUESTIONS

1. The two possible states of operation of a thyristor are fully conducting and fully nonconducting.
2. a. A junction diode is a two-layer device. An example is a rectifier diode (or a switching diode).
   b. A BJT is a three-layer device. An example is an NPN or PNP transistor.
   c. A thyristor is a four-layer device. An example is an SCR (or GCS, LASCR, or SCS).
3. The conditions required for turning on an SCR are (1) forward-biasing voltage between the cathode and anode, and (2) forward-biasing voltage between the cathode and gate. The SCR is turned off only by removing or reversing the cathode-anode voltage.
4. Applying a voltage to the gate (G) that is positive relative to the cathode (K) biases Q_2 on. The collector current for Q_2 forward biases Q_1 to turn on Q_1. The collector current for Q_1, in turn, maintains the base current for Q_2. The two transistors (equivalent of an SCR) are thus locked on.

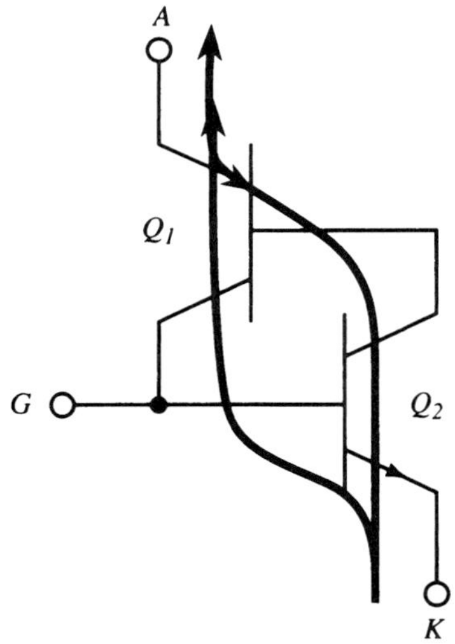

5. Holding current is the least amount of forward-conducting current a thyristor can conduct and remain conducting.
6. Forward-blocking voltage is the maximum amount of forward-bias voltage the device can withstand without breaking down and conducting. The peak-reverse voltage is the maximum amount of reverse-bias voltage that can be applied without breaking down the device. The forward-blocking voltage and peak-reverse voltage rating for a thyristor are frequently the same value (but with opposite polarities).

7. GCS = gate-controlled switch
   LASCR = light-activated SCR
   SCR = silicon-controlled rectifier
   SCS = silicon-controlled switch
8. Back to back

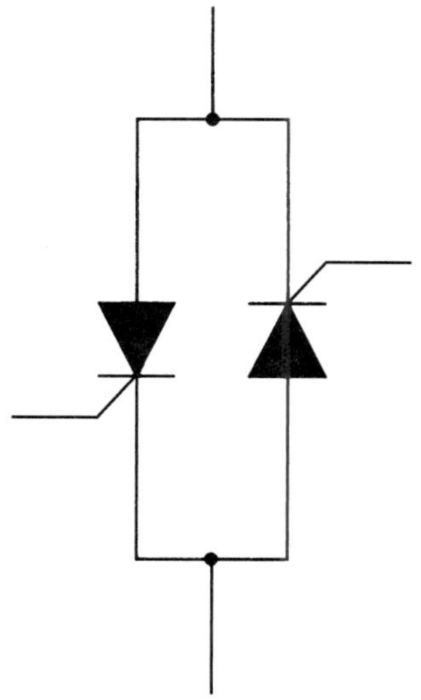

9. The most important difference between an SCR and a GCS is that a GCS can be both gated on and gated off. An SCR can only be gated on with gate pulses.
10. The SCS has two gates that can be used for triggering the device on. An SCR has only one gate for this purpose.
11. A LASCR can be switched on by applying a positive gate pulse or by shining a light onto its lens assembly.
12. a. Diac
    b. SCR, GCS, LASCR, triac
    c. SCS
    d. Diac, triac
13. A bidirectional device is one that is capable of conducting current in both directions. A common resistor is bidirectional because it can conduct current in both directions. A junction diode is not bidirectional because it passes current in only one direction.
14. A diac begins conducting when the applied voltage exceeds the forward-blocking or reverse breakdown rating. A diac stops conducting when the applied voltage goes to zero.
15. Since a diac is switched on only by exceeding the forward-blocking voltage and peak-reverse voltage, these ratings are extremely important to the theory of operation of a diac (a bidirectional device).
16. A triac symbol has a gate element, whereas a diac symbol does not.
17. A triac is gated on by applying a voltage between the two anode connections (polarity is not relevant) and applying a forward-biasing voltage to the gate. A triac is turned off when the applied voltage is reduced to zero or reversed in polarity.
18. An SCR can conduct in only one direction (from cathode to anode), whereas a triac can conduct in both directions.
19. An ac phase-control circuit is used for switching on a circuit at an adjustable point along each half-cycle of the ac waveform. Such circuits are used for controlling the amount of ac power applied to motors, heating elements, and lamps.

20.

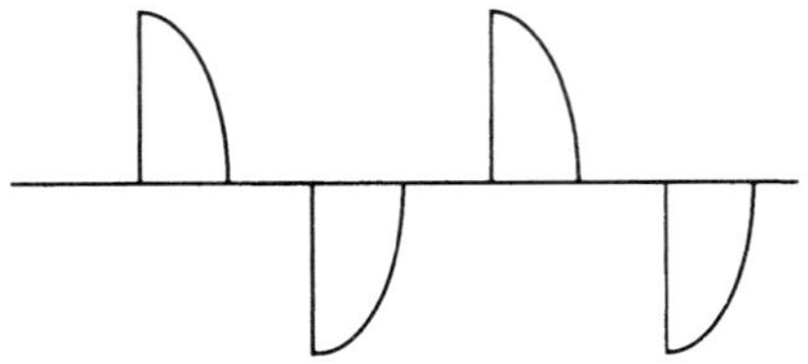

ANALYSIS QUESTIONS

1. An SCR can be considered a power amplifier because it uses a small gate voltage and current to switch on relatively large amounts of anode current and voltage. It can be considered a nonlinear amplifier because the output (load) waveform can be significantly different from the input (gate) waveform.
2. Student research and results will vary.
3. The only similarity between the action of a diac and a zener diode is that they begin conducting only after the applied voltage reaches a certain level. They are quite different in the way they are turned off. The zener diode stops conducting when the applied voltage falls below the zener level; however, the diac does not stop conducting until the applied voltage drops to nearly zero.
4. The UJT (unijunction transistor) is a negative-resistance device that has been commonly used as a low-frequency *RC* oscillator. The spiked, sawtooth output waveform is suitable for triggering SCRs and other gated thyristors. Phase control is accomplished by powering the UJT with the unfiltered output from a full-wave rectifier circuit. Oscillation (and the occurrence of the first trigger pulse) thus begins at a fixed point on each half-cycle. Varying the *RC* time constant varies the trigger interval.

CHAPTER 31

PRACTICE PROBLEMS 1

$R = 325\ \Omega$

PRACTICE PROBLEMS 2

Closed switches are S1, S3, S4, S6, and S7.

45.5 mA

IN-PROCESS LEARNING CHECK 1

1. An LED emits light energy when an electron in the *conduction* band falls to the *valence* band.
2. The choice of *semiconductor material* determines the color of the light from the LED.
3. The arrow on the LED symbol points *away from* the diode.
4. An LED emits light when it is *forward* biased.
5. The formula for calculating the value of the resistor to be used in series with an LED and its dc voltage source is $R = (V_S - V_D)/I_F$, where:
   V_S is *the voltage of the source*
   V_D is *the voltage across the diode*
   I_F is *the forward diode current*
6. The reverse-breakdown voltage of an LED is generally *lower* than that of a typical rectifier diode, and the forward voltage drop of an LED is *higher* than that of a rectifier diode.
7. Light falling onto the depletion region of a reverse-biased diode *increases* the amount of reverse leakage current.
8. The arrow on a photodiode symbol points *toward* the diode.
9. A photodiode is connected into the circuit so that it is *reverse* biased.
10. In an opto-isolator, an *LED* is the source and a *photodiode* is the detector of light energy.

USING EXCEL

$V_S = 9$ V

$V_D = 1.2$ V

$I_D = 24$ mA

Therefore, $R = 325\ \Omega$

REVIEW QUESTIONS

1. The conditions required to get an LED to emit light are:
   a. The LED must be forward biased.
   b. The LED must be conducting correctly.
2. 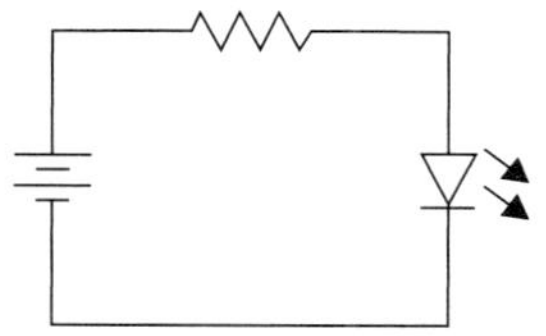
3. The light falling onto the photodiode junction causes a reverse-leakage current to flow. An increase in the light intensity results in an increase of the reverse current.
4. 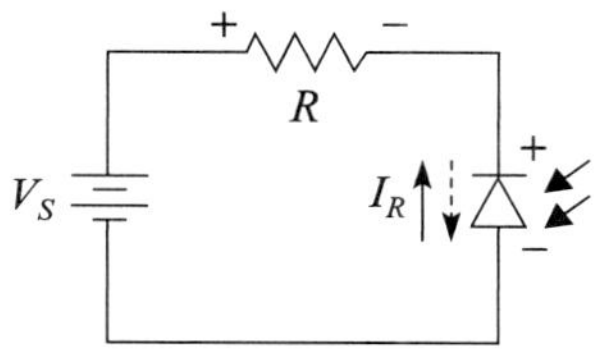

5. For the LED symbol, the arrows point away from the "diode" symbol. For the photodiode symbol, the arrows point into the "diode" symbol.
6. LED and photodiode
7. The light energy emitted from the laser diode is more intense than the light energy emitted from the LED.

PROBLEMS

1. a. 25 mA
   b. 25 mA
   c. 113 mW
2. $R = (9 - 1.5)/150$ mA
   $R = 5\ \Omega$
3. Switches S1, S2, S3, and S9
   36.4 mA

ANALYSIS QUESTIONS

1. Laser diodes provide the source of coherent light for scanning the bar code.
2. (Depends on parts selected for comparisons)

Answers to Laboratory Projects to accompany Foundations of Electronics, 5e (Electron Flow Version) and Foundations of Electronics: Circuits and Devices, 5e (Electron Flow Version) and Foundations of Electronics: Circuits and Devices, 2e (Conventional Flow Version)

Includes answers to:

- Procedures
- Story Behind the Numbers
- Summary

SECTION III

PART 1

USE AND CARE OF METERS

PROJECT 1 Voltmeters

In this project you will begin learning some of the standard procedures for safely using meters to make measurements. To reinforce your thinking, we will take this opportunity to stress key points of meter care in brief form. Think of and properly apply these key factors *every time you use a meter*!

Safety Hints

1. Use the proper METER MODE/FUNCTION (dc or ac, volts, amperes/mA, or ohms).
2. On nonautoranging meters, be sure the RANGE is high enough for what you will measure.

   NOTE ➤ If not sure, START with the HIGHEST range switch position and work down until the reading is easiest to interpret.
3. Be sure to OBSERVE POLARITY when measuring dc.
4. When MEASURING VOLTAGE, be sure meter is connected IN PARALLEL with the two points having the potential difference to be measured.
5. Use PERSONAL SAFETY cautions! (Power off when connecting test leads or holding only one lead with other hand in pocket, and so on.)

PROCEDURE

1. Use the precautions listed above and use a DMM (multimeter) to measure the voltage of a dry cell (e.g., flashlight, battery, and so on).

OBSERVATION Cell voltage measures ____**1.5**____ V.

CONCLUSION Mode/function used was (*dc, ac*) ___**dc**___ volts. The red test lead was connected to the (+, −) ___**+**___ terminal of the cell. The black test lead was connected to the (+, −) ___**−**___ terminal of the cell. If not autoranging, the range selector switch was in the ____**(depends on meter used)**____ voltage range position.

2. If a variable-voltage power supply is available, use the voltmeter to monitor the power supply's output voltage terminals and carefully adjust the power supply to 5-V, 10-V, and 15-V output settings, in that order. Have the instructor check your setting each time.

OBSERVATION 5-V setting OK. ➤ **Instructor initial:**______________

10-V setting OK. ➤ **Instructor initial:**______________

15-V setting OK. ➤ **Instructor initial:**______________

CONCLUSION What mode/function was the meter set in? **dc**. The red test lead was connected to the (+, −) + output terminal of the power supply. The black test lead was connected to the (+, −) − output terminal of the power supply. What meter voltage range setting was used? **(depends on meter used)**.

Lab Projects pages 3–4

PROJECT 2 Ammeters

In this project, you should be aware of the following precautions and procedures related to connecting meters and making current measurements.

Safety Hints

1. Turn POWER OFF circuit into which ammeter will be connected.
2. BREAK THE CIRCUIT in the appropriate place and INSERT THE METER IN SERIES, observing the following rules.
3. Use the proper METER MODE/FUNCTION (dc current mode).
4. On nonautoranging meters be sure the RANGE is high enough for the current to be measured. (Start with the HIGHEST range and work down, if appropriate.)
5. Be sure to OBSERVE PROPER POLARITY when measuring dc.
6. After meter is connected properly, TURN POWER ON AND TAKE READING.
7. Observe appropriate PERSONAL SAFETY rules.

PROCEDURE

1. Connect the circuit shown in Figure 2–1.

2. Set up the DMM to measure currents in the range of 0–1 mA dc.

OBSERVATION Mode/Function switch = **dc**. Range switch = **(depends on meter used)**.

3. Follow the safety rules outlined above. Break the circuit at point *A* and insert the meter.

OBSERVATION Check: (✔ *Yes*, ___ *No*) Proper polarity observed? **yes**.

Connected in series? **yes**. Proper range? **yes**.

CONCLUSION Current will enter the meter through the (*black, red*) **black** lead and will exit the meter through the (*black, red*) **red** lead.

4. Connect the 1.5-V cell to the circuit as shown and read the ammeter.

OBSERVATION Measured current is: **0.15 mA** mA.

CONCLUSION What change(s) would need to be made in the meter setup if *V* were to be greatly increased or if *R* were to be greatly decreased? **Change range selector to higher range if meter is not autoranging.**

PROCEDURE

1. Disconnect the 1.5-V cell from the previous circuit in Figure 2–1. Set up the DMM to read current in the range of 0 to 10 mA. Have your instructor check your setup.

   OBSERVATION Setup OK. ➤ **Instructor initial:**______________

2. With power OFF, connect a variable-voltage power supply (VVPS) to the circuit so the direction of current through the circuit will be the same as it was when the 1.5-V cell was connected, Figure 2–2.

   OBSERVATION Setup OK. ➤ **Instructor initial:**______________

3. Set the voltage control on the VVPS to the zero output setting. Then, turn power supply on and SLOWLY adjust output until you measure 5 mA through the circuit.

   OBSERVATION Procedure OK. ➤ **Instructor initial:**______________

4. What is the voltage applied to the circuit when the VVPS is adjusted so that there is current flow of 2.5 mA?

   STOP **CAUTION: Remember the safety rules.**

   OBSERVATION Voltage measures ____25____ V. Current measures ____2.5____ mA.

PROJECT 3 Ohmmeters

In this project, you should be aware of the following precautions and procedures about using the ohmmeter to make resistance measurements.

Safety Hints

1. Turn POWER OFF and/or DISCONNECT circuit from power source.
2. ISOLATE COMPONENT being measured from the rest of the circuit, whenever possible, to prevent "sneak" paths.
3. Use the proper METER MODE/FUNCTION (dc and ohms).
4. On nonautoranging meters, be sure the RANGE is appropriate for the range of resistance anticipated.
5. Connect test probes across component or circuit to be tested and make the measurement.
6. BE SURE TO TURN SELECTOR SWITCH OFF of the "ohms" range to a high-voltage range position (e.g., 1,000 V) or to the "OFF" position, if your meter has it, when finished using the ohmmeter.

PROCEDURE

1. Obtain a 1,000-Ω and a 10,000-Ω resistor. Use proper procedures; measure and record the resistance value of each resistor using the ohmmeter.

OBSERVATION The 1,000-Ω resistor measures ≈ **1,000** Ω

The 10,000-Ω resistor measures ≈ **10,000** Ω.

CONCLUSION Did the two resistors measure exactly 1,000 Ω and 10,000 Ω, respectively? **no**.

What could have caused any differences? **Resistor tolerances; meter error/tolerances**.

2. Obtain a kit of five unknown resistor values from your instructor. Using the ohmmeter, measure and record their values as numbered.

OBSERVATION Measured values are:

R #1 = **will depend on** Ω.
R #2 = **the kit supplied** Ω.
R #3 = ____ Ω.
R #4 = ____ Ω.
R #5 = ____ Ω.

➤ **Instructor initial:** ____

CONCLUSION When the value of R to be measured is unknown, it is generally best to start by trying to measure on the (*highest, lowest*) **highest** R range and by then working (*up, down*) **down** through the other ranges until the R value can be easily read. **(If meter is not autoranging.)**

3. If available, obtain several circuits from your instructor, some of which have continuity, some of which do not. Use the ohmmeter to identify the circuits that have continuity and those that do not have continuity.

OBSERVATION Continuity check: If there is continuity, fill in the blank with "yes," if not, fill in the blank with "no."

Circuit #1 **will depend on**. Circuit #4 ______.
Circuit #2 **circuits supplied**. Circuit #5 ______.
Circuit #3 **by instructor**.

CONCLUSION When there was continuity, the ohmmeter reading indicated a (*high, low*) **low** resistance reading.
When there was NO continuity, the ohmmeter indicated **0 L or °** ohms of resistance.

PART 1 Summary

Complete the following review questions, indicating the appropriate response by placing a check in the box next to the correct answer.

1. When preparing to measure an unknown dc voltage, the mode/function and range switches should be set at
   - ☐ dc mode, lowest V range
   - ☐ ac mode, lowest V range
   - ☑ dc mode, highest V range
   - ☐ ac mode, highest V range

2. The red test lead on a multimeter should be connected to the
   - ☑ positive input jack on the meter
   - ☐ negative input jack on the meter
   - ☐ neither of these

3. When measuring dc voltage, the meter's black test lead is normally connected to the more ___________ point of the component or circuit being measured.
   - ☑ negative
   - ☐ positive

4. Precautions for preparing to measure current include
   - ☐ Circuit off; correct polarity; correct range; connect meter in parallel
   - ☑ Circuit off; correct polarity; correct range; connect meter in series
   - ☐ Circuit on; correct polarity; correct range; connect meter in parallel
   - ☐ Circuit on; correct polarity; correct range; connect meter in series

5. Precautions when using an ohmmeter to measure the resistance value of a resistor include
   - ☐ Power off; correct range; correct polarity; turn off ohms mode when through measuring *R*
   - ☑ Power off; correct range; turn off ohms mode when through measuring *R*
   - ☐ Power on; correct range; correct polarity; turn off ohms mode when through measuring *R*

6. Turning the meter off the ohms mode when through may prevent
   - ☑ Possibility of battery drainage; possibility of meter damage
   - ☐ Using wrong range; meter damaging a circuit

Lab Projects page 11

PART 2

OHM'S LAW

PROJECT 4 Resistor Color Code Review and Practice

CHART 4-1

1st Color	2nd Color	3rd Color	4th Color	Ohms Value	Tolerance Percent
Red	Violet	Yellow	Gold	**270 kΩ**	**5%**
Brown	Black	Green	None	**1 MΩ**	**20%**
Orange	White	Black	Gold	**39 Ω**	**5%**
Yellow	Violet	Orange	Silver	**47 kΩ**	**10%**
Gray	Red	Brown	Gold	**820 Ω**	**5%**
Green	Brown	Black	Gold	**51 Ω**	**5%**
Blue	Red	Brown	Gold	**620 Ω**	**5%**
Green	Blue	Orange	Silver	**56 kΩ**	**10%**

CHART 4-2

Ohms Value	Colors		
100 Ω	**Brown**	**Black**	**Brown**
12 Ω	**Brown**	**Red**	**Black**
1.0 Ω	**Brown**	**Black**	**Gold**
13 kΩ	**Brown**	**Orange**	**Orange**
2 MΩ	**Red**	**Black**	**Green**
91 Ω	**White**	**Brown**	**Black**

NOTE ➤ These charts are for practice only and therefore may have some values called for which are not standard available resistor values.

PROCEDURE

1. List the 10 colors used in the resistor color code to represent 0,1,2,3,4,5,6,7,8, and 9.

OBSERVATION Colors used in the resistor color code are as follows:

0 = **Black**. 5 = **Green**.
1 = **Brown**. 6 = **Blue**.
2 = **Red**. 7 = **Violet**.
3 = **Orange**. 8 = **Grey**.
4 = **Yellow**. 9 = **White**.

2. List the other colors in the color code generally used to indicate resistor tolerance or used as special multipliers.

OBSERVATION Special colors are:

gold and **silver**.

CONCLUSION The color used to indicate 5% tolerance is **gold**.
The color used to indicate 10% tolerance is **silver**.
The color used to indicate a 0.1 multiplier is **gold**.
The color used to indicate a 0.01 multiplier is **silver**.

3. Fill in the resistance and tolerance values on Chart 4-1, as appropriate.
4. Fill in the colors that can be used to indicate the values in Chart 4-2, as appropriate.
5. Obtain a set of 10 resistors having assorted values and tolerances and use the color code to determine their values and tolerances, as appropriate.

OBSERVATION Values and tolerances of resistors in the kit are as follows:
Answers depend on resistors supplied.

R #1 = ______ Ω ____ % tolerance. *R* #6 = ______ Ω ____ % tolerance.
R #2 = ______ Ω ____ % tolerance. *R* #7 = ______ Ω ____ % tolerance.
R #3 = ______ Ω ____ % tolerance. *R* #8 = ______ Ω ____ % tolerance.
R #4 = ______ Ω ____ % tolerance. *R* #9 = ______ Ω ____ % tolerance.
R #5 = ______ Ω ____ % tolerance. *R* #10 = ______ Ω ____ % tolerance.

➤ **Instructor initial:**__________

6. For precision resistors, a five-band color-coding system is frequently used. List the meaning of each band in the observation column.

OBSERVATION Band #1 = **first value digit**. Band #4 = **decimal multiplier**.
Band #2 = **second value digit**. Band #5 = **tolerance**.
Band #3 = **third value digit**.

7. For the fifth band on these precision resistors, list the meaning of each color listed in the Observation section, as appropriate.

OBSERVATION Brown = ± 1 %. Blue = ± 0.25 %.
Red = ± 2 %. Violet = ± 0.1 %.
Green = ± 0.5 %.

8. Fill in the resistance and tolerance values for the five-band precision resistors listed in Optional Chart 4-3, as appropriate.

Optional Chart

CHART 4-3 Special Precision Resistors

1st Color	2nd Color	3rd Color	4th Color	5th Color	Resistance Value (First 4 bands)	Tolerance Percent (5th band)
Brown	Brown	Black	Red	Brown	11 kΩ	1%
Orange	Blue	Black	Red	Red	36 kΩ	2%
Brown	Brown	Black	Orange	Brown	110 kΩ	1%
White	Brown	Black	Orange	Red	910 kΩ	2%
Red	Yellow	Black	Gold	Green	24 Ω	0.5%
Brown	Brown	Black	Silver	Blue	1.1 Ω	0.25%

Lab Projects pages 15–17

PROJECT 5 Relationship of *I* and *V* with *R* Constant

As you perform this project, remember that Ohm's Law states that $I = \frac{V}{R}$.

PROCEDURE

1. Connect the initial circuit as shown in Figure 5-1.
2. Adjust V_A to obtain 1/2 scale deflection on the 1-mA range.

   OBSERVATION Current is **0.5** mA.

3. Measure the V_A.

   OBSERVATION V_A measures **5** V.

4. Use Ohm's Law ($I = V/R$) and calculate I from the measured value of V and the indicated R value.

   OBSERVATION $V_A =$ **5** V. $R =$ **10 k** Ω.

   CONCLUSION I calculated = **0.5** mA.

5. Increase V_A to twice its original value and note the new current reading.

   OBSERVATION $V_A =$ **10** V. I now = **1** mA.

   CONCLUSION Doubling V_A caused I to: **double**. From this we conclude that with R constant (unchanged), current is directly proportional to **V**.

6. Reduce V_A to 2 volts and note the new current reading.

   OBSERVATION I now = **0.2** mA.

   CONCLUSION Reducing V_A to 2 volts caused the current to (*increase, decrease*) **decrease** proportionately. This shows again that current stays "in step" (is directly proportional) with voltage when R is unchanged. This means that if R is held constant and V is increased, I will (*increase, decrease*) **increase**; if V is decreased, I will (*increase, decrease*) **decrease**.

Lab Projects pages 19–20

PROJECT 6 Relationship of *I* and *R* with *V* Constant

PROCEDURE

1. Connect the initial circuit as shown in Figure 6-1.
2. Adjust V_A to obtain 1 mA of current.

   OBSERVATION Current is ___1___ mA.

3. Measure V_A and **be sure *not* to change V_A for the rest of the steps in this section.**

   OBSERVATION V_A measures ___10___ V.

4. Use Ohm's Law ($R = V/I$) and calculate R from the measured values of V and I.

   OBSERVATION $V =$ ___10___ V. $I =$ ___1___ mA.

   CONCLUSION R calculated = ___10 k___ Ω.

5. Remove the 10-kΩ resistor from the circuit, replace it with a 100-kΩ resistor, and note the new current reading.

   OBSERVATION R now = ___100 k___ Ω. I now = ___0.1___ mA.

   CONCLUSION Keeping V_A constant at ___10___ volts and increasing R by 10 times to a value of ___100 k___ ohms caused the current to (*increase, decrease*) ___decrease___ to one ___tenth___ of its original value. From this we conclude that I is inversely proportional to R. This means that if R is increased, I will (*increase, decrease*) ___decrease___ proportionately; if R is decreased, I will (*increase, decrease*) ___increase___ by the same factor as R was decreased.

6. Remove the 100-kΩ resistor from the circuit, replace it with a 47-kΩ resistor, and record the new current reading.

   NOTE ➤ Keep V_A the same as it was.

   OBSERVATION $V_A =$ ___10___ V. $I =$ ___0.21___ mA.
   $R =$ ___47 k___ Ω.

   CONCLUSION The circuit resistance for this step is *approximately* (*1/4, 1/2*) ___1/2___ that of the previous step; the applied voltage is the same; and the resulting current is approximately (*2, 4*) ___2___ times that of Step 5. This again tends to prove that I is (*directly, inversely*) ___inversely___ proportional to R, with V held constant.

Lab Projects pages 21–23

Optional Step

7. Remove the 47-kΩ resistor from the circuit and replace it with a 27-kΩ resistor. Record the new current reading.

OBSERVATION V_A = **10** V. I = **0.37** mA.
R = **27** Ω.

CONCLUSION Did the circuit current change from the previous step? **yes**. Was the R larger or smaller than in the previous step? **smaller**. In your own words, explain what this data proves: **As R decreases, I increases. As R increases, I decreases. I and R are inversely proportional.**

PROJECT 7 Relationship of Power to *V* with *R* Constant

PROCEDURE

1. Connect the initial circuit as shown in Figure 7-1.
2. Adjust V_A to obtain 0.5 mA of current and measure V_A.

   OBSERVATION I = ____**0.5**____ mA. V_A = ____**5**____ V.

3. Use the $V \times I$ power formula and calculate the power dissipated by R.

   OBSERVATION V = ____**5**____ V. I = ____**0.5**____ mA.

   CONCLUSION P calculated = ____**2.5**____ mW.

4. Change V_A to obtain 1 mA of current. Measure V_A and calculate P using the $V \times I$ formula.

   OBSERVATION V now = ____**10**____ V. I now = ____**1**____ mA.

   CONCLUSION P calculated now = __**10**__ mW. Doubling V caused the current to increase by how many times? __**2**__. Thus, the product of $V \times I$ increased __**4**__ times (when we doubled the voltage and kept R constant). From this we conclude that power is proportional to the ____**square**____ of the voltage when R is not changed.

5. Calculate the power for the measured values of step 4 using the $P = V^2/R$ formula.

   OBSERVATION V = ____**10**____ V. R = ____**10 k**____ Ω.

   CONCLUSION P calculated = ____**10**____ mW.

6. Change V_A to 2.5 volts. Measure the appropriate voltage and current values and calculate P by both of the above formulas.

   OBSERVATION V now = ____**2.5**____ V. I now = ____**0.25**____ mA.

   CONCLUSION P calculated by both methods is approximately __**0.625**__ mW. Compared to the 10-volt V_A condition, we now have (*1/2, 1/4, 1/8*) __**1/4**__ the V_A and (*1/4, 1/8, 1/16, 1/32*) __**1/16**__ the power dissipation. This again illustrates that P is related to the (*square*, $\sqrt{\ }$) __**square**__ of the voltage when R remains unchanged.

Optional Steps

7. Assume a V_A value of 4 volts. Calculate and predict what the circuit current and power would be for that value of applied voltage.

   OBSERVATION Predicted I = ____**0.4**____ mA. Predicted P = ____**1.6**____ mW.

Lab Projects pages 25–27

8. Change the circuit V_A to 4 volts; measure I, and calculate P.

OBSERVATION Measured I = ____**0.4**____ mA. Calculated P = ____**1.6**____ mW.

CONCLUSION Was the current with 4 volts applied voltage higher or lower than when there were 2.5 volts applied? ____**higher**____. Was the calcualted power higher or lower? ____**higher**____. How many times higher or lower? ____**2.56**____. Is the change coherent with the concept of power being proportional to V^2? ____**yes**____.

PROJECT 8 Relationship of Power to *I* with *R* Constant

PROCEDURE

1. Connect the initial circuit as shown in Figure 8-1.
2. Adjust V_A to 7 volts and note the current.

OBSERVATION $I =$ **0.7** mA.

3. Use the measured values of V and I and calculate P by the formula $P = V \times I$.

OBSERVATION $V =$ **7** V. $I =$ **0.7** mA.

CONCLUSION P calculated = **4.9** mW.

4. Decrease V_A until I is 1/2 its original value and calculate P using both the $V \times I$ and $I^2 R$ formulas.

OBSERVATION V now = **3.5** V. I now = **0.35** mA.

CONCLUSION P calculated by both methods is approximately **1.225** mW. This is (*1/8, 1/4, 1/2*) **1/4** the power dissipated when I was double the value of the current for this step. From this we can conclude that P is proportional to I^2 when R in the circuit is unchanged. This means that if the circuit current has doubled, the power dissipation has increased (*1, 2, 3, 4*) **4** times; or, if circuit current were decreased to one-third its original value, the power must decrease to (*1/3, 1/6, 1/9*) **1/9** its original value.

Optional Step

5. Change V_A until it is approximately one-third the original 7-volt value (approximately 2.33 volts). Make appropriate measurements and calculations to fill in the blanks in the Observation section.

OBSERVATION V_A now = **2.3** V. $V \times I$ now = **0.529** mW.
I now = **0.23** mW V^2/R now = **0.529** mW.

CONCLUSION Is the power approximately one-ninth that when 7 volts was applied? **yes**. Does this verify that power is related to the square of the circuit current with R constant? **yes**.

PART 2 Story Behind the Numbers

Data Table

Given Values		Measured Current
V applied Value	Value of *R* Used	Value of Circuit Current (in mA)
10 V	10 kΩ	**1.0**
10 V	27 kΩ	**0.37**
10 V	47 kΩ	**0.21**
20 V	10 kΩ	**2.0**
20 V	27 kΩ	**0.74**
20 V	47 kΩ	**0.43**

Graph 1

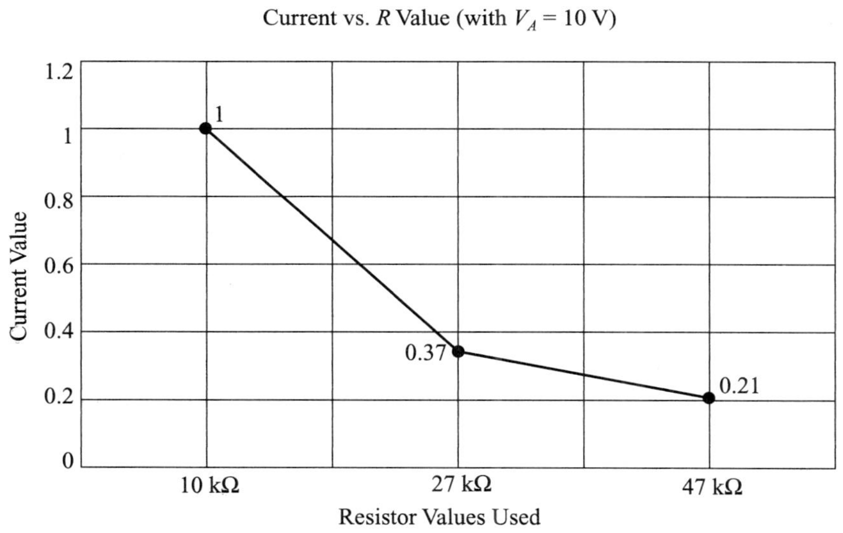

Graph 2

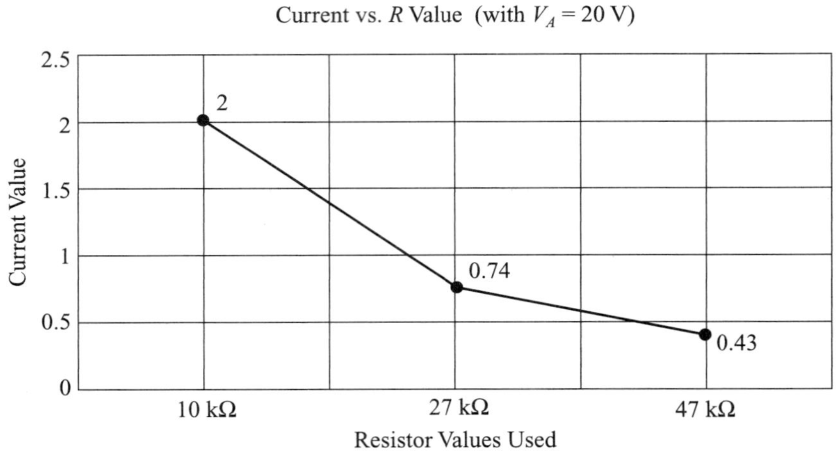

Answers to Analysis Questions

1. Current is inversely related to resistance for any applied voltage.

2. As voltage was doubled, current doubled through a given resistance value. For example, with 10 V applied and an R value of 10 kΩ, current was 1 mA. With 20 V applied and an R value of 10 kΩ, current was 2 mA … etc.

3. At 10 V applied the 10-kΩ R dissipated 10 mW. At 20 V applied it dissipated 40 mW. At 10 V applied the 27-kΩ R dissipated 3.7 mW. At 20 V applied it dissipated 14.8 mW.

4. Observed that when voltage was doubled power quadrupled for a given R value. That is, power is proportional to the square of the voltage, as is shown by the formula $P = V^2/R$. For example, when R was 10 kΩ, R dissipated 10 mW with 10 V applied, but dissipated 40 mW with 20 V applied.

5. The data collected reveal that current is directly related to voltage for a given resistance. That is, if voltage doubles, current will double, etc. Also, the data show that current is inversely proportional to resistance value. That is, for a given voltage, if resistance doubles, current will halve, etc. Finally, the data show that power is proportional to voltage squared for a given resistance value. That is, for a given R value, if voltage is doubled, the power dissipated by that R will be four times the original value.

Lab Projects pages 31–33

PART 2 Summary

Complete the following review questions, indicating the appropriate response by placing a check in the box next to the correct answer.

1. If *V* increases and *R* remains the same, then *I* will
   - ☑ increase
   - ☐ decrease
   - ☐ remain the same
2. If *I* increases and *R* remains the same, then *V* must have
   - ☑ increased
   - ☐ decreased
   - ☐ remained the same
3. If *R* increases and *V* remains the same, then *I* will
   - ☐ increase
   - ☑ decrease
   - ☐ remain the same
4. If *V* is doubled and *R* is halved, then *I* will
   - ☐ double
   - ☐ halve
   - ☑ quadruple
   - ☐ remain the same
5. If *V* is doubled, and *R* remains the same, then *P* will
   - ☐ double
   - ☐ halve
   - ☑ quadruple
   - ☐ remain the same
6. If *I* is halved, and *R* remains the same, then *P* will
   - ☐ double
   - ☐ halve
   - ☐ quadruple
   - ☑ decrease to one-quarter
   - ☐ remain the same
7. If *V* is doubled and *R* is halved, then *P* will
   - ☐ decrease 4 times
   - ☑ increase 8 times
   - ☐ increase 16 times
   - ☐ remain the same
8. Increasing the voltage applied to a circuit will cause
   a. Current to
   - ☑ increase
   - ☐ decrease
   - ☐ remain the same

   b. Resistance to
   - ☐ increase
   - ☐ decrease
   - ☑ remain the same

   c. Power dissipated to
   - ☑ increase
   - ☐ decrease
   - ☐ remain the same

Lab Projects pages 35–36

9. Decreasing the resistance in a circuit will cause
   a. Current to
      - ☑ increase
      - ☐ decrease
      - ☐ remain the same
   b. Voltage applied to
      - ☐ increase
      - ☐ decrease
      - ☑ remain the same
   c. Power dissipated to
      - ☑ increase
      - ☐ decrease
      - ☐ remain the same

10. In an electrical circuit
    a. current is directly proportional to
       - ☑ V
       - ☐ R
    b. and inversely proportional to
       - ☐ V
       - ☑ R
    c. while power is proportional to the square of the
       - ☑ V
       - ☐ R

Lab Projects pages 35–36

PART 3

SERIES CIRCUITS

PROJECT 9 Total Resistance in Series Circuits

PROCEDURE

1. Connect the initial circuit shown in Figure 9-1.

   STOP **CAUTION: Do not connect power to the circuit for this project!**

2. Use an ohmmeter and measure the total resistance of the circuit.

   OBSERVATION R total = **30 k** Ω.

3. Use an ohmmeter, measure the resistance of each individual resistor, and record your observations.

   OBSERVATION R_1 = **10 k** Ω. R_3 = **10 k** Ω.
   R_2 = **10 k** Ω. **DMM readings will vary slightly.**

4. Add the resistances of the individual resistors and note the result.

   OBSERVATION $R_1 + R_2 + R_3$ = **30 k** Ω.

   CONCLUSION The total resistance of a series circuit equals the (*product, sum*) **sum** of all the individual resistances; therefore, R_T = $\boldsymbol{R_1 + R_2 + R_3 \ldots}$.

5. Predict what the new R_T would be if R_1 were changed to a 27-kΩ resistor.

   OBSERVATION Predicted R_T = **47 k** Ω.

6. Change R_1 to 27 kΩ and measure the new R_T.

   OBSERVATION New R_T = **47 k** Ω.

   CONCLUSION Changing any element's resistance in a series circuit while the rest of the elements are unchanged will cause the circuit's total resistance to (*remain the same, change*) **change**. If any element's R increases, then R_T will (*increase, decrease*) **increase**. If any element's R decreases, then R_T will (*increase, decrease*) **decrease**.

Optional Step

7. Change the circuit so that R_1 = 47 kΩ, R_2 = 100 kΩ, and R_3 = 100 kΩ. Predict R_T's value; then measure R_T to verify your prediction.

OBSERVATION Predicted R_T = **247** kΩ. Measured R_T = **247** kΩ.

CONCLUSION Did R_T equal the sum of the individual resistances in this case? **yes**. What might cause a difference in predicted and measured values? ***R* tolerance, Meter tolerance**

Lab Projects pages 39–40

PROJECT 10 Current in Series Circuits

PROCEDURE

1. Connect the initial circuit shown in Figure 10-1.
2. Apply 9.5 volts to the circuit and note the current.

OBSERVATION Number of paths for current in circuit = ___**1**___.
Current = ___**0.7**___ mA.

3. Swap positions of R_1 and the current meter and note the current reading.

OBSERVATION Current reading is now ___**0.7**___ mA.

4. Swap positions of the current meter and each of the remaining resistors and note the current reading each time.

OBSERVATION Current reading in every case was ___**0.7**___ mA.

CONCLUSION No matter where the current meter was placed in the circuit, the current reading was the same. This indicates that the current through all parts of a series circuit is the ___**same**___ current.

5. Change R_1 to a 10-kΩ resistor. Move the meter to various spots in the circuit and note the current reading.

OBSERVATION Current reading in all cases was ___**0.47**___ mA.

CONCLUSION Increasing any R in a series circuit affects the current through all parts of the circuit. Would changing R by decreasing it cause I to change? ___**yes**___. Increase or decrease? ___**increase**___.

6. Now change V_A to 19 volts. Move the meter to various spots in the circuit and note the current reading.

OBSERVATION Current reading in all cases was ___**0.94**___ mA.

CONCLUSION Changing V_A for the series circuit caused I to change through all parts of the circuit.

It changed by (*a different, the same*) ___**the same**___ amount in all parts because the current through all parts of a series circuit is (*a different, the same*) ___**the same**___ current. This is because there is only ___**one**___ path for current through series elements.

Lab Projects pages 41–42

PROJECT 11 Voltage Distribution in Series Circuits

PROCEDURE

1. Connect the initial circuit shown in Figure 11-1.
2. Apply 9.5 volts to the circuit and measure each of the individual voltage drops. Also calculate the circuit current.

OBSERVATION V_A = **9.5** V. V_{R_3} = **6.75** V.
V_{R_1} = **0.25** V. I calculated = **0.25** mA.
V_{R_2} = **2.5** V.

CONCLUSION Since the I is the same through all the resistors, the voltage drop across any given resistor is directly related to its R compared to the total circuit (*number of Rs, resistance*) **resistance**.

3. Add all the individual voltage drops and note the sum.

OBSERVATION $V_{R_1} + V_{R_2} + V_{R_3}$ equals **9.5** V or V **applied**.

CONCLUSION In essence, Kirchhoff's Voltage Law states that the arithmetic sum of voltage drops around any circuit closed loop must equal V applied. Does it? **yes**.

4. Calculate what fraction of the applied voltage is dropped by each of the resistors. Express your answer as a fraction. (Example: 1/38, 10/38, etc.)

OBSERVATION *Example:*
V_{R1} = **1/38th** V_A. V_{R3} = **27/38ths** V_A.
V_{R_2} = **10/38ths** V_A.

CONCLUSION Each resistor dropped the same fraction of V applied as its **resistance** value is of the total **resistance**.

5. Compute the fractional relationship of V_{R_1} to V_{R_2} and V_{R_3}. Express answers as fractions.

OBSERVATION *Example:*
V_{R_1} = **1/10th** V_{R_2}. V_{R_1} = **1/27th** V_{R_3}.

CONCLUSION Because I is the same through all elements in a series circuit and since V = ***I*** × ***R***, the voltage drops across the resistors are related to each other by the same factor as their ***Rs***.

6. Predict what value V_{R_2} would be if V_A were 19 volts. Change V_A to 19 volts and measure V_{R_2}.

OBSERVATION V_{R_2} predicted = **5** V. V_{R_2} measured = **5** V.

CONCLUSION V_{R_2} is now what fraction of V_A? **10/38ths**. Is this the same fraction as when V_A = 9.5 volts? **yes**. Changing V (*does, does not*) **does not**

change the distribution percentages of V_A. If one of the *R*s were changed, would the distribution percentages change? ___yes___. We may conclude that in series circuits, the largest *R* will drop the (*least, most*) ___most___ voltage, and the smallest *R* the (*least, most*) ___least___ voltage.

7. Use the voltage-divider rule and calculate V_{R_1} and V_{R_3} assuming $V_T = 19$ V.

OBSERVATION V_{R_1} calculated = ___0.5___ V. V_{R_3} calculated = ___13.5___ V.

Table 2

8. Measure V_{R_1} and V_{R_3} with 19 V applied to circuit.

OBSERVATION V_{R_1} measured = ___0.5___ V. V_{R_3} measured = ___13.5___ V.

CONCLUSION Do the measured values for V_{R_1} and V_{R_3} confirm the calculations using the voltage-divider rule? (*yes, no*) ___yes___.

Optional Steps

9. Turn off the power supply and replace R_3 with a 100-kΩ resistor.

10. Adjust voltage applied to the circuit to approximately 28 volts and measure the resistor voltage drops.

OBSERVATION V_{R_1} = ___0.26___ V. V_{R_3} = ___26___ V.
V_{R_2} = ___2.6___ V.

CONCLUSION Is V_{R_3} equal to about 10 times V_{R_2}? ___yes___? Is V_{R_2} about 10 times greater than V_{R_1}? ___yes___. Is the ratio of V_{R_3}'s voltage to V_{R_1}'s voltage about equal to their *R* ratio? ___yes___? What might cause the ratios discussed to not be exactly 10:1 in each case?
___***R* tolerance, Meter tolerance**___

PROJECT 12 Power Distribution in Series Circuits

PROCEDURE

1. Connect the initial circuit shown in Figure 12-1.

2. Apply 19 volts (V_A) to the circuit. Measure the current and the individual voltage drops and calculate the power dissipated by each of the resistors.

OBSERVATION $I =$ **0.5** mA. $V_{R_2} =$ **5.0** V.
$V_{R_1} =$ **0.5** V. $V_{R_3} =$ **13.5** V.

CONCLUSION P_{R_1} calculated = **0.25** mW.
P_{R_2} calculated = **2.5** mW.
P_{R_3} calculated = **6.75** mW.

We may conclude that since the I is the same through all resistors, the I^2R or (*voltage, power*) **power** dissipated by each resistor is directly related to its (*size, resistance*) **resistance** value. Furthermore, the power distribution throughout the circuit is the same as the (*I, R*) ***R*** distribution. This means that the largest value R will dissipate the (*most, least*) **most** amount of power; the smallest R, the (*most, least*) **least** power.

3. Add all the individual power dissipations and note the sum. Also, calculate P_T by the formula: $P_T = V_T \times I_T$.

OBSERVATION $P_{R_1} + P_{R_2} + P_{R_3}$ equals **9.5** mW. $V_T \times I_T =$ **9.5** mW.

CONCLUSION The total power in a series circuit is equal to the (*product, sum*) **sum** of all the individual power dissipations.

4. What ratio does P_{R_1} have to P_{R_3}? Express as a ratio. (For example: 1:10, 2:5, etc.)

OBSERVATION Ratio = **1:27**.

CONCLUSION The ratios of power dissipated by two resistors in a series circuit is the same as their **resistance** ratio.

5. Indicate what would happen to P_T and to the individual power dissipations if V_A were cut in half.

OBSERVATION $P_T =$ **one-fourth** as much as before. Individual Ps would also be **one-fourth** the original.

CONCLUSION For a given R, decreasing V to one-half will also cause the circuit current to (*increase, decrease*) **decrease** to **1/2**. Therefore, the product of $V \times I$ will be (*1/2, 1/4*) **1/4** the original value. If Rs remain unchanged but V is changed, the percentage of P_T dissipated by any given R

Lab Projects pages 47–49

will (*change, not change*) ______**not change**______ but the actual value of power dissipated will (*change, not change*) ______**change**______.

Optional Step

6. Change the V_A to a value of 9.5 volts. Measure and calculate values, as required to fill in the blanks in the Observation section.

OBSERVATION

V_A = **9.5** V.
I = **0.25** mA.
P_T = **2.37** mW.
V_{R_1} = **0.25** V.
P_{R_1} = **0.06** mW.
V_{R_2} = **2.5** V.
P_{R_2} = **0.625** mW.
V_{R_3} = **6.75** V.
P_{R_3} = **1.69** mW.

CONCLUSION When V_A was reduced to half its original value, current decreased to ______**one-half**______ its original value and total power dissipated by the circuit decreased to ______**one-fourth**______ its original value. Does the power dissipated by each resistor also change by this same factor? ___**yes**___. Does this verify the data in step 5? ___**yes**___.

PROJECT 13 Effects of an Open in Series Circuits

PROCEDURE

1. Connect the initial circuit as shown in Figure 13-1.

2. Adjust V applied to 19 volts. Measure and record the current and individual voltage drops.

OBSERVATION $I =$ **0.5** mA. $V_{R_2} =$ **5.0** V.
$V_{R_1} =$ **0.5** V. $V_{R_3} =$ **13.5** V.

3. To simulate an R becoming open in this series circuit, remove R_2 and leave the circuit open between R_1 and R_3. Measure and record the circuit current and the individual voltage drops.

OBSERVATION $I =$ **0** mA. $V_{R_2} =$ **19** V. (across open)
$V_{R_1} =$ **0** V. $V_{R_3} =$ **0** V.

CONCLUSION Since in a series circuit there is only one path for current flow, opening any element within the series circuit will cause (*continuity, discontinuity*) **discontinuity**. The R_T of the circuit then appears to be infinitely (*high, low*) **high**. The voltage drop across R_1 was **0** volts because, with zero current, the $I \times R$ drop must be **0**. We conclude that if any part of a series circuit opens, R_T will (*increase, decrease*) **increase** to **∞**; I_T will (*increase, decrease*) **decrease** to **0**. The voltage drops across the unopened elements will (*increase, decrease*) **decrease** to **0**, and the potential difference across the open portion of the circuit will (*increase, decrease*) **increase** to **V applied**.

4. Assume R_2 were replaced in the circuit and R_3 removed. List the predicted results in the Observation section.

OBSERVATION $I =$ **0** mA. $V_{R_2} =$ **0** V.
$V_{R_1} =$ **0** V. $V_{R_3} =$ **19** V.

5. Make the change suggested in step 4 above, make appropriate measurements, and note the results.

OBSERVATION $I =$ **0** mA. $V_{R_2} =$ **0** V.
$V_{R_1} =$ **0** V. $V_{R_3} =$ **19** V.

CONCLUSION Do the results of these measurements verify the conclusions of step 3? **yes**. The main difference noted in the parameters for this step compared to step 3 is that V applied now appears across R **3**, rather than R **2**.

PROJECT 14 Effects of a Short in Series Circuits

PROCEDURE

1. Connect the initial circuit as shown in Figure 14-1.
2. Adjust V applied to 20 volts. Measure and record the current and individual voltage drops.

OBSERVATION $I =$ **0.425** mA. $V_{R_2} =$ **4.25** V.
$V_{R_1} =$ **4.25** V. $V_{R_3} =$ **11.48** V.

3. To simulate an R "shorting out," remove R_2 and replace it with a jumper wire. Measure and note the circuit I and individual voltage drops with the shorted element replacing R_2.

OBSERVATION $I =$ **0.54** mA. $V_{R_2} =$ **0** V.
(across short)

$V_{R_1} =$ **5.4** V. $V_{R_3} =$ **14.6** V.

CONCLUSION Shorting out R_2 has caused the circuit R_T to (*increase, decrease*) **decrease** to **37 k** ohms. This then caused I_T to (*increase, decrease*) **increase**. The new (*higher, lower*) **higher** current caused the $I \times R$ drops across the unshorted elements (Rs) in the circuit to (*increase, decrease*) **increase**. The R between the shorted terminals is effectively **0** ohms. Therefore, the $I \times R$ drop across the shorted element or section of a series circuit will (*increase, decrease*) **decrease** to **0**. We conclude that if any part of a series circuit shorts, R_T will (*increase, decrease*) **decrease**;

I_T will (*increase, decrease*) **increase**; the voltage drops across the

unshorted elements will (*increase, decrease*) **increase**; and the V across the shorted element or section of the circuit will (*increase, decrease*) **decrease** to **0**. The total power supplied to the circuit will (*increase, decrease*) **increase** with the shorted condition.

Lab Projects pages 55–56

PART 3 Story Behind the Numbers

Table 1

Identifiers	Circuit 1 Measured Values	Circuit 2 Measured Values
R_1	10,000 Ω	10,000 Ω
R_2	10,000 Ω	27,000 Ω
R_3	10,000 Ω	10,000 Ω
R_4	10,000 Ω	47,000 Ω
R_T	40,000 Ω	94,000 Ω
V_{R_1}	5 V	2.13 V
V_{R_2}	5 V	5.75 V
V_{R_3}	5 V	2.13 V
V_{R_4}	5 V	10.01 V
V_T	20 V	20 V
I_T	0.5 mA	0.213 mA

Table 2

Identifiers	Calculated *P* Values for Circuit 1	Calculated *P* Values for Circuit 2
P_{R_1}	2.5 mW	0.45 mW
P_{R_2}	2.5 mW	1.22 mW
P_{R_3}	2.5 mW	0.45 mW
P_{R_4}	2.5 mW	2.13 mW
P_T	10 mW	4.25 mW

Lab Projects pages 57–60

Graph 1

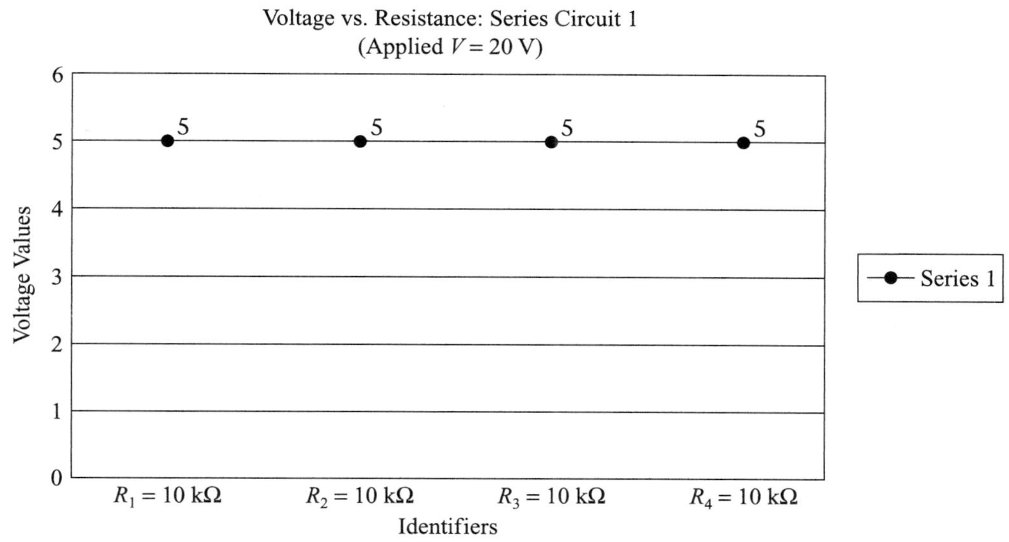

Graph 2

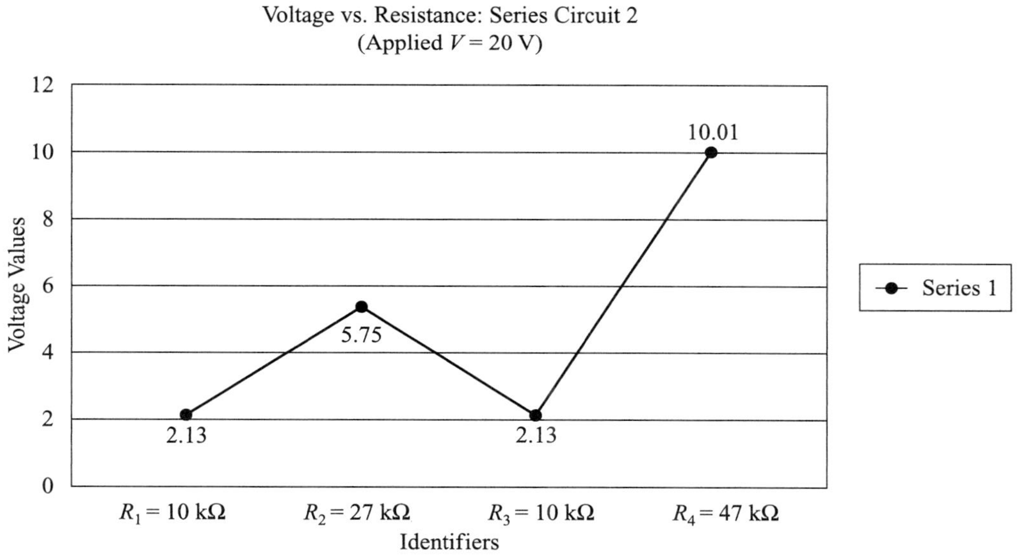

Answers to Analysis Questions

1. Yes, the current is the same since there is only one path for current. Verification example: All the 10-kΩ *R*s had the same voltage drops with the *R* tolerance ratings.

2. Yes, each was proportional to its resistance value and to other resistor values in series with it. Also, in Circuit 1, R_1 was one-fourth of total *R* and dropped one-fourth of V_A; R_2 was one-fourth of total *R* and dropped one-fourth of V_A, etc. Likewise, in Circuit 2, the proportionality statements held true.

Lab Projects pages 57–60

3. In circuit 2, R_4 (47 kΩ) dissipated the most power, and both R_1 and R_3 dissipated the least power. This is true because in the series circuit all resistors have the same current through them. Therefore, the I^2R value would be greatest for the largest resistor value, and least for the smallest resistor value.

4. No. The *R*s did not precisely match the color code. This is because resistors cannot typically be manufactured with zero percent tolerance. Most resistors are rated by the manufacturer as having a plus and minus tolerance rating from color-coded value.

5. No. In Circuit 1 the color-coded value would be 40 kΩ and measured R_T was 39.8 kΩ. In Circuit 2 the color-coded value would be 94 kΩ and measured R_T was 95.8 kΩ.

6. Assuming 5% tolerance resisistors, the answers would be:

   Minimum R_T possible if all resistors were *below* color-coded value by the amount of their tolerances: Circuit 1 minimum possible R_T __**38**__ kΩ. Circuit 2 minimum possible R_T __**89.3**__ kΩ.

   Maximum R_T possible if all resistors were *above* color-coded value by the amount of their tolerances: Circuit 1 maximum possible R_T __**42**__ kΩ. Circuit 2 maximum possible R_T __**98.7**__ kΩ.

7. (a) $I_T = 0$, (b) $P_T = 0$, (c) V across opened $R = V_A$, V across other *R*s = 0

8. (a) R_T decreases, (b) I_T increases, (c) P_T increases

Lab Projects pages 57–60

PART 3 Summary

Complete the following review questions, indicating the appropriate response by placing a check in the box next to the correct answer.

1. The total resistance in a series circuit is equal to
   - ☑ the sum of all the *R*s
   - ☐ the largest *R* minus smallest *R*
   - ☐ neither of these

2. In a series circuit there is
   - ☑ only one path for current
   - ☐ as many paths as there are components

3. The current in a series circuit is
   - ☑ the same through all parts
   - ☐ different through each component

4. The highest voltage drop in a series circuit appears across
   - ☐ the smallest *R*
   - ☑ the highest *R*
   - ☐ the average value *R*

5. If R_1 is 10 times larger in value than R_2,
   - ☑ V_1 = 10 times V_2
   - ☐ V_1 = one-tenth V_2
   - ☐ neither answer is correct

6. In a series circuit, the smallest *R* will dissipate
   - ☐ the most power
   - ☐ no power
   - ☑ the least power
   - ☐ none of these

7. In a series circuit, the applied voltage equals
   - ☐ the largest *V* drop minus the smallest *V* drop
   - ☑ the sum of all *V* drops

8. In a series circuit, the percentage of the circuit applied voltage dropped across any given resistor is directly related and dependent upon the ratio of the given *R* to
   - ☑ *R* total
   - ☐ *I* total
   - ☐ *V* total
   - ☐ *P* total

9. In a series circuit, if part of the circuit becomes open, then
   a. R_T will
      - ☑ increase
      - ☐ decrease
      - ☐ remain the same

   b. V_A will
      - ☐ increase
      - ☐ decrease
      - ☑ remain the same

   c. I_T will
      - ☐ increase
      - ☑ decrease
      - ☐ remain the same

Lab Projects pages 61–62

d. The voltage across the opened portion of the circuit will

- ☑ increase
- ☐ remain the same
- ☐ decrease

e. The voltage drops across the unopened elements will

- ☐ increase
- ☐ remain the same
- ☑ decrease

10. In a series circuit, if part of the circuit becomes shorted, then

a. R_T will

- ☐ increase
- ☐ remain the same
- ☑ decrease

b. V_A will

- ☐ increase
- ☑ remain the same
- ☐ decrease

c. I_T will

- ☑ increase
- ☐ remain the same
- ☐ decrease

d. The voltage across the shorted portion of the circuit will

- ☐ increase
- ☐ remain the same
- ☑ decrease

e. The voltage drops across the other components in the circuit will

- ☑ increase
- ☐ remain the same
- ☐ decrease

In summarizing, whether you use these questions, or some of your own, the instructor should be sure the students understand the following basic facts:

1. Total resistance in a series circuit is simply the sum of all the individual resistances.
2. Voltage and power distribution is directly related to the R distribution.
3. V_T = the sum of all V drops in series.
4. P_T = the sum of all the P dissipations.
5. The effects of opens or shorts in series circuits on R_T, I_T, P_T, and individual component parameters.

Lab Projects pages 61–62

Optional Troubleshooting Exercises

To be optionally used to challenge students after they have completed the "Series Circuits" group of projects and the "Summary." The instructor may wish to add more problems to be solved than the three examples shown here.

Series Circuit Troubleshooting Challenge

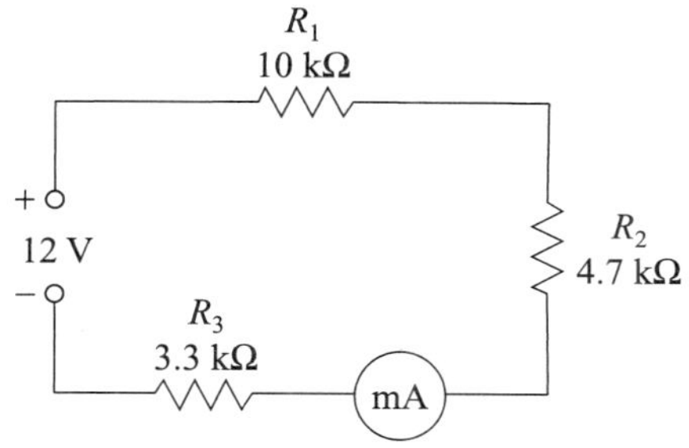

Circuit Diagram

The instructor should set up this series circuit with *one* of the following faults in it, and then have students troubleshoot and find the problem. Have students log each step they use so you can evaluate, comment on, or grade their performance when they have finished the troubleshooting procedure. Instruct students to use the SIMPLER sequence in performing this exercise.

NOTE ➤ Only put in *one* of these probems at a time for students to troubleshoot. You may want to repeat the exercise and give them a second or third problem, as you desire.

1. Trouble is: Resistor R_1 somewhat greater in value than its color-coded value.
   (Starting point symptom to be given to students is: *Circuit current is lower than normal*)
2. Trouble is: Resistor R_2 open.
   (Starting point symptom to be given to students is: *No circuit current*)
3. Trouble is: Resistor R_3 shorted.
   (Starting point symptom to be given to students is: *Higher than normal circuit current*)

PART 4

PARALLEL CIRCUITS

PROJECT 15 Equivalent Resistance in Parallel Circuits

PROCEDURE

1. Connect the initial circuit as shown in Figure 15-1.
2. Measure R_e at points A and B.

   OBSERVATION R_e = **10 k** Ω.

3. Use the popular product-over-the-sum formula and calculate R_e if a 10-kΩ R were inserted between points C and D on the trainer.

   $R_e = R_1 \times R_2 \div (R_1 + R_2)$

   OBSERVATION R_e calculated = **5 k** Ω.

   CONCLUSION When two resistors of equal value are connected in parallel, R_e is equal to one-half the R of (*one, both*) **one** branch(es).

   After calculating R_e, measure it with a DMM at points A and B with a second 10-kΩ R inserted at points C and D.

   OBSERVATION R_e measured = **5 k** Ω.

4. Assume a third 10-kΩ R is to be added at points E and F and calculate R_e. Use the reciprocal method, or

   $R_e = R_{e}^{1} \times R_3 \div (R_{e}^{1} + R_3)$

   OBSERVATION R_e calculated = **3.3 k** Ω.

   CONCLUSION When three resistors of equal value are connected in parallel, R_e is equal to (*1/2, 1/3, 1/4*) **1/3** the R of one branch.

   After calculating, connect a third 10-kΩ R between points E and F and measure R_e to verify your calculations.

   OBSERVATION R_e measured = **3.3 k** Ω.

Lab Projects pages 67–71

CONCLUSION In parallel circuits, the total resistance of the circuit is always less than the (*highest, lowest*) ____**lowest**____ branch R. If two or more *unequal* Rs are in parallel, can we divide one branch R by the number of branches to find R_e? ____**no**____.

5. Change R_2 to a 47-kΩ R and R_3 to a 100-kΩ R. Use the assumed voltage method of solving for R_e as follows:
   a. Assume 47 volts applied.
   b. Solve for each branch I.
   c. Calculate I_T.
   d. Find R_e by Ohm's Law.

$$C = \frac{1}{2\pi f X_C}; \text{ and } f = \frac{1}{2\pi C X_C}.$$

OBSERVATION I_{R_1} = ____**4.7**____ mA. I_T = ____**6.18**____ mA.
I_{R_2} = ____**1.0**____ mA. R_e = ____**7.6 k**____ Ω.
I_{R_3} = ____**0.48**____ mA.

After calculating R_e, use the DMM and measure R_e to verify your calculations.

OBSERVATION R_e measured = ____**7.6 k**____ Ω.

CONCLUSION The assumed voltage method of finding R_e is sometimes easier to use than the product-over-the-sum method, especially when several branches of unequal values are involved. Many times, if an appropriate value of V applied is assumed, one can solve for the branch currents in his/her head, then add the branch currents easily to solve for (R_T, I_T) ____I_T____. Then all that remains is a simple (*multiplication, division*) ____**division**____ problem to find R_e.

NOTE ➤ Using a calculator, the reciprocal method is also very easy!

6. With R_1 = 10 kΩ, R_2 = 47 kΩ, and R_3 = 100 kΩ as shown in Figure 15-2, what will the parameters (electrical circuit values) be if we apply 7.6 volts? (Use 7.6 as the assumed voltage and calculate as before.)

OBSERVATION I_{R_1} = ____**0.76**____ mA. I_T = ____**1.0**____ mA.
I_{R_2} = ____**0.16**____ mA. R_e = ____**7.6 k**____ Ω.
I_{R_3} = ____**0.076**____ mA.

CONCLUSION Changing V applied does not change (R_T, I_T) ____R_T____. The voltage that is "assumed" does make a difference in difficulty of computing the final result (R_e). From the preceding steps, it would seem wise to choose a value of assumed voltage that is easily divided by the resistance values for solving branch currents, since no matter what voltage is assumed, it does not change the actual circuit (I, R) ____**R**____.

Lab Projects pages 67–71

7. With the circuit described in step 6 connected, insert the milliammeter to read I_T. Apply 7.6 volts to the circuit and measure I_T. After measuring, REMOVE the power supply and current meter from the circuit. Replace the current meter with a jumper wire.

OBSERVATION $I_T =$ **1.0** mA.

CONCLUSION The assumed voltage method of calculating R_e (*has been, has not been*) **has been** verified.

8. A practical problem that sometimes confronts a technician is what value R must be put in parallel with the existing Rs in order to arrive at a desired equivalent resistance. A simple formula that helps solve this is:

$$R_u = \frac{R_k \times R_e}{R_k - R_e}$$

where: R_u is R unknown, R_k is R known, and R_e is the desired resultant equivalent R. Assume you have a 27-kΩ R and want an R_e of 17 kΩ. Use the above formula, solve for R_u, then connect the circuit on the matrix using an available resistor as close to the calculated R_u as possible and measure R_e to verify results.

OBSERVATION $R_u =$ **45.9 k** Ω.

CONCLUSION The formula for solving for the unknown needed parallel resistance to arrive at a given equivalent resistance seems to work, because the results of the practical circuit were close to the theoretical result. It has been proven again that R_e turns out to be (*less, more*) **less** than the least resistance branch.

PROJECT 16 Current in Parallel Circuits

NOTE ➤ Due to using standard-value resistors, some approximating is called for in this project to simply demonstrate concepts.

PROCEDURE

1. Connect the initial circuit as shown in Figure 16-1.
2. Replace the appropriate jumper with the current meter to read I_1. Apply 5 volts to the circuit and note I_1. Then move the meters and jumpers as required to measure I_2, I_3, and I_T.

OBSERVATION I_1 = **0.5** mA. I_3 = **0.106** mA.
I_2 = **0.185** mA. I_T = **0.791** mA.

CONCLUSION R_2 is roughly **3** times larger in resistance than R_1. The current through R_2 is roughly **1/3** the current through R_1. R_3 is roughly **5** times larger in R value than R_1 and its current is roughly **1/5** of the current through R_1. From this we conclude that the current through parallel branches is (*directly, inversely*) **inversely** proportional to the branch Rs. Also, we observe from the measured currents that total circuit current equals the (*product, sum*) **sum** of the **branch** currents.

3. Change R_2 from 27 kΩ to 100 kΩ. Measure and record all the circuit currents one at a time. Make V applied 5 volts in each case.

OBSERVATION I_1 = **0.5** mA. I_3 = **0.106** mA.
I_2 = **0.05** mA. I_T = **0.656** mA.

CONCLUSION When R_2 was changed from 27 kΩ to 100 kΩ, keeping the same V_A, did the current through R_1 or R_3 change? **no**. Did I_T change? **yes**. If so, was the change in I_T the same as the change in I_2? **yes**. This again proves that I_T is simply the sum of all the **branch** currents. Since R_2 is 10 times larger than R_1, its current should be **1/10th** of I_1. Is it? **yes, close**.

4. Refer again to the circuit diagram, and make the appropriate statements in the Conclusion section that will verify Kirchhoff's Current Law, which states that the current away from any point in a circuit must equal the current to that point.

CONCLUSION The currents going away from point A, (D) in the circuit are labeled as **I_1** and **I_2** and **I_3**. These equal the current coming to point A, (D) and this current is designated by the symbol **I_T**. The current *to and* *away from* point *B* is labeled as I_3 on the diagram. The currents to and away from point *C* include I_2 and I_3, and the currents to and away from point *D* include I_1 and I_2 and I_3.

Lab Projects pages 73–74

PROJECT 17 Voltage in Parallel Circuits

PROCEDURE

1. Connect the initial circuit as shown in Figure 17-1.
2. Apply 10 volts V_A to the circuit and measure all the circuit voltages. Record in the Observation section.

OBSERVATION V_A = 10 V. V_{R_2} = 10 V.
V_{R_1} = 10 V. V_{R_3} = 10 V.

CONCLUSION We can conclude from the measurements that the voltage across all the resistors in parallel is the **same**. Also, in a simple, purely parallel circuit configuration, the voltage across each branch equals *V* **applied**.

3. Change the value of R_2 from 10 kΩ to 100 kΩ and measure all the circuit voltages. Record in the Observation section.

OBSERVATION V_A = 10 V. V_{R_2} = 10 V.
V_{R_1} = 10 V. V_{R_3} = 10 V.

CONCLUSION Did all the voltages measure the same as before? **yes**. This indicates that changing the value of branch *R*s in a parallel circuit (*does, does not*) **does not** alter the branch voltages. As a matter of fact, the branch voltages are not separate voltages, but really all the **same** *V*. Does changing the value of a branch *R* change the circuit total current? **yes**. Does it change the current through the unchanged *R* branch(es)? **no**. Does it change the current through the changed *R* branch? **yes**.

4. Add a fourth branch to the parallel circuit by inserting a 10-kΩ resistor between points *A* and *B*. Measure the circuit voltages with 10 volts applied.

OBSERVATION V_A = 10 V. V_{R_3} = 10 V.
V_{R_1} = 10 V. V_{R_4} = 10 V.
V_{R_2} = 10 V.

CONCLUSION From the results of this step, we conclude that adding additional branches to a parallel circuit (*does, does not*) **does not** alter the voltage across the parallel branches. (This assumes that the source of voltage is capable of meeting the current requirements of the circuit without its output voltage being altered.)

Lab Projects pages 75–76

PROJECT 18 Power Distribution in Parallel Circuits

PROCEDURE

1. Connect the initial circuit as shown in Figure 18-1.
2. Apply 10 volts to the circuit, replace the appropriate "branch jumpers" one at a time with the milliammeter in order to measure each of the branch currents (I_1, I_2, and I_3), and record the results in the Observation section.

OBSERVATION I_1 = **1.0** mA. I_3 = **0.21** mA.
I_2 = **0.37** mA.

CONCLUSION Using the law that total current in a parallel circuit is equal to the (*product, sum*) **sum** of the **branch** currents, the total current for the circuit must be **1.58** mA. Since the total power dissipated in any resistive circuit can be calculated as V_T times **I_T**, then the total power dissipated by this circuit and being supplied by the source is **15.8** mW.

3. Use the appropriate power formula(s) and calculate the individual power dissipations of R_1, R_2, and R_3. Record your answers.

OBSERVATION P_{R_1} = **10** mW. P_{R_3} = **2.1** mW.
P_{R_2} = **3.7** mW.

CONCLUSION Does the sum of $P_{R_1} + P_{R_2} + P_{R_3}$ equal P_T? **yes**. Which resistor dissipates the most power? **R_1**. This resistor is the (*largest, smallest*) **smallest** value R in the circuit. Which resistor dissipated the least power? **R_3**. This resistor is the (*largest, smallest*) **largest** value R in the circuit. From this we conclude that in a parallel circuit the smaller the branch R, the (*lesser, greater*) **greater** power it will dissipate, because all Vs are the same, and the Is are inversely proportional to the branch Rs. Therefore, the $V \times I$ product will be greater if the resistor is of (*low, high*) **low** value.

4. Change R_2 from a 27-kΩ to a 10-kΩ resistor. Measure the branch currents and calculate the individual power dissipations and the P_T with 10 volts V applied.

OBSERVATION I_1 = **1.0** mA. P_{R_2} = **10** mW.
I_2 = **1.0** mA. P_{R_3} = **2.1** mW.
I_3 = **0.21** mA. P_T = **22.1** mW.
P_{R_1} = **10** mW.

CONCLUSION

Changing the resistance of one branch in a parallel circuit will cause the power dissipated by that branch to change and the ____**total**____ power to change. But the power dissipated by the other branches will remain the same as long as their resistance values are not changed.

PROJECT 19 Effects of an Open in Parallel Circuits

PROCEDURE

1. Connect the initial circuit as shown in Figure 19-1.

2. Apply 20 volts to the circuit and note the measured total current. Use Ohm's Law and calculate R_T and the branch currents.

OBSERVATION $I_T =$ **0.825** mA. $I_2 =$ **0.2** mA.
$R_T =$ **24.2 k** Ω. $I_3 =$ **0.20** mA.
$I_1 =$ **0.425** mA.

CONCLUSION R_T is less than the (*highest, lowest*) **lowest** resistance branch. I_T is equal to the **sum** of the branch currents. The ratios of the branch currents to each other is inverse to the ratios of their **resistances** to each other.

3. Remove R_1 from the circuit to simulate an open in that branch. Apply 20 volts to the circuit and measure I_T, I_1, I_2, and I_3. Calculate R_T.

OBSERVATION $I_T =$ **0.4** mA. $I_2 =$ **0.2** mA.
$R_T =$ **50 k** Ω. $I_3 =$ **0.2** mA.
$I_1 =$ **0** mA.

CONCLUSION It can be concluded that if any branch of a parallel circuit opens, R_T will (*increase, decrease*) **increase**; therefore, I_T will (*increase, decrease*) **decrease** since V remained the same. Did the current through R_2 and R_3 change when R_1 opened? **no**. The reason R_T increased was that when R_1 opened there was one less **current** path. The current through the opened branch (*increased, decreased*) **decreased** to **0** mA; whereas, the current through the unopened branches (*increased, decreased, remained the same*) **remained the same**. The voltage across all branches (*did, did not*) **did not** change when the R_1 branch was opened.

4. Use the proper formula(s) and calculate the branch power dissipations and P_T for the normal circuit condition of step 2.

OBSERVATION $P_T =$ **16.5** mW. $P_{R_2} =$ **4.0** mW.
$P_{R_1} =$ **8.5** mW. $P_{R_3} =$ **4.0** mW.

5. Use the proper formula(s) and calculate the branch power dissipations and P_T for the circuit conditions when R_1 was open in step 3.

OBSERVATION $P_T =$ **8.0** mW. $P_{R_2} =$ **4.0** mW.

$P_{R_1} =$ **0.0** mW. $P_{R_3} =$ **4.0** mW.

CONCLUSION Opening one branch of a parallel circuit (*does, does not*) **does not** affect the power dissipated by the other branches. However, total *P* will (*increase, decrease*) **decrease** by the amount of power that was being dissipated by the opened branch previous to its opening.

The instructor might call attention to the fact that in any kind of circuit, an open causes R_T to increase and I_T and P_T to decrease.

PROJECT 20 Effects of a Short in Parallel Circuits

PROCEDURE

1. Connect the initial circuit as shown in Figure 20-1.

   CAUTION: Do not use the power supply or milliammeter for this demonstration or project.

2. Use the DMM and measure R_e at points A and B.

   OBSERVATION R_e = **24.2 k** Ω.

3. To simulate R_1 becoming shorted, remove R_1 and replace it with a jumper wire; then measure R_e again.

   OBSERVATION R_e = **0.0** Ω.

   CONCLUSION A short across one branch of a parallel circuit causes R_e to (*increase, decrease*) **decrease** effectively to **0** Ω. If we were using a power supply that could supply infinite current, the current through R_2 would be **0** mA; through R_3 would be **0** mA; and through the shorted branch would be ∞ mA. Since $V = I \times R$, what is the IR drop across the shorted branch? **0** V. In parallel, the voltage across all branches in parallel is the **same** voltage; therefore, V_2 and V_3 for the conditions described above would be **0** V.

4. Replace R_1 into the circuit and simulate a short in branch R_2 by replacing R_2 with a jumper wire. Measure R_e again.

   OBSERVATION R_e = **0.0** Ω.

   CONCLUSION A short of *any* branch of a parallel circuit will cause R_e to (*increase, decrease*) **decrease** and I_T to (*increase, decrease*) **increase** (before the power supply fuse blows). Also, current through the "unshorted" branches would (*increase, decrease*) **decrease**, and current through the "shorted" branch would (*increase, decrease*) **increase**. The voltage across all branches would (*increase, decrease*) **decrease**. It should be noted that in the strictest sense, if any branch of a parallel circuit is "shorted," all branches are shorted. However, only one branch "contains" the actual shorted element that is "shunting" the whole circuit with the undesired low resistance path, whose resistance approaches (° , *0*) **0** Ω.

The instructor may note the fact that a short in ANY kind of circuit causes R_T to decrease and I_T and P_T to increase. For parallel circuits, I_T increases to the capacity of the power source and may blow fuses, breakers, burn out components, or even cause a fire.

Lab Projects pages 83–84

PART 4 Story Behind the Numbers

Table 1

Identifiers	Initial Circuit Theoretical (Calculated) Values	Initial Circuit Measured Values	Modified Circuit (R_3 Removed) Measured Values
V_A	25.00 V	25.00 V	—
I_{R_1}	0.93 mA	0.93 mA	—
I_{R_2}	0.53 mA	0.53 mA	—
I_{R_3}	0.25 mA	0.25 mA	—
I_{R_4}	0.25 mA	0.25 mA	—
I_T^*	1.96 mA	1.96 mA	1.71 mA
R_T	12.75 kΩ	12.75 kΩ	14.62 kΩ

Table 2

Identifiers	Using Initial Circuit Theoretical (Calculated) Values	Using Initial Circuit Measured Values	Using Modified Circuit (R_3 Removed) MeasuredValues
P_{R_1}	23.25 mW	23.25 mW	—
P_{R_2}	13.25 mW	13.25 mW	—
P_{R_3}	6.25 mW	6.25 mW	—
P_{R_4}	6.25 mW	6.25 mW	—
P_T	49 mW	49 mW	42.75 mW

Lab Projects pages 85–87

Graph 1

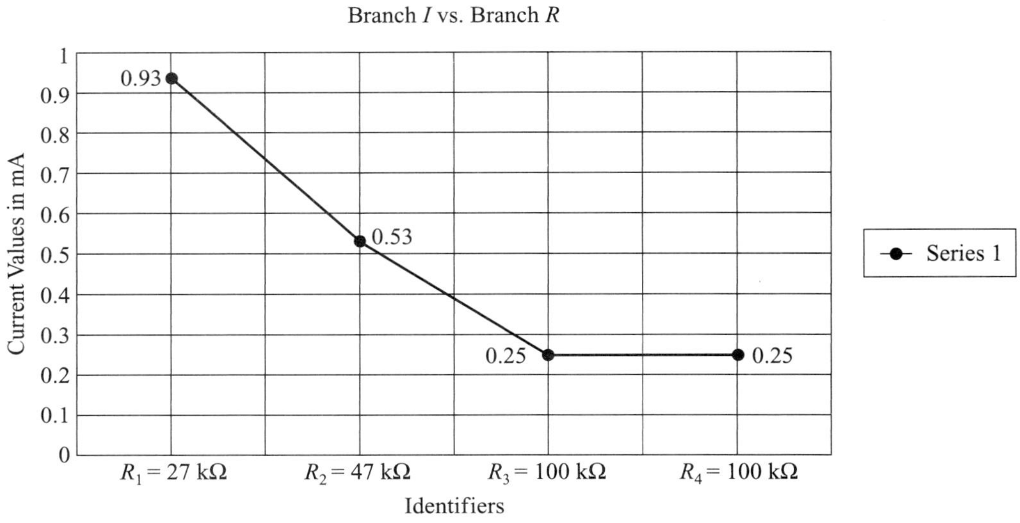

Graph 2

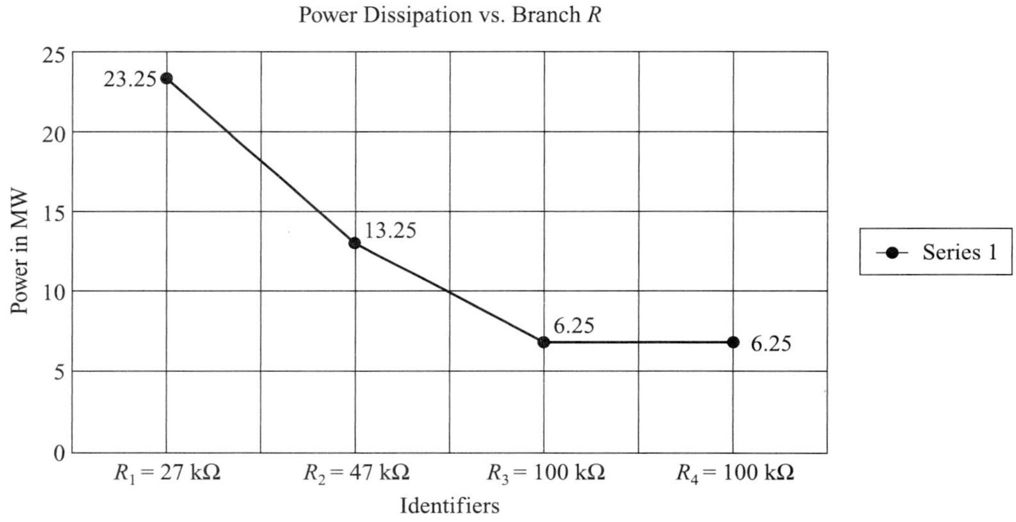

Answers to Analysis Questions

1. Branch currents are inversely related to each branch resistance value for any given applied voltage.

2. Branch power dissipations were inversely related to branch resistance values, since each branch had the same V, and $P = V \times I$. The power dissipated by the 27-kΩ resistor was approximately double that dissipated by the 47 kΩ branch; the power dissipated by the 47-kΩ branch was approximately double that dissipated by the 100-kΩ branches, etc.

3. The approximate ratio was inverse to the resistors; that is, 100/47 or about 2.12:1. For example, the power dissipated by the 47-kΩ *R* was about 13.25 mW. The power dissipated by one of the 100-kΩ branch *R*s was about 6.25 mW, which is about a 2.12:1 ratio.

4. Total current decreased and total power decreased since *V* was the same and *I* total decreased. When one branch was opened, there was one less branch to conduct current; thus *R* total increased, and with the same applied voltage, total current therefore decreased.

5. Since the standard rule indicates 2 times the power expected to be dissipated, the minimum power rating for R_1 should be 2 × 23.25 mW, or 46.5 mW. Anything above 1/8 W would do; however, 1/4 W or larger ratings are more commonly available and will work.

Lab Projects pages 85–87

PART 4 Summary

Complete the following review questions, indicating the appropriate response by placing a check in the box next to the correct answer.

1. A greater change of R_e occurs if the resistance that is added in shunt with the existing circuit has a resistance value that is
   - ☐ high
   - ☑ low
2. Adding another resistor in parallel with an existing circuit will cause the circuit's total current to
   - ☑ increase
   - ☐ decrease
   - ☐ remain the same
3. Adding another resistor in parallel with an existing parallel circuit will cause the currents through the original branches to
   - ☐ increase
   - ☐ decrease
   - ☑ remain the same
4. In a simple two-branch parallel circuit consisting of a 2-Ω and a 3-Ω resistor in parallel, three-fifths of the total circuit current will flow through
   - ☐ the 3-Ω *R*
   - ☑ the 2-Ω *R*
   - ☐ neither *R*
5. The total resistance of the circuit in question 4 is
   - ☐ 8 Ω
   - ☐ 2 Ω
   - ☑ 1.2 Ω
   - ☐ 2.1 Ω
   - ☐ 3 Ω
   - ☐ none of these
6. In a circuit consisting of 100-kΩ, 47-kΩ, and 1-kΩ branches, the total circuit resistance must be
   - ☐ more than 1 kΩ
   - ☑ less than 1 kΩ
   - ☐ more than 100 kΩ
   - ☐ more than 47 kΩ
7. The important fact to remember about parallel circuits is that the voltage across all parallel branches is
   - ☐ divided equally
   - ☐ different
   - ☑ the same
   - ☐ none of these
8. When one branch of a parallel circuit opens
   a. V_A will
   - ☐ increase
   - ☐ decrease
   - ☑ remain the same

   b. I_T will
   - ☐ increase
   - ☑ decrease
   - ☐ remain the same

   c. The unopened branch currents will
   - ☐ increase
   - ☐ decrease
   - ☑ remain the same

Lab Projects pages 89–90

d. The opened branch currents will

- ☐ increase
- ☑ decrease
- ☐ remain the same

e. R_T will

- ☑ increase
- ☐ decrease
- ☐ remain the same

9. If any branch of a parallel circuit "shorts," it will cause

- ☑ excessive current
- ☐ excessive voltage
- ☐ excessive resistance
- ☐ none of these

10. In a parallel circuit, the resistor that dissipates the most power is

- ☑ the lowest value *R*
- ☐ the highest value *R*
- ☐ neither of these

Lab Projects pages 89–90

Optional Troubleshooting Exercises

To be optionally used to challenge students after they have completed the "Parallel Circuits" group of projects and the "Summary." The instructor may wish to add more problems to be solved than the three examples shown here.

Parallel Circuit Troubleshooting Challenge

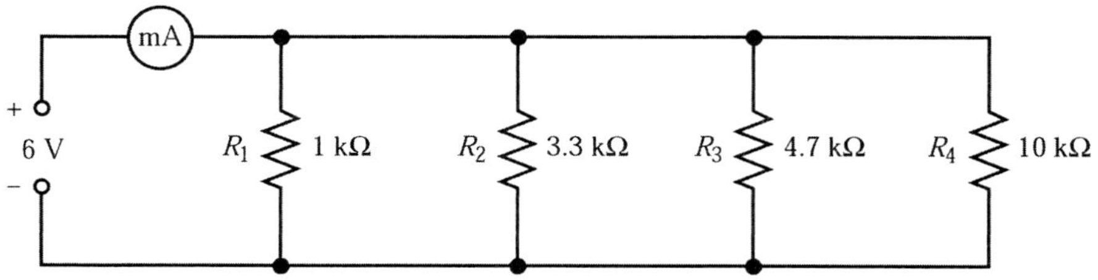

Circuit Diagram

The instructor should set up this parallel circuit with *one* of the following faults in it, and then have students troubleshoot and find the problem. Have students log each step they use so you can evaluate, comment on, or grade their performance when they have finished the troubleshooting procedure. Instruct students to use the SIMPLER sequence in performing this exercise.

NOTE ➤ Only put in *one* of these problems at a time for students to troubleshoot. You may want to repeat the exercise and give them a second or third problem, as you desire.

1. Trouble is: Resistor R_3 is open.
   (Starting point symptom to be given to students is: *Circuit current is too low*)
2. Trouble is: Broken current path (bad jumper or broken wire) between R_3 and R_4.
   (Starting point symptom to be given to students is: *Circuit current is too low*)
3. Trouble is: Resistor R_3 is much lower in value than its color-coded value.
   (Starting point symptom to be given to students is: *Circuit current is somewhat high*)

PART 5

SERIES-PARALLEL CIRCUITS

PROJECT 21 Total Resistance in Series-Parallel Circuits

NOTE ➤ The term R_T generally refers to total circuit resistance, whereas the term R_e (R equivalent) may refer to the total resultant resistance of specific parallel resistors in just a portion of the circuit. If the total circuit is a purely parallel circuit, the term R_e is the same as R_T, for that particular case.

PROCEDURE

1. Connect the initial circuit as shown in Figure 21-1.
2. Calculate the circuit R_T from the indicated resistance values as follows: Add $R_1 + R_e$ of R_2 and R_4 (in parallel) + R_3.

OBSERVATION R_T = ____16 k____ Ω.

CONCLUSION Is the total resistance of this circuit less than the least resistor? ___No___. Is the total resistance of this circuit equal to the sum of all the individual resistors? ___No___.

3. Apply just enough voltage to the circuit to obtain 1 mA of current and measure V_A, V_{R_1}, V_{R_2}, V_{R_3}, and V_{R_4}.

OBSERVATION V_A = ____16____ V. V_{R_3} = ____10____ V.
V_{R_1} = ____1____ V. V_{R_4} = ____5____ V.
V_{R_2} = ____5____ V.

CONCLUSION Since there is ___1___ mA of total circuit current and V_A is ___16___ V, then by Ohm's Law, R_T must be ___16 k___ Ω. Which resistors in this circuit are in parallel with one another? ___R_2___ and ___R_4___. Is the voltage across these two Rs the same? ___Yes___. Which resistor(s) are in series with the main line (carry I_T)? ___R_1 and R_3___.

4. DISCONNECT the *power supply and milliammeter* from the circuit and measure R_T (at appropriate points in the circuit) with an ohmmeter.

OBSERVATION R_T = ____16 k____ Ω.

CONCLUSION It can be concluded that in series-parallel circuits the total resistance is equal to the sum of all the components in (*series, parallel*) **series** with the main line (components through which I_T passes) plus the equivalent resistance of parallel elements, whose R_e is effectively in (*series, parallel*) **series** with the main line components.

5. Using the logic just discussed, calculate what the circuit R_T would be if a 10-kΩ resistor were added to the circuit between points *A* and *B*. After calculating R_T, measure it with a DMM to verify your thinking (meter and power supply still not connected).

OBSERVATION R_{calc} = **11 k** Ω. R_{meas} = **11 k** Ω.

CONCLUSION R_T was calculated by adding the resistance of *R* **1** + the R_e of *R* **2** and *R* **4** + the R_e of *R* **3** and *R* **5**. Would R_T change if the jumper were removed that runs between the top of R_3 and the top of R_5? **Yes**.

Optional Steps

6. Assume that the 10-kΩ resistor is left in the circuit between points *A* and *B*. Also assume that the circuit is to be rewired so that all four 10-kΩ resistors will be in parallel with each other, and that combination is to still be in series with R_1 (the 1-kΩ resistor). Predict the total resistance for the conditions just described.

OBSERVATION $R_{predicted}$ = **3.5** kΩ.

CONCLUSION Is the predicted resistance higher or lower than that measured in step 5? **lower**. Why? **4 *Rs* now in parallel** .

7. Actually rewire the circuit so that the four 10-kΩ resistors are in parallel with each other, and in series with the 1-kΩ resistor; then, *without any power applied to the circuit,* measure and record the circuit total resistance.

OBSERVATION R_{meas} = **3.5** kΩ.

CONCLUSION Did the measured value reasonably agree with your predicted value in the preceding step? **Yes**. If a 1-kΩ resistor had been placed across points *A* and *B* rather than the 10-kΩ resistor, would total resistance have been higher or lower? **lower**. Logic and the measurements of this project seem to indicate that a low *R* value placed in parallel with an existing S-P circuit will affect the circuit's total resistance (*more, less*) **more** than a larger *R* value being placed in parallel with the existing circuit.

Lab Projects pages 95–97

8. Actually replace R_5 (across points A and B) with a 1-kΩ resistor; then, measure total resistance to see if it verifies your conclusions.

OBSERVATION R_{meas} = 1.77 kΩ.

CONCLUSION Are the conclusions verified? **Yes**.

To provide more practice for students, the instructor may continue changing Rs and circuit configurations, then the students may predict the new R_es. Verify by measurement for each change.

Lab Projects pages 95–97

PROJECT 22 Current in Series-Parallel Circuits

PROCEDURE

1. Connect the initial circuit as shown in Figure 22-1.
2. Apply just enough voltage to the circuit to obtain 0.5 mA of I_T. Measure all the individual voltage drops and calculate the current through each component as required to fill in the blanks in the Observation section.

OBSERVATION

V_{R_1} = **0.5** V.
V_{R_2} = **11.5** V.
V_{R_3} = **5** V.
V_A = **17** V.
I_{R_1} = **0.5** mA.
I_{R_2} = **0.115** mA.
I_{R_3} = **0.5** mA.
V_{R_4} = **3.85** V.
V_{R_5} = **3.85** V.
V_{R_6} = **3.85** V.
I_{R_4} = **0.385** mA.
I_{R_5} = **0.385** mA.
I_{R_6} = **0.385** mA.

CONCLUSION The components that are in series with the main line and through which total current flows are: R **1** and R **3**. The components that are in series with each other, but not the main line are: **R_4**, **R_5**, and **R_6**. What value of resistance is in parallel with R_2? **30 k** Ω. Does the current divide through the parallel branches inverse to the resistance relationships? **Yes**. It should be noted that the current and voltage distribution in a series-parallel circuit for any given component is dependent upon its location in the circuit. However, the series "sections" of the circuit can be analyzed by the same rules as used in **series** circuits, and the parallel sections of the circuit can be analyzed by the rules for **parallel** circuits. To finalize the circuit analysis, you then combine the results of the sectional analysis.

3. Analyze the circuit and predict what will happen to all the currents in the circuit if R_5 is changed to a 27-kΩ resistor. Will they: (*increase, decrease,* or *remain the same*)? Note your predictions in the Observation section.

OBSERVATION

I_{R_1} will **decrease**.
I_{R_2} will **increase**.
I_{R_3} will **decrease**.
I_{R_4} will **decrease**.
I_{R_5} will **decrease**.
I_{R_6} will **decrease**.
I_T will **decrease**.

CONCLUSION Increasing the value of any resistor in a series-parallel circuit (or any other circuit) will cause I_T to (*increase, decrease*) **decrease**. It is possible for current to increase through a component in parallel with the changed value R because a larger percentage of I_T (even though I_T has decreased some) will pass through the unchanged branch.

4. To verify your predictions of the previous step 3 in Figure 22-1, change R_5 from a 10-kΩ resistor to a 27-kΩ resistor as shown in the diagram in Figure 22-2. Measure all the circuit voltages with 17 volts applied to the circuit. Calculate the current through each component using Ohm's Law. Fill in the blanks as required in the Observation section. Compare these results with those of the previous step 2 in Figure 22-1 to see if the current(s) increased or decreased.

OBSERVATION

V_{R_1} =	**0.395**	V.	I_{R_1} =	**0.395**	mA.
V_{R_2} =	**12.65**	V.	I_{R_2} =	**0.125**	mA.
V_{R_3} =	**3.95**	V.	I_{R_3} =	**0.395**	mA.
V_{R_4} =	**2.7**	V.	I_{R_4} =	**0.27**	mA.
V_{R_5} =	**7.3**	V.	I_{R_5} =	**0.27**	mA.
V_{R_6} =	**2.7**	V.	I_{R_6} =	**0.27**	mA.
V_A =	**17**	V.	I_T =	**0.395**	mA.

CONCLUSION Do the measurements agree with the predictions? **Yes**. Another way of analyzing why I_{R_2} increased even though I_T decreased is to consider the circuit to be a "simple" series circuit of R_1 in series with the R_e of R_2 in parallel with R_4, R_5, R_6, and R_3. Thus, the "equivalent" series circuit is $R_1 + R_e + R_3$. Because R_5 was increased from 10 kΩ to 27 kΩ, R_e will increase. Now R_e is a bigger percentage of R_T, and thus more of V_A will be dropped across R_e and hence R_2. Since R_2 has not itself changed value, yet V_{R_2} is higher, the current through R_2 must also be higher.

5. Remove the current meter. Change R_5 from 27 kΩ to 1 kΩ (which is lower than the original circuit's 10 kΩ). As time permits, predict the circuit parameter "trends" (as opposed to specific values). After predicting, make any measurements required to verify or correct the predictions.

OBSERVATION The currents that will increase are: **all except I_{R_2}**.

The currents that will decrease are: **none except I_{R_2}**.

CONCLUSION Decreasing the value of any resistor in a series-parallel circuit (or any other circuit) will cause I_T to (*increase, decrease*) **increase**. The current through R **2** decreased because with a higher I_T the $I \times R$ drops increased across R **1** and R **3**, which means less of V_A was left to be dropped by R **2**. Since this resistor has not changed value and it dropped less V, then the current through it must have (*increased, decreased*) **decreased**.

Optional Step

6. Use the circuit of Figure 22-2; however, change R_1 from 1 kΩ to 12 kΩ. Use 17 volts applied voltage. Measure and calculate, as appropriate, to fill in blanks and draw conclusions.

OBSERVATION $I_T =$ **0.31** mA. $I_{R_4} =$ **0.21** mA.
$I_{R_2} =$ **0.1** mA.

CONCLUSION Does R_1 carry total current? **Yes**. Did the circuit total current change when R_1 was changed? **Yes**. Did the value of current through R_2 change? **Yes**. Did the percentage of total current through the R_2 branch change when R_1 was changed? **No**. (Be careful when thinking about this question!) Did the percentage of total current carried by the branch with R_4 in it change when R_1 was changed? **No**. This indicates that changing a series component that carries total current in a series-parallel circuit will cause a change in total circuit current, but (*will, will not*) **will not** affect the division of current through parallel branches in the circuit.

Lab Projects pages 99–101

PROJECT 23 Voltage Distribution in Series-Parallel Circuits

PROCEDURE

1. Connect the initial circuit as shown in Figure 23-1.

2. Assume an I_T of 0.5 mA and calculate all the circuit voltages. Note results of calculations.

NOTE ➤ Solve for the R_e of branch R_2 in parallel with branch ($R_3 + R_4$) so you can use the current-divider rule(s) to find currents through these particular branch resistors. This will then allow you to find their individual resistor voltage drops, using the appropriate $I \times R$ values.

OBSERVATION V_{R_1} = 0.5 V. V_{R_4} = 1.56 V.
V_{R_2} = 3.43 V. V_{R_5} = 5.0 V.
V_{R_3} = 1.87 V.

CONCLUSION According to Kirchhoff's Voltage Law, the applied voltage to obtain these results would be 8.93 V.

3. Apply just enough V_A to obtain 0.5 mA of I_T and measure all the circuit voltages.

OBSERVATION V_{R_1} = 0.5 V. V_{R_4} = 1.56 V.
V_{R_2} = 3.43 V. V_{R_5} = 5.0 V.
V_{R_3} = 1.87 V. V_A = 8.93 V.

CONCLUSION Did the measurements verify the calculations of step 2? Yes. Did the largest value R drop the most voltage? No. Because of its location in the circuit, only a small portion of the total current passes through this component. The important consideration when analyzing voltage distribution in a series-parallel circuit is the electrical location of each component being considered. I_T only passes through those components that are in series with the (*largest branch, main line*) main line or source. Parallel sections of the circuit must be analyzed using parallel circuit rules. However, the "net R_e" of a parallel section can be considered as being in series with (*largest branch, main line*) main line components.

4. Predict what will happen to V_{R_2} if R_3 is replaced with a jumper wire short. After making the prediction, make the suggested circuit change and note the results.

OBSERVATION V_{R_2} will decrease. V_{R_2} did decrease.

A CONCLUSION Decreasing the R in the parallel branch causes total current to (*increase, decrease*) ___**increase**___, which causes the components that are in series with the main line $I \times R$ drops to (*increase, decrease*) ___**increase**___. Thus, less of V_A is dropped across R_2.

The instructor may further illustrate these concepts by changing a main line series component and having the students note the similarities and contrasts of the effects on voltage distribution, as compared to a change in R of a parallel branch.

PROJECT 24 Power Distribution in Series-Parallel Circuits

PROCEDURE

1. Connect the initial circuit as shown in Figure 24-1.
2. Apply just enough V_A to obtain 0.5 mA of current through the meter.

OBSERVATION V_A = **14** V. (approximately)

CONCLUSION R_T must equal V_T/I_T or **28** kΩ.

3. Measure all voltage drops and calculate the power dissipated by each component.

OBSERVATION

V_{R_1} = **5** V. P_{R_2} = **2.7** mW.
V_{R_2} = **8.5** V. P_{R_3} = **0.25** mW.
V_{R_3} = **0.5** V. P_{R_4} = **1.55** mW.
V_{R_4} = **8.5** V. P_T = **7** mW.
P_{R_1} = **2.5** mW.

CONCLUSION The power dissipated by each component in this circuit is directly proportional to its **voltage** drop and the **current** through it. This statement is true for any circuit. For the parallel sections of the circuit, the larger the R, the (*more, less*) **less** the power dissipated by that component. For the components in series with the main line, the larger the R, the (*more, less*) **more** the power dissipated by that component. Which component in the circuit dissipated the most power? **R_2**. Notice this is not the largest R in the circuit, as it would have to be in a simple series circuit if it dissipated the most power. Also, it is not the smallest R, as it would have to be in a simple parallel circuit if it dissipated the most power. The key to the power distribution for series-parallel circuits is again the electrical (*size, location*) **location** of the component being considered in the circuit.

4. Swap the positions of R_2 and R_4. Make measurements and determine if the powers dissipated by each of the Rs have changed. (Measure voltages and compare to step 3 results.)

OBSERVATION Results are: (*same, different*) **same**.

CONCLUSION It is not the "physical" location that is important but rather the **electrical** location of components that determines the distribution of power (and other parameters) in a series-parallel circuit.

Further practice can be given by changing values of components in different electrical locations, or sections of the circuit.

PROJECT 25 Effects of an Open in Series-Parallel Circuits

PROCEDURE

1. Connect the initial circuit as shown in Figure 25-1.
2. Increase V applied until 1 mA of current is indicated by the milliammeter. Measure and note the circuit voltages.

OBSERVATION V_A = **25** V. V_{R_3} = **5** V.
V_{R_1} = **10** V. V_{R_4} = **10** V.
V_{R_2} = **5** V.

3. Calculate the current through each component using Ohm's Law.

OBSERVATION I_{R_1} = **1** mA. I_{R_3} = **0.5** mA.
I_{R_2} = **0.5** mA. I_{R_4} = **1** mA.

CONCLUSION Total current flows through R **1** and R **4**. The current divides equally between R **2** and R **3** as they are **equal** resistances and are in **parallel**, thus they have the same voltage drop.

4. Remove R_1 from the circuit to simulate R_1 "opening." Measure the circuit voltages.

OBSERVATION V_A = **25** V. V_{R_3} = **0** V.
V_{R_1} = **25** V. V_{R_4} = **0** V.
(across the open)
V_{R_2} = **0** V.

CONCLUSION Opening a component in series with the source caused I_T to (*increase, decrease*) **decrease** to **0** mA. The voltage drops across all the good components decreased to **0** V. The potential difference across the open became equal to V **applied**. What electrical law does this verify? **Kirchhoff's** Voltage Law.

5. Replace R_1 into the circuit. Now remove R_3 to simulate it "opening." Measure the circuit voltages.

OBSERVATION V_A = **25** V. V_{R_3} = **8.3** V.
(across the open)
V_{R_1} = **8.3** V. V_{R_4} = **8.3** V.
V_{R_2} = **8.3** V.

▲ CONCLUSION Opening a parallel branch in a series-parallel circuit causes I_T to (*increase, decrease*) ____**decrease**____ but not to ____**0**____ mA. The voltage drops across the components in series with the main line (*increase, decrease*) ____**decrease**____, whereas, the voltage across the components directly in parallel with the open (*increase, decrease*) ____**increase**____. No matter whether an open occurs in the series or parallel sections of the series-parallel circuit, the circuit total resistance will (*increase, decrease*) ____**increase**____.

Optional Step

6. Use the circuit of Figure 25-1. With the power off, remove R_4 from the circuit to simulate R_4 opening. Turn the power back on and measure and record V_A and the voltage across the open terminals where R_4 was previously connected.

▲ OBSERVATION V_A measures ____**25**____ V. V_{R_4} measures ____**≅ 25**____ V.

▲ CONCLUSION Did the voltage across the opened resistor exactly equal V_A? ____**close**____. If not, was it higher or lower than V_A? ____**lower**____. Explain what might cause this to be true. ____**Meter across open**____

________________________________.

The instructor should reemphasize that an open occurring IN ANY TYPE OF CIRCUIT will cause R_T to increase and I_T to decrease. To analyze what happens to the parameters for any given component in a series-parallel circuit, consider its electrical location.

Lab Projects pages 107–109

PROJECT 26 Effects of a Short in Series-Parallel Circuits

PROCEDURE

1. Connect the initial circuit as shown in Figure 26-1.
2. Increase V applied until 0.5 mA of current is indicated by the milliammeter. Measure and note the circuit voltages.

OBSERVATION V_A = **12.5** V. V_{R_3} = **2.5** V.
V_{R_1} = **5** V. V_{R_4} = **5** V.
V_{R_2} = **2.5** V.

3. Calculate the current through each component using Ohm's Law.

OBSERVATION I_{R_1} = **0.5** mA. I_{R_3} = **0.25** mA.
I_{R_2} = **0.25** mA. I_{R_4} = **0.5** mA.

CONCLUSION Total current flows through R_1 and R **4**. One- (*third, half, fourth*) **half** of total current flows through R_2 and one-half through R *(1, 2, 3, 4)* **3**.

4. Simulate R_1 shorting by replacing it with a jumper wire. Measure the circuit voltages and record.

OBSERVATION V_A = **12.5** V. V_{R_3} = **4.17** V.
V_{R_1} = **0** V. V_{R_4} = **8.33** V.
(across the short)
V_{R_2} = **4.17** V.

CONCLUSION Shorting a component in series with the source caused I_T to (*increase, decrease*) **increase** the voltage drops across all the "unshorted" or normal components to (*increase, decrease*) **increase** and the voltage across the shorted element to (*increase, decrease*) **decrease** to **0** V. Do Kirchhoff's Voltage and Current Laws still hold true when there is a short in the circuit? **Yes**.

5. Replace R_1 into the circuit. Now replace R_3 with a short. Measure and record the circuit voltages.

OBSERVATION V_A = **12.5** V. V_{R_3} = **0** V.
V_{R_1} = **6.25** V. V_{R_4} = **6.25** V.
V_{R_2} = **0** V.

▲ CONCLUSION

Shorting a parallel branch caused I_T to (*increase, decrease*) ____**increase**____ to ____**0.625**____ mA. The voltage across the shorted branch and any branch in parallel with it (*increases, decreases*) ____**decreases**____ to ____**0**____ V. The voltages across all other components (*increased, decreased*) ____**increased**____ because total current (*increased, decreased*) ____**increased**____. A short anywhere in any type of circuit will cause the circuit total resistance to (*increase, decrease*) ____**decrease**____. Therefore, the circuit total current will (*increase, decrease*) ____**increase**____.

The instructor may mention to the students that in many instances when a short occurs in an electrical circuit, the extra voltage across the remaining components may damage them and cause the circuit to act abnormally and cause a fire. When choosing to short out other components for further illustration, be careful not to exceed the meter current range and DO NOT short out the power supply.

Lab Projects pages 111–112

PART 5 Story Behind the Numbers

Data Table

V_A to be 24 V for all circuits used in this project.

Identifiers	Initial Circuit Color-Coded & Calculated Values	Initial Circuit Color-Coded & Measured Values (approx.)	Modified Circuit 1 (R_1 Replaced by a Jumper) Color-Coded & Measured Values (approx.)	Modified Circuit 2 (R_3 Replaced by a Jumper) Color-Coded & Measured Values (approx.)	Resitor Identifiers for Graphs *x*-axis
*R_T in kΩ	24	24	14	18.25	Initial Circuit
*R_1 in kΩ	10	10	NA (0)	10	R_1 = 10 kΩ
*R_2 in kΩ	47	47	47	47	R_2 = 47 kΩ
*R_3 in kΩ	10	10	10	NA (0)	R_3 = 10 kΩ
*R_4 in kΩ	10	10	10	10	R_4 = 10 kΩ
	Calculated Values	Measured Values	Measured Values	Measured Values	
I_T in mA	1 (approx)	1	1.71	1.315	Modified S-P Circuit 1
I_{R_1} in mA	1	1	NA	1.315	R_1 = Jumpered
I_{R_2} in mA	0.29	0.298	0.51	0.231	R_2 = 47 kΩ
I_{R_3} in mA	0.7	0.7	1.2	NA	R_3 = 10 kΩ
I_{R_4} in mA	0.7	0.7	1.2	1.085	R_4 = 10 kΩ
V_A in V	23.63	24	24	24	
V_{R_1} in V	10	10	NA	13.15	
V_{R_2} in V	13.63	14	24	10.85	
V_{R_3} in V	7	7	12	NA	
V_{R_4} in V	7	7	12	10.85	
P_T in mW	24	24	41.04	31.56	Modified S-P Circuit 2
P_{R_1} in mW	10	10	NA	17.29	R_1 = 10 kΩ
P_{R_2} in mW	4.17	4.17	12.24	2.49	R_2 = 47 kΩ
P_{R_3} in mW	4.9	4.9	14.4	NA	R_3 = Jumpered
P_{R_4} in mW	4.9	4.9	14.4	11.77	R_4 = 10 kΩ
MultiSIM Measured Values					
I_{RL_1} (mA)		—	—	—	
I_{RL_2} (mA)		—	—	—	
I_{RL_3} (mA)		—	—	—	
Point A (V)		—	—	—	
Point B (V)		—	—	—	
Point C (V)		—	—	—	

NOTE:** ***DO NOT have power supply connected to circuit when making *R* measurements for the gray boxes!

Lab Projects pages 113–117

Graph 1

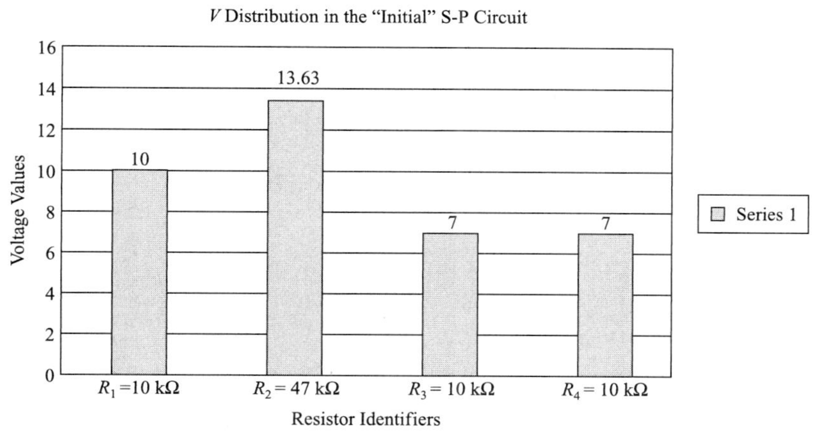

Graph 2

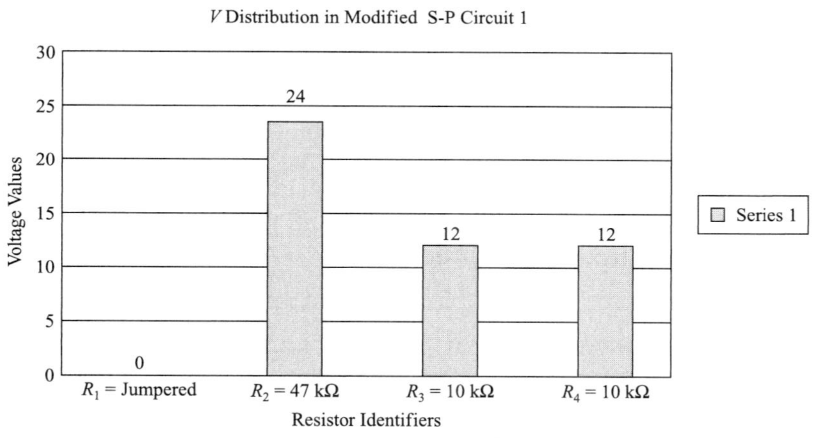

Graph 3

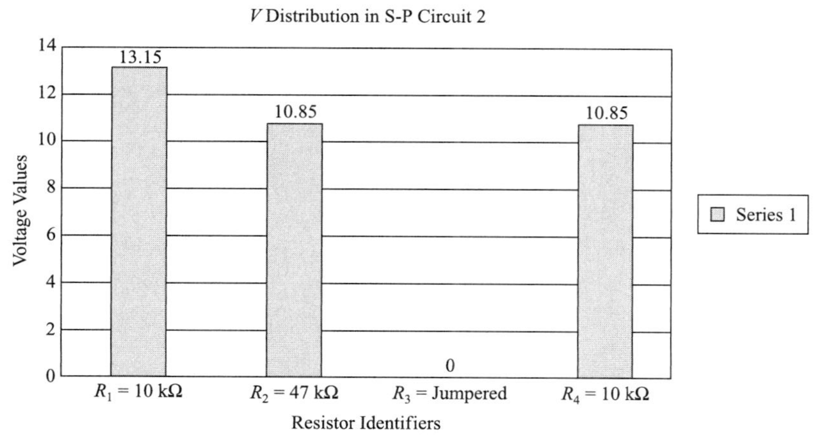

Lab Projects pages 113–117

Answers to Analysis Questions

1. R_1

2. Meter connected between points *A* and *B*.

3. Current through R_2

4. Current through resistors R_3 and R_4

5. Because the current passing through R_1 is only about 3 times greater than that through R_2 but R_2 is nearly 5 times greater in resistance than R_1, the *IR* drop across R_2 is greater than the *IR* drop across R_1.

6. Because they have the same *R* value and the same current through them (since they are in series with each other), the *IR* drops are equal. This would be true for any value of applied voltage. As long as their *R* values are the same and the current through each of them is the same current, their *IR* drops will be equal.

7. In series-parallel circuits, the location of any given component determines what portion of total current will pass through it. In this case, total current passes through R_1, but only a portion of total current passes through R_3 and R_4. Even though the *R* values are equal, the *IR* parameters are different due to the differences in the *I* values.

8. Factors include: location in circuit; *R* value; voltage applied to circuit

9. Circuit modification 1, wherein R_1 was shorted out, had the greatest impact. This is because R_1 was in series with the source and greatly affected voltage applied to the remainder of the circuit, which in turn greatly affected currents through the components, thus affecting *IR* drops and power dissipations.

10. In Initial Circuit, R_2 felt about 13.63/24 or approximately 56.7% of V_A. In Modified Circuit 1, R_2 felt 24/24 or 100% of V_A. In Modified Circuit 2, R_2 felt 10.85/24, or about 45% of V_A.

11. In Initial Circuit, R_2 felt about 13.63/24 or approximately 56.7% of V_A. In Modified Circuit 1, R_2 felt 24/24 or 100% of V_A. In Modified Circuit 2, R_2 felt 10.85/24, or about 45% of V_A.

Lab Projects pages 113–117

PART 5 Summary

Complete the following review questions, indicating the appropriate response by placing a check in the box next to the correct answer.

1. The total resistance (R_T) of a series-parallel circuit is always greater than the sum of the Rs through which total current flows.
   - ☑ True
   - ☐ False

2. Total current passes through every component in a series-parallel circuit.
   - ☐ True
   - ☑ False

3. Considering only the components in series with the main line, the resistor that will dissipate the most power is
   - ☑ the largest R
   - ☐ the smallest R
   - ☐ neither of these

4. Considering only the components in the parallel sections of the series-parallel circuit, the component that will dissipate the most power is
   - ☐ the largest R
   - ☑ the smallest R
   - ☐ neither of these

5. In a series-parallel circuit, does the largest resistor always have the largest $I \times R$ drop?
   - ☐ Yes
   - ☑ No

6. In a series-parallel circuit, if a component in series with the source (in the main line) opens, the voltage across the other components will
   - ☐ increase
   - ☑ decrease
   - ☐ remain the same

7. If a resistor in a parallel section of a series-parallel circuit opens, the voltage across the unopened components will
   - ☑ increase in some cases and decrease in others
   - ☐ decrease in all cases
   - ☐ increase in all cases

8. If a component in series with the main line shorts, the voltage across all the other components will
   - ☑ increase
   - ☐ decrease
   - ☐ remain the same

Lab Projects pages 119–120

9. If a resistor in a parallel section of a series-parallel circuit shorts, the voltage across all the other components will
   - [x] increase in some cases and decrease in others
   - [ ] decrease in all cases
   - [ ] increase in all cases

10. To analyze a series-parallel circuit, it is necessary to use both series circuit rules and parallel circuit rules.
   - [x] True
   - [ ] False

Lab Projects pages 119–120

Optional Troubleshooting Exercises

To be optionally used to challenge students after they have completed the "Series-Parallel Circuits" group of projects and the "Summary." The instructor may wish to add more problems to be solved than the three examples shown here.

Series-Parallel Circuit Troubleshooting Challenge

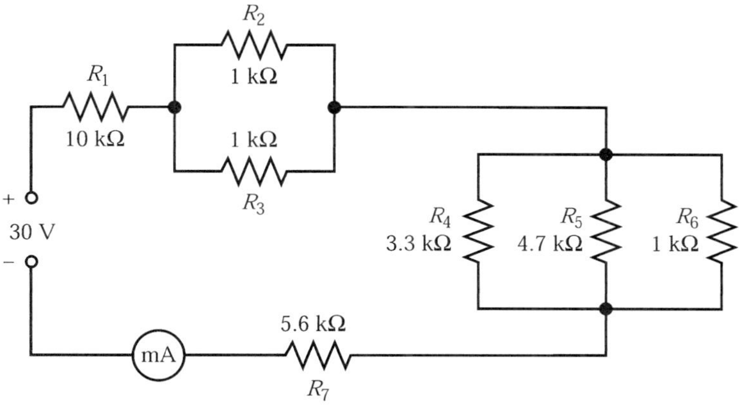

Circuit Diagram

The instructor should set up this series-parallel circuit with *one* of the following faults in it, and then have students troubleshoot and find the problem. Have students log each step they use so you can evaluate, comment on, or grade their performance when they have finished the troubleshooting procedure. Instruct students to use the SIMPLER sequence in performing this exercise.

NOTE ➤ Only put in *one* of these problems at a time for students to troubleshoot. You may want to repeat the exercise and give them a second or third problem, as you desire.

1. Trouble is: Resistor R_6 is open.
   (Starting point symptom to be given to students is: *Voltage across* R_1 is lower than normal)

2. Trouble is: Resistor R_1 is much higher in value than it is color coded value.
   (Starting point symptom to be given to students is: *Voltage across* R_6 is lower than normal)

3. Trouble is: Resistor R_1 is shorted.
   (Starting point symptom to be given to students is: *Voltage across* R_5 is higher than normal)

Instructions for Troubleshooting Exercises can be found on page xi of the Introduction.

PART 6

BASIC NETWORK THEOREMS

PROJECT 27 Thevenin's Theorem

SPECIAL NOTE:

For our purposes, Thevenin's theorem might be stated as follows: The current through a load resistor (R_L) connected between any two points on an existing resistive network (circuit) can be calculated by dividing the voltage at those two points, prior to connecting R_L, by the sum of R that would be measured at those two points (with the voltage source(s) replaced by shorts) + R_L. Therefore: $I_L = V_{TH} / R_{TH} + R_L$.

In essence, Thevenin's theorem tells us that the entire network of resistors and voltage source(s) can be replaced by an equivalent circuit of a voltage source, called V_{TH}, and a single series resistor, called R_{TH}.

V_{TH} = the open-circuit voltage at the two points (V without R_L connected). R_{TH} = resistance seen when looking back into the circuit at the two points, with any sources replaced by shorts (and without R_L connected).

In order to quickly calculate I_L (current through R_L) when R_L is to be connected to any existing network (complicated or otherwise), we then consider R_L as being connected across the Thevenin equivalent circuit so the circuit consists of V_{TH} supplying voltage to R_{TH} and R_L in series. Therefore: $I_L = V_{TH}/R_{TH} + R_L$.

PROCEDURE

1. Connect the initial circuit as shown in Figure 27-1.

2. Apply the voltage distribution characteristics of series circuits and determine what the voltage would be between points A and B with a source V of 20 volts.

OBSERVATION V_{A-B} = **10** V.

CONCLUSION In a series circuit consisting of two equal value resistors, each resistor will drop (*1/3, 1/2, 1/4*) **1/2** of V applied. If the load resistor, R_L, is to be connected to points A and B, then V_{TH} = **10** V.

3. Assume the source were zero ohms and calculate R_{TH}.

NOTE ➤ R_1 and R_2 would then be effectively in parallel.

OBSERVATION R_{TH} = **5 k** Ω.

CONCLUSION Therefore, the Thevenin equivalent circuit would be drawn as follows:

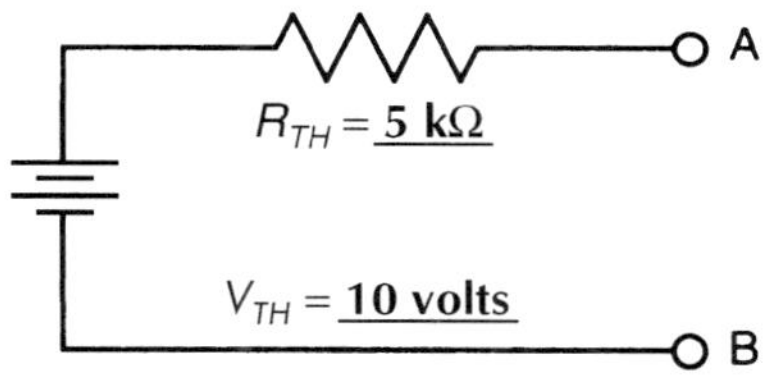

4. If R_L is to be a 10-kΩ resistor, calculate I_L and V_L.

OBSERVATION I_L = **0.67** mA. V_L = **6.7** V.

CONCLUSION $V_L = I_L \times$ R_L.

5. Actually apply 20 volts of V applied. Measure and record circuit parameters as indicated in the Observation section.

OBSERVATION V_A = **20** V. V_{R_2} = **13.3** V.
V_L = **6.7** V.

CONCLUSION With R_L connected, total circuit current is **1.3** mA. If R_L were disconnected, I_T would = **1** mA and V_2 would = **10** V.

6. Using the Thevenin equivalent circuit, as appropriate, calculate and project what the values of V_L and I_L would be if R_L were changed to a value of 4.7 kΩ.

OBSERVATION V_L calculated = **4.84** V. I_L calculated = **1.03** mA.

7. Change R_L to a value of 4.7 kΩ and make measurements needed to compare measurements with your projections.

OBSERVATION V_L measured = **4.84** V.

CONCLUSION Did the measured and projected values for V_L reasonably compare? **Yes**. Were the calculations made easier by virtue of the Thevenin equivalent circuit approach? **Yes**.

8. Connect the circuit as shown in Figure 27-2.

9. Calculate V_{TH} and R_{TH} (assume 13 volts V applied).

OBSERVATION V_{TH} = **10** V. R_{TH} = **2.3 k** Ω.

CONCLUSION Therefore, the Thevenin equivalent circuit would be drawn as follows:

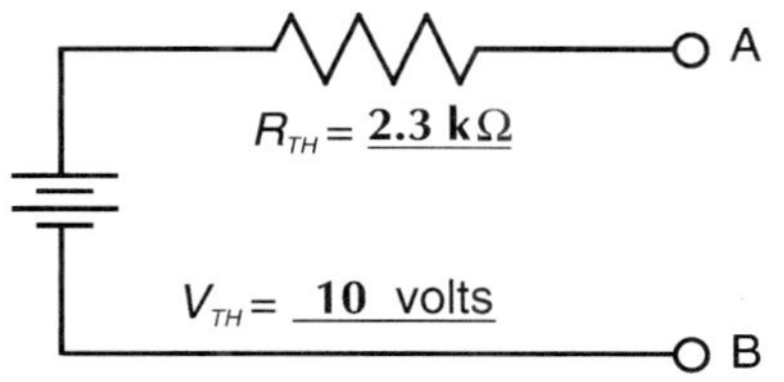

10. If R_L is to be a 10-kΩ resistor, calculate I_L and V_L.

OBSERVATION I_L = **0.813** mA. V_L = **8.13** V.

CONCLUSION $I_L = V_{TH} / R_L + R_{TH}$ = **0.813** mA.
$V_L = I_L \times R_L$ = **0.813 mA** × 10 kΩ = **8.13** V.

11. Apply 13 volts to the circuit. Measure the circuit parameters as indicated in the Observation section.

OBSERVATION V_A = **13** V. V_{R_4} = **1.63** V.
V_L = **8.13** V.

CONCLUSION The main advantage of using Thevenin's theorem for solving I_L is not that it particularly simplifies the initial calculations for a single value of R_L, but rather, that if R_L were going to be changed several times, V_{TH} and **R_{TH}** could be used for each case. This means that each new change of R_L does not necessitate solving the whole network problem each time. All that is necessary is to substitute the new value of R_L in the formula: $I_L = V_{TH}$ / **$R_{TH} + R_L$**.

The students may be given more practice by changing R_L in either or both of the illustrated circuits. The students can calculate and predict parameters, change R_L as appropriate on the actual circuit, and make measurements that verify Thevenin's theorem. Once V_{TH} and R_{TH} are known, solving circuit parameters for any value of R_L boils down to simply solving a two-resistor series circuit problem.

NOTE ➤ The instructor may mention that Thevenin's approach assumes a constant voltage source, that is, a source whose terminal voltage will remain constant regardless of load current.

Lab Projects pages 125–127

PROJECT 28 Norton's Theorem

SPECIAL NOTE:

For our purposes, Norton's theorem might be stated as follows: Any two-terminal network of resistors and source(s) may be represented by a single current source shunted by a single resistance. The size of the current source is determined by calculating the amount of current that will flow through the two terminals if they are shorted. The size of the resistance (shunt R) is equal to the resistance that would be seen looking back into the network from the two points (when they are not shorted and without an R_L).

In essence, the above statement means:

- I_N (equivalent Norton current source) = current that would flow through a short at points A and B if A and B were shorted.
- R_N = resistance seen when looking back into the circuit at the two points (to which R_L will be connected), with any sources replaced by shorts. (In effect, R_N is the same as R_{TH} would be.)

In order to quickly calculate I_L (current through R_L) when R_L is to be connected to points A and B, we then consider R_L as being connected across the Norton equivalent circuit so the circuit consists of I_N supplying current to R_N and R_L in parallel and the current dividing according to the rules of parallel circuits. Therefore: $I_L = (R_N/R_N + R_L) \times I_N$.

PROCEDURE

1. Connect the initial circuit as shown in Figure 28-1.
2. Assume a short across points A and B and determine what the current through this short would be if the source voltage is to be 15 volts.

OBSERVATION I_N = ____1.5____ mA.

CONCLUSION In effect, R ____1____ is shorted out and the circuit total resistance is ____10 k____ Ω.

3. Assume the source were zero ohms and calculate R_N.

NOTE ➤ R_1 and R_2 would then be effectively in parallel.

OBSERVATION R_N = ____5 k____ Ω.

CONCLUSION The Norton equivalent circuit would therefore be drawn as follows:

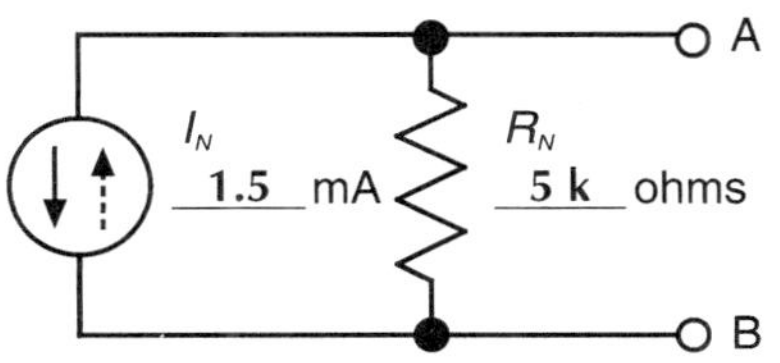

Lab Projects pages 129–131

4. If R_L is to be a 10-kΩ resistor, calculate I_L and V_L. (Use the formula for I_L shown in the Special Note.)

OBSERVATION I_L = **0.5** mA. V_L = **5** V.

CONCLUSION $V_L = I_L \times$ **R_L**.

5. Actually apply 15 volts to the circuit and measure the circuit parameters to verify predicted results.

OBSERVATION V_A = **15** V. V_{R_2} = **10** V.
V_L = **5** V.

CONCLUSION Does Norton's theorem actually work in analyzing and predicting I_L? **Yes**.

6. Using the Norton equivalent circuit, as appropriate, calculate and project what the values of V_L and I_L would be if R_L were changed to a value of 5.6 kΩ.

OBSERVATION V_L calculated = **3.96** V. I_L calculated = **0.707** mA.

7. Change R_L to a value of 5.6 kΩ and make measurements needed to compare measurements with your projections.

OBSERVATION V_L measured = **3.96** V.

CONCLUSION Did the measured and projected values for V_L reasonably compare? **Yes**. Were the calculations made easier by virtue of the Norton equivalent circuit approach? **Yes**.

8. Connect the initial circuit as shown in Figure 28-2.

9. Calculate I_N and R_N (assume 20 volts V_A).

OBSERVATION I_N = **1** mA. R_N = **6.66 k** Ω.

CONCLUSION $I_N = V_A /$ **20 k** Ω = **1** mA.
R_N = 10 kΩ × **20 k** /10 kΩ + **20 kΩ** = **6.66 kΩ**.

10. If R_L is to be 1 kΩ, calculate I_L and V_L for a V_A of 20 volts.

OBSERVATION I_L = **0.87** mA. V_L = **0.87** V.

CONCLUSION $I_L = \left(\frac{R_N}{R_N + R_L}\right) \times I_N$ = **0.87** mA.

11. Apply 20 volts to the circuit. Measure parameters and verify the predictions made by calculations using Norton's theorem.

OBSERVATION V_A = **20** V. V_{R_3} = **9.56** V.
V_L = **0.87** V.

CONCLUSION The predictions made using Norton's theorem (*were, were not*) **were** verified by the measurements.

12. If R_L were to be changed to a 10-kΩ resistor, use Norton's theorem and predict the parameters called for in the Observation section (assume 20 volts V_A).

OBSERVATION I_N = **1** mA. V_{R_1} = **8** V.
R_N = **6.66 k** Ω. V_{R_2} = **4** V.
I_L = **0.4** mA. V_{R_3} = **8** V.
V_L = **4** V.

CONCLUSION Changing the value of R_L that will be used (*does, does not*) **does not** change the Norton equivalent circuit. The Norton equivalent circuit for the circuit shown on the setup diagram can therefore be drawn as follows:

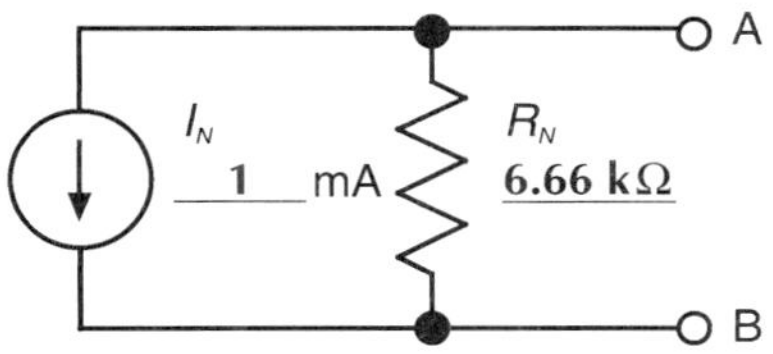

The instructor may wish to tell the students that once I_N and R_N are known, solving circuit parameters for any value of R_L simplifies into solving a two-resistor parallel circuit problem of current division.

NOTE ➤ The instructor may also mention that Norton's approach assumes a constant-current source, that is, a source whose output current will remain constant regardless of voltage.

PROJECT 29 Maximum Power Transfer Theorem

SPECIAL NOTE:

For our purposes, the maximum power transfer theorem might be stated as: Maximum power will be transferred from the generator (or source) to the load (R_L) when the resistance of the load equals the resistance of the generator (or source).

PROCEDURE

1. Connect the initial circuit as shown in Figure 29-1.
2. Apply 10 volts to the circuit. Measure V_L and calculate P_L.

OBSERVATION V_L = **5** V. P_L = **25** mW.

CONCLUSION This circuit is solved as a simple (*series, parallel*) **series** circuit.

3. Assume that R_1 represents the resistance of the generator (R_g). Calculate the power transfer efficiency as follows:

$$\text{Efficiency} = (P_{out}/P_{in}) \times 100 \text{ where } P_{out} = P_L \text{ and } P_{in} = P_T$$

OBSERVATION P_{out} = **25** mW. Efficiency = **50** %.
P_{in} = **50** mW.

CONCLUSION When $R_L = R_g$, (*1/3, 1/4, 1/2*) **1/2** the total power provided by the source is transferred to the load (R_L).

4. Change R_L to 10 kΩ. Measure V_L and calculate P_L and efficiency (use 10 volts V_A).

OBSERVATION V_L = **9.1** V. P_{in} = **9.1** mW.
P_L = **8.28** mW. Efficiency = **91** %.
P_{out} = **8.28** mW.

CONCLUSION Making R_L larger than R_g caused efficiency to (*increase, decrease*) **increase**. However, P_L was (*less, more*) **less** than when $R_L = R_g$. The reason for this is that, with the higher R_L, the circuit total resistance was larger. Thus the total power supplied to the circuit by the source was (*less, more*) **less** than when $R_L = R_g$, even though P_L was a (*larger, smaller*) **larger** percentage of P_T, the power transferred to the 10-kΩ load was (*less, more*) **less** than to the 1-kΩ load.

5. Change R_L to a 100-Ω resistor. Measure V_L and calculate P_L and efficiency (use 10 volts V_A).

OBSERVATION V_L = **0.9** V. P_{in} = **91** mW.
P_L = **8.19** mW. Efficiency = **9** %.
P_{out} = **8.19** mW.

▲ CONCLUSION

When R_L is smaller than R_g, the efficiency is (*more, less*) ______**less**______ than 50% as when $R_L = R_g$. Total power produced by the source with R_L smaller than R_g is greater than for maximum power transfer conditions. However, a smaller percentage of the power is dissipated by R ___L___.

PART 6 Story Behind the Numbers

Data Table

Identifiers	Color-Coded & Calculated Values	Measured Values (Approximate)
For Thevenin's Theorem		
$R_1 = 1\ k\Omega$	1	—
$R_2 = 10\ k\Omega$	10	—
$R_3 = 1\ k\Omega$	1	—
$R_4 = 1\ k\Omega$	1	—
$V_A = 13$ V	—	13
$R_{thevenin}$ (kΩ)	2.3	2.3
$V_{thevenin}$ (V)	10	10
R_L (color code) (kΩ)	27	27
I_{RL} (mA)	0.341	0.341
V_{RL} (V)	9.215	9.215
Thevenin Equivalent Circuit: V_{TH}= V ; R_{TH}= k Ω in series with R_L= k Ω		
For Norton's Theorem		
$R_1 = 10\ k\Omega$	10	—
$R_2 = 10\ k\Omega$	10	—
$R_3 = 10\ k\Omega$	10	—
R_4 = not in circuit	—	—
$V_A = 20$ V	—	20
R_{norton} (kΩ)	6.66	6.66
I_{norton} (mA)	1	1
R_L (color code) (kΩ)	1	1
I_{RL} (mA)	0.87	0.87
V_{RL} (V)	0.87	0.87

Lab Projects pages 135–139

Data Table (*continued*)

Identifiers	Color-Coded & Calculated Values	Measured Values (Approximate)
Norton Equivalent Circuit: I_N= mA; R_N = k Ω in parallel with R_L = k Ω		
For Maximum Power Transfer Theorem		
R_g (R_1) = 10 kΩ	10	—
R_{L1} = 1 kΩ	1	—
R_{L2} = 4.7 kΩ	4.7	—
R_{L3} = 10 kΩ	10	—
R_{L4} = 27 kΩ	27	—
R_{L5} = 47 kΩ	47	—
V_A = 20 V	—	20
V_{L1} (V)	1.82	1.82
V_{L2} (V)	6.39	6.39
V_{L3} (V)	10.00	10
V_{L4} (V)	14.59	14.59
V_{L5} (V)	16.49	16.49
P_{L1} (mW)	3.31	—
P_{L2} (mW)	8.70	—
P_{L3} (mW)	10.00	—
P_{L4} (mW)	7.89	—
P_{L5} (mW)	5.79	—

Graph

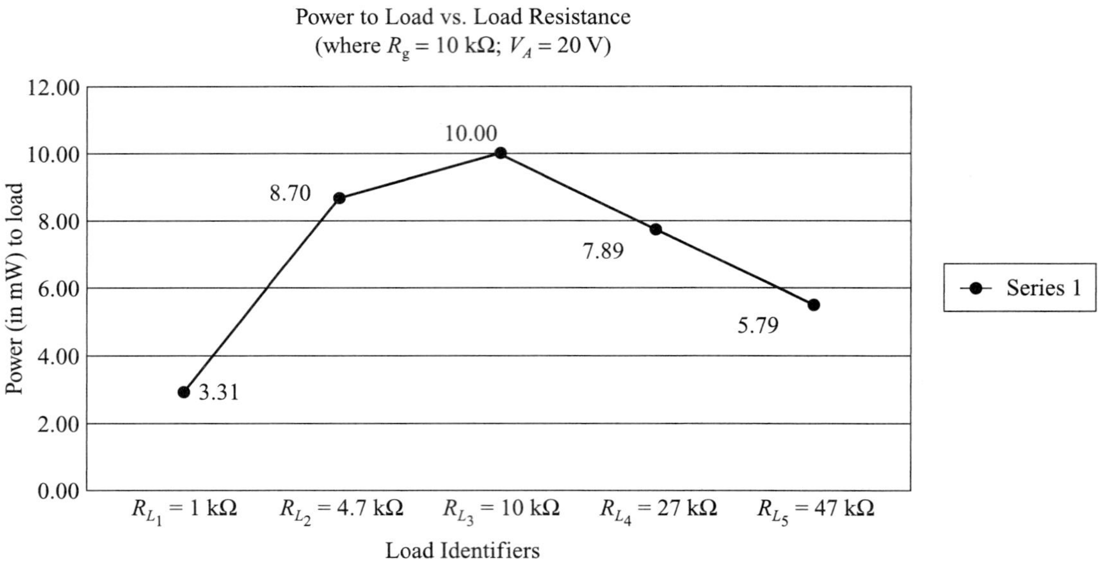

Answers to Analysis Questions

1. V_{TH}
2. Circuit resistor values or network circuit configuration connections
3. It makes it easy to determine load parameters for changes in load resistance; that is, a simple two-resistor series circuit analysis is all that is needed.
4. No. Use the existing Thevenin equivalent circuit parameters and perform the two-resistor series circuit analysis with the new load resistance value.
5. The equivalent circuit is a voltage source and a series resistance called R_{TH}. The type of circuit analysis is a simple two-resistor series circuit analysis.
6. I_N would change.
7. Resistor values, or configuration of network
8. The equivalent circuit consists of a constant-current generator with a resistor having the value of R_N in parallel with the constant-current source. Parallel circuit analysis is used to find load current and load voltage.
9. No.
10. The ease of calculating load parameters for different load values by using a simple two-resistor parallel circuit analysis.
11. R_L was 10 kΩ, which was equal to R_g.
12. *Note*: You will need to calculate the total power supplied by the source (P_{in}) for each case and the power delivered to the load (P_{out}) for each case. Then, use the formula: Efficiency = $P_{out}/P_{in} \times 100$.

Lab Projects pages 135–139

(Eff. with R_{L1} = approx. (3.3/36) × 100 = 9.16%; with R_{L2} = (8.7 /27.2) × 100 = 31.9%; with R_{L3} = (10/20) × 100 = 50%; with R_{L4} = (7.89/10.8) × 100 = 73%; with R_{L5} = (5.79/7) × 100 = 82.7%)

13. 50%
14. Because less total power was provided and available for transfer
15. The source was delivering more power than the maximum power transfer situation. The larger part or share of the power was being consumed by the source internal resistance (R_g).

Lab Projects pages 135–139

PART 6 Summary

Complete the following review questions, indicating the appropriate response by placing a check in the box next to the correct answer.

1. To find R_{TH}, we look back into the network from the two points to which R_L will be connected and assume the source(s) as
   - ☐ open
   - ☑ shorted
   - ☐ neither of these
2. When solving for R_{TH}, we assume that R_L is
   - ☐ connected to the circuit
   - ☑ not connected to the circuit
3. When solving for R_{TH}, we assume the part of the circuit connected to the two points to which R_L will be connected to be
   - ☐ open
   - ☐ shorted
   - ☑ neither of these
4. A Thevenin equivalent circuit consists of
   - ☑ a voltage source and series R
   - ☐ a current source and shunt R
   - ☐ neither of these
5. To find R_N, we look back into the network from the two points to which R_L will be connected and assume the source(s) as
   - ☐ open
   - ☑ shorted
   - ☐ neither of these
6. When solving for I_N, we assume the part of the circuit connected to the two points to which R_L will be connected to be
   - ☐ open
   - ☑ shorted
   - ☐ neither of these
7. A Norton equivalent circuit consists of
   - ☐ a voltage source and series R
   - ☑ a current source and shunt R
   - ☐ neither of these
8. Maximum power transfer from source to load occurs when
   - ☐ R_L is greater than R_g
   - ☐ R_L is less than R_g
   - ☑ $R_L = R_g$
   - ☐ all are the same
9. Highest efficiency occurs when
   - ☑ R_L is greater than R_g
   - ☐ R_L is less than R_g
   - ☐ $R_L = R_g$
   - ☐ makes no difference
10. At maximum power transfer, the efficiency is
    - ☐ 100%
    - ☐ 25%
    - ☐ 10%
    - ☑ 50%
    - ☐ none of these

Lab Projects pages 141–142

If time allows, give the students some problems with specific values involved. Be sure the students understand the following basic concepts:

1. According to Thevenin's theorem, an equivalent circuit comprised of a single voltage source and a single series resistor may be used to represent any two-terminal network.
2. By Norton's theorem, an equivalent circuit comprised of a single current source and a single shunt resistance may be used to represent any two-terminal network.
3. The maximum power transfer theorem says that maximum power is transferred from source to load when $R_L = R_g$.
4. Norton's theorem is often used in analyzing transistor circuits.
5. Thevenin's theorem is a powerful tool for simplifying complex network problems.
6. The maximum power transfer theorem is used in designing transmitters, determining loading factors for power plants, and for various other practical applications.

Lab Projects pages 141–142

PART 7

NETWORK ANALYSIS TECHNIQUES

PROJECT 30 Loop/Mesh Analysis

SPECIAL NOTE:

Steps to use the Loop/Mesh analysis approach are summarized as follows:

1. Label the assumed mesh currents on the diagram. (Use the arbitrary clockwise direction convention.)
2. Assign mathematical polarities to sources. If assumed mesh current enters positive *(negative)* terminal of source, consider source a positive polarity. If mesh current enters negative *(positive)* terminal of source, consider source a negative polarity.
3. Write Kirchhoff's equations for each loop's voltage drops. Assume voltages across resistors in a given loop positive when the voltage drop is caused by its "own mesh" current and negative when caused by an adjacent mesh's assumed current.
4. Solve the resulting equations, using simultaneous equation methods.
5. Verify results using Ohm's Law and Kirchhoff's Law.

PROCEDURE

1. Draw appropriate arrows and label the meshes on the diagram of Figure 30-1.

OBSERVATION Number of meshes labeled were **2**.

CONCLUSION The convention used to draw the arrows for assumed mesh currents was to use a (*CW, CCW*) **CW** direction.

2. Assign appropriate polarities to use for sources when performing Loop/Mesh calculations.

OBSERVATION V_{S_1} = (+, –) **+ *(–)***. V_{S_2} = (+, –) **– *(+)***.

CONCLUSION The reason V_{S_1} was assigned (+, –) **+ *(–)*** polarity is because mesh current enters its (+, –) **+** terminal. V_{S_2} was assigned a (+, –) **– *(+)*** polarity because mesh current enters its (+, –) **–** terminal.

3. Write Kirchhoff's equations for each loop's voltage drops.

OBSERVATION Loop A: **$9{,}500I_A - 5{,}600I_B = 15$ V *(–15V)***.
Loop B: **$-5{,}600I_A + 12{,}400I_B = -12$ V *(12V)***.

Lab Projects pages 145–147

4. Solve equations from step 3 by using simultaneous equation techniques.

OBSERVATION Equation 1: **$21{,}033I_A - 12{,}400I_B = 33.21$ *(−33.21)*** .
Equation 2: **$-5{,}600I_A + 12{,}400I_B = -12$ *(12)*** .
$I_A =$ **1.377 mA *(– 1.377 mA)*** . $I_B =$ **−0.346 mA *(0.346 mA)*** .

5. Use mesh currents and solve for the voltage drops across each resistor.

OBSERVATION $V_{R_1} =$ **5.37** V. $V_{R_3} =$ **2.35** V.
$V_{R_2} =$ **9.65** V.

CONCLUSION Does the sum of $V_{R_1} + V_{R_2}$ equal source V_{S_1}'s value? **Yes**. Does the sum of $V_{R_3} + V_{R_2}$ equal source V_{S_2}'s value? **Yes**.

6. Connect the circuit of Figure 30-1. (Be sure to observe source polarities.)

7. Adjust VVPS 1 so that $V_{S_1} = 15$ V.

8. Adjust VVPS 2 so that $V_{S_2} = 12$ V.

9. Measure voltage drops across each resistor and indicate polarity of each of their voltage drops with respect to point *A* on the diagram.

OBSERVATION $V_{R_1} = (+, -)$ **−5.37** V. $V_{R_3} = (+, -)$ **+2.35** V.
$V_{R_2} = (+, -)$ **+9.65** V.

CONCLUSION Which resistor's voltage drop represents the result of both meshs' currents? **R_2**.

10. Compare the voltages measured in step 9 with your earlier calculations in step 5.

OBSERVATION Step 5 $V_{R_1} =$ **5.37** V. Step 5 $V_{R_3} =$ **2.35** V.
Measured $V_{R_1} =$ **5.37** V. Measured $V_{R_3} =$ **2.35** V.
Step 5 $V_{R_2} =$ **9.65** V.
Measured $V_{R_2} =$ **9.65** V.

CONCLUSION Actual circuit operating parameters (*do, do not*) **do** verify the theoretical approach.

11. Calculate the current through each resistor and indicate the direction of current through each *R* by drawing an appropriate arrow *(or dotted line arrow)* on the diagram next to each *R*.

OBSERVATION $I_{R_1} =$ **1.37** mA. $I_{R_3} =$ **0.345** mA.
$I_{R_2} =$ **1.72** mA.

CONCLUSION (*Ohm's, Kirchhoff's*) **Ohm's** Law was used to find current values. Are the calculated currents the same in every case as the "assumed mesh" currents? **No**. Why not? **Mesh currents don't branch**
.

PROJECT 31 Nodal Analysis

SPECIAL NOTE:

The Nodal analysis approach steps are summarized as follows:

1. Find the major nodes and select the reference node.
2. Designate currents into and out of major nodes.
3. Write Kirchhoff's Current Law equations for currents into and out of major nodes. (Use V/R statements to represent each current.)
4. Develop equations to solve for voltage across resistor(s) associated with the reference node, and solve for V and I for reference node component(s).
5. Using data accumulated, solve for voltages and currents through other components in the circuit.
6. Verify results using Kirchhoff's Voltage Law.

PROCEDURE

1. Use arrows *(or dotted line arrows for conventional flow)* and designate currents into and out of the major node on Figure 31-1.

OBSERVATION *On the diagram:*

I_1 is current through R ___**1**___. I_3 is current through R ___**2**___.

I_2 is current through R ___**3**___.

2. Write Kirchhoff's current equation for the loop with the 15-V source.

OBSERVATION $I_1 + I$ ___**2**___ $= I$ ___**3**___.

CONCLUSION Converting these currents into V/R statements, we can say that $V_{R_1}/R_1 +$ V_R ___**3**___ $/R$ ___**3**___ equals V_R ___**2**___ $/R$ ___**2**___.

The voltage and current of special interest, due to being associated with the reference node, is V_R __**2**__ and current I __**3**__.

3. Write equations for V across R_2 using loop voltage and current statements.

OBSERVATION

For 15-V loop:

$V_{R_1} + V_{R_2} = 15$ V

$V_{R_1} = 15 -$ ___$\boldsymbol{V_{R_2}}$___.

For 12-V loop:

$V_{R_3} + V_{R_2} = 12$ V

$V_{R_3} = 12 -$ ___$\boldsymbol{V_{R_2}}$___.

CONCLUSION Voltage across R_2 is a result of current I ___**3**___.

Translated into V/R current statements:

15 V $- V_{R_2}/10$ kΩ plus 12 V $- V_{R_2}/1$ kΩ equals $V_{R_2}/5$ kΩ.

4. Perform appropriate math to solve for V_{R_2}.

OBSERVATION V_{R_2} = **10.38** V.

5. Use Ohm's law and solve for I_{R_2}.

OBSERVATION I_{R_2} = **2.076** mA.

6. Solve for V_{R_3}, I_{R_3}, V_{R_1}, and I_{R_1}.

OBSERVATION V_{R_3} = 12 − **10.38** = **1.62** V.
I_{R_3} = **1.62** V ÷ **1** kΩ.
equals **1.62** mA.

V_{R_1} = 15 − **10.38** = **4.62** V.
I_{R_1} = **4.62** V ÷ **10** kΩ.
equals **0.462** mA.

7. Verify results of calculations.

OBSERVATION $V_{R_1} + V_{R_2} = V_{S_1}$
4.62 + **10.38** = 15 V.

$V_{R_2} + V_{R_3} = V_{S_2}$
10.38 + **1.62** = 12 V.

CONCLUSION Did the calculated values using the Nodal approach appropriately check out with Kirchhoff's Law(s)? **Yes**.

8. Connect the circuit shown in Figure 31-1. (Be sure to observe source polarities.)

9. Adjust VVPS 1 so that V_{S_1} equals 15 volts. Adjust VVPS 2 so that V_{S_2} equals 12 volts.

10. Measure voltage drops across the resistors.

OBSERVATION V_{R_1} measures **4.62** V.
V_{R_2} measures **10.38** V.
V_{R_3} measures **1.62** V.

CONCLUSION Do the measured values of *V*s reasonably agree with your earlier calculations? **Yes**.

11. Using the measured values of *V* and the color-coded values of resistors, calculate current through each resistor.

OBSERVATION I_{R_1} = **0.462** mA.
I_{R_2} = **2.076** mA.
I_{R_3} = **1.62** mA.

CONCLUSION Do the current values calculated on the basis of actual circuit measurements reasonably agree with your previous theoretical calculations? **Yes**.

PROJECT 32 Wye-Delta and Delta-Wye Conversions

SPECIAL NOTE:

The process of converting wye-to-delta and delta-to-wye networks enables selection of appropriate circuit component values that will provide the same electrical load to sources when converting from one of these circuit configurations to the other. For our purposes, we will use the nomenclatures R_1, R_2, and R_3 to designate wye configuration resistive components and R_A, R_B, and R_C to designate delta configuration resistive components. See Figure 32-1.

To find values of delta components when the wye component values are known, use:

$$R_A = \frac{(R_1R_2)+(R_2R_3)+(R_3R_1)}{R_1}$$

$$R_B = \frac{(R_1R_2)+(R_2R_3)+(R_3R_1)}{R_2}$$

$$R_C = \frac{(R_1R_2)+(R_2R_3)+(R_3R_1)}{R_3}$$

To find wye component values when the delta component values are known, use:

$$R_1 = \frac{R_B R_C}{R_A + R_B + R_C}$$

$$R_2 = \frac{R_C R_A}{R_A + R_B + R_C}$$

$$R_3 = \frac{R_A R_B}{R_A + R_B + R_C}$$

PROCEDURE

1. Connect the wye portion of the circuit shown in Figure 32-1, using the component values shown.

OBSERVATION R_1 = 3.3 kΩ. R_3 = 5.6 kΩ.
R_2 = 4.7 kΩ.

2. Using an ohmmeter, measure and record the resistance readings from point *A* to *B*, from point *B* to *C*, and from point *A* to *C*.

OBSERVATION A to B = 10.3 kΩ. A to C = 8.0 kΩ.
B to C = 8.9 kΩ.

CONCLUSION When measuring from A to B, you are actually measuring the series resistance of *R* 2 and *R* 3. The resistance from point *B* to point *C* equals the series resistance of *R* 3 and *R* 1. The resistance from point *A* to point *C* equals the series resistance of *R* 2 and *R* 1.

3. Use the appropriate formulas and solve for the values of resistors that would form a delta network that is equivalent to the wye network used in steps 1 and 2.

OBSERVATION $R_A =$ 18.28 kΩ. $R_C =$ 10.77 kΩ.
$R_B =$ 12.83 kΩ.

CONCLUSION The equivalent delta resistor values are (*greater, smaller*) greater in value than the wye network resistor values.

4. Connect a delta circuit using the approximate (available) resistance values calculated in step 3.

NOTE ➤ You may have to use more than one resistor for each delta leg in order to use standard, available resistor values.

CONCLUSION Is it necessary to use more than one standard resistor for each leg, in this case? Yes.

5. Use the space in the Observation section and draw a schematic showing your circuit setup, which shows the *R*s used for each leg. Label the *R*s making up R_A, R_B, R_C, as appropriate. Also label test points *A, B,* and *C* on your diagram.

6. Use an ohmmeter and measure the *R* values found between points *A* to *B*, *B* to *C*, and *A* to *C*.

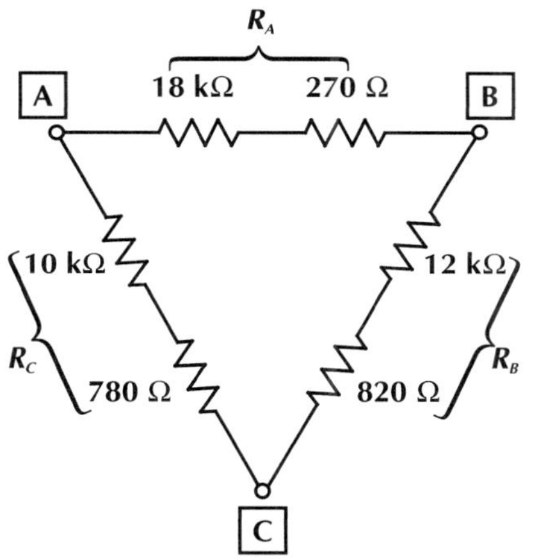

OBSERVATION A to B measures 10.3 kΩ. A to C measures 8.0 kΩ.
B to C measures 8.9 kΩ.

CONCLUSION Do the measured values for this equivalent delta network come close to the values measured between the same test points in the wye network? (*yes, no*) Yes. If not, what could cause some of the differences? $R_{tolerance}$.

When measuring *R* between points *A* and *B*, you are measuring the equivalent total *R* of R_A in parallel with what components? $R_B + R_C$. When measuring

R between points B and C, you are measuring the equivalent total R of R_B in parallel with what components? ____$R_C + R_A$____. When measuring R between points A and C, you are measuring the equivalent total R of R_C in parallel with what components? ____$R_A + R_B$____.

Lab Projects pages 153–155

PART 7 Story Behind the Numbers

Procedure

1. The circuit diagram above will be used for *both* Mesh analysis and Nodal analysis portions of this project.

2. First, you should fill in the required data prior to performing a Mesh analysis of the above circuit.

   - Label the assumed mesh currents on the diagram. (Use the arbitrary clockwise direction convention.)
   - Assign mathematical polarities to the sources. If assumed mesh current enters the positive terminal of the source, consider the source a positive polarity. If the mesh current enters the negative terminal of the source, consider the source a negative polarity. *(NOTE: Use reverse polarities if considering conventional current flow directions.)*

3. Write Kirchhoff's equations for each loop's voltage drops.

   Loop A: $(10{,}000I_A + 10{,}000I_A) - 10{,}000I_B = 15\text{ V}$ *(–15 V)*

   Equation 1: $20{,}000I_A - 10{,}000I_B = 15\text{ V}$ *(–15 V)*

   Loop B: $-10{,}000I_A + (10{,}000I_B + 10{,}000I_B) = -12\text{ V}$ *(12 V)*

   Equation 2: $-10{,}000I_A + 20{,}000I_B = -12\text{ V}$ *(12 V)*

4. Use simultaneous equations and solve the two equations above to find I_A.

   $40{,}000I_A - 20{,}000I_B = 30\text{ V}$ *(–30 V)*
   $-10{,}000I_A + 20{,}000I_B = -12\text{ V}$ *(12 V)*
   $30{,}000I_A = 18\text{ V}$ *(–18 V)*
   $I_A = 0.6\text{ mA}$ *(–0.6 mA)*

5. Substitute the value of I_A into Equation 2 (in step 3) and solve for the value of I_B.

   $-10{,}000(0.6\text{ mA})$ *(–0.6 mA)* $+ 20{,}000I_B = -12\text{ V}$ *(12 V)*
   $-6 + 20{,}000I_B = -12\text{ V}$ *(12 V)*
   $+20{,}000I_B = -6$ *(6)*
   $I_B = 0.3\text{ mA}$

6. Use the mesh current values you found and solve for the voltage drops across each resistor in the circuit.

   $I_A =$ 0.6 mA $I_B =$ 0.3 mA

   $V_{R_1} = 0.6\text{ mA} \times 10\text{ k}\Omega = 6\text{ V}$

   $V_{R_2} = (0.6\text{ mA} + 0.3\text{ mA}) \times 10\text{ k}\Omega = 9\text{ V}$

   $V_{R_3} = 0.3\text{ mA} \times 10\text{ k}\Omega = 3\text{ V}$

7. With the circuit connected as shown, measure V_{R_1}, V_{R_2}, and V_{R_3} to see if the measured values correlate well with the theoretically calculated ones from your Mesh analysis.

8. Find the major nodes and select the ground reference point as the *reference node.*

Lab Projects pages 157–159

9. Use arrows *(or dotted line arrows)* to show currents into and out of the major nodes. Let the current through R_1 be identified as I_1. Let the current through R_2 be identified as I_3. Let the current through R_3 be identified as I_2.

HINT ➤ Be careful. Renumber identifier I_3 properly while solving the network.

10. On the diagram: I ___1___ $+ I$ ___2___ $= I$ ___3___.

11. Write Kirchhoff's current equation for the loop with the 15-V source.

NOTE ➤ Convert each current into an appropriate V/R statement: e.g., I_{R_1} would be stated as V_{R_1}/R_1, and so on.

$$V_{R_1}/R_1 + V_{R_3}/R_3 = V_{R_2}/R_2$$

12. Write equations for the voltage across R_2 using loop voltage and current statements.

For 15-V loop: $V_{R_1} + V_{R_2} = 15$ V; $V_{R_1} = 15 -$ ___**V_{R_2}**___.

For 12-V loop: $V_{R_3} + V_{R_2} = 12$ V; $V_{R_3} = 12 -$ ___**V_{R_2}**___.

13. Perform appropriate math to solve for V_{R_2}. Then use Ohm's Law to solve for I_2.

$V_{R_2} = 9$ V $I_{R_2} = 0.9$ mA

14. Solve for:

V_{R_3} ___**12 – 9 = 3 V**___, I_{R_3} ___**3 V/10 k = 0.3 mA**___,

V_{R_1} ___**15 – 9 = 6 V**___, and I_{R_1} ___**6 V/10 k = 0.6 mA**___.

15. With the circuit connected as shown, measure: V_{R_1} ___**6 V**___, V_{R_2} ___**9 V**___, and V_{R_3} ___**3 V**___ to see if the measured values correlate well with the theoretically calculated ones from your Nodal analysis.

16. After completing all the calculations and measurements data called for thus far, answer the Analysis Questions and produce a brief Technical Lab Report to complete the project.

Answers to Analysis Questions

1. Yes. Kirchhoff's Laws/Ohm's Law
2. Assumed "mesh" currents do not branch as actual circuit currents do.
3. Single window-frame definition
4. The answer is dependent upon the student's opinion.
5. Yes. Current would decrease to half, since *R*s had doubled and *V* was not changed.

Lab Projects pages 157–159

PART 7 Summary

Complete the following review questions, indicating the appropriate response by placing a check in the box next to the correct answer.

1. When using the Loop/Mesh analysis technique, assume that mesh currents
   - ☐ should branch at circuit junction points
   - ☑ should not branch at circuit junction points

2. The arbitrary convention regarding directions of assumed mesh currents is that
   - ☑ they should be shown circulating in a clockwise direction through each loop
   - ☐ they should be shown circulating in a counterclockwise direction through each loop

3. Voltage drops across resistors caused by their own assumed mesh currents are
   - ☐ considered negative, when used in mesh equations
   - ☑ considered positive, when used in mesh equations
   - ☐ polarity assignment is not necessary

4. The convention for source polarity assignment(s) indicates that
   - ☐ if the mesh current returns to the source at its positive terminal, the source will be considered negative in equations
   - ☑ if the mesh current returns to the source at its negative terminal, the source will be considered negative in equations
   - ☐ polarity assignments are not necessary for sources

5. A major node, for Nodal analysis purposes, is a junction where
   - ☐ two components join
   - ☑ three components join
   - ☐ four components join
   - ☐ none of the above

6. To qualify as a "reference node" in the Nodal analysis approach
   - ☐ any node or junction point can be called the reference node
   - ☐ any major node may be designated as the reference node
   - ☑ only the major node with the most junctions can be designated as the reference node

7. The Nodal approach is based on
   - ☐ Kirchhoff's Voltage Law
   - ☑ Kirchhoff's Current Law
   - ☐ Ohm's Law only

8. Delta networks may also sometimes be called
   - ☐ Tee networks
   - ☐ Vee networks
   - ☑ Pi networks
   - ☐ L networks

9. When converting from a wye network to a delta network,
   - ☑ the numerator in the mathematical equation stays the same for finding all three equivalent delta resistors
   - ☐ the denominator in the mathematical equation stays the same for finding all three equivalent delta resistors
   - ☐ neither the numerator nor the denominator is the same for the three equations used to find the equivalent delta resistors

Lab Projects pages 161–162

10. When converting from a delta network to a wye network,
    - ☐ the numerator in the mathematical equation stays the same for finding all three equivalent delta resistors
    - ☑ the denominator in the mathematical equation stays the same for finding all three equivalent delta resistors
    - ☐ neither the numerator nor the denominator is the same for the three equations used to find the equivalent delta resistors

11. When measuring resistance between outside terminal points, which network configuration results in measuring the series resistance of two resistors?
    - ☑ wye network
    - ☐ delta network

12. When measuring resistance between outside terminal points, which network configuration results in measuring the equivalent resistance of parallel resistors?
    - ☐ wye network
    - ☑ delta network

Lab Projects pages 161–162

PART 8

THE OSCILLOSCOPE

PROJECT 33 Basic Operation: Familiarization

SPECIAL NOTE:

The oscilloscope is a versatile test instrument that can visually display the relationship between (1) two electrical quantities and (2) an electrical quantity and time. In this project, you will become acquainted with the basic jacks and controls, and in some cases, screen menus used in obtaining visual waveforms with the scope.

Due to the wide variety of oscilloscopes in use, both in training facilities and in industry, it is not feasible to give precise instructions for every situation. The spectrum of scopes being used may be from the simplest, triggered, dual-trace oscilloscopes to the most sophisticated digital phosphor scopes with automated measurement and math functions appearing on the screen as alphanumeric readouts. Additionally, some scopes may have computer interfacing capability.

Basically, we will try to give generic enough instructions that you can adapt them to the particular scopes you are using at your training facility. It will be important for you, as the student, to acquire the appropriate operator's manual from your instructor for the type of scope you are using to aid you in performing the various tasks in this project and several following projects.

PROCEDURE

1. Obtain the available scope and the appropriate scope operator's manual. Referring to the oscilloscope and appropriate operating manual, locate and point out to your instructor as many of the listed jacks and controls as possible.

NOTE ➤ Different brands and models may use different names for some of these items. For each item listed, write the name your scope uses that is equivalent to the name we have on our list.

OBSERVATION *Identify the following:*

- Power

  ______________________(equiv.)

- Vertical Gain Control(s)

  ______________________(equiv.)

  ______________________(equiv.)

- Intensity
  ____________________(equiv.)
- Focus
  ____________________(equiv.)
- Ground Jack
  ____________________(equiv.)
- AC-GND-DC switch
  ____________________(equiv.)
- Vertical Input Jack(s)
  ____________________(equiv.)
  ____________________(equiv.)
- Vertical Position
  ____________________(equiv.)
- Vertical Mode(s)
  ____________________(equiv.)
  ____________________(equiv.)
- Horizontal Position
  ____________________(equiv.)
- Horiz. Time Base Control(s)
  ____________________(equiv.)
  ____________________(equiv.)
- Trigger Mode(s)
  ____________________(equiv.)
  ____________________(equiv.)
- Trigger Source(s)
  ____________________(equiv.)
  ____________________(equiv.)
- Trigger Level
  ____________________(equiv.)

CONCLUSION

Which controls are used to center the display on the scope screen?

V and H position controls

Which controls adjust the speed of the horizontal trace?

Time/Division and Sweep Variable controls

Which controls help keep the waveform from moving?

Trigger controls (Sync.)

Which controls adjust the height of the waveform for a given signal input?

V Volts/Division and Variable controls

Lab Projects pages 165–167

PROJECT 34 Basic Operation: Controls Manipulation

PROCEDURE

1. Connect the function generator or audio generator to the oscilloscope as shown in Figure 34-1. Adjust the generator frequency to 100 Hz.

NOTE ➤ Refer to the signal generator operator manual, as required, to learn how to set the output frequencies of the generator.

2. Using the various controls, obtain as many of the waveforms shown on the opposite page as possible. Demonstrate waveforms A, B, and D for your instructor.

NOTE ➤ If you have trouble achieving stable waveforms (that is, "stopping the waveform(s)"), ask the instructor to help you adjust the trigger control(s), as appropriate.

OBSERVATION *Demonstration of waveforms:*

A ______________________ D ______________________

B ______________________

➤ Instructor initial: ____________

CONCLUSION The main controls manipulated to achieve waveforms were the **position** controls and the **volts/division** controls.

To obtain the waveform shown in D required changing the **time/division** controls.

3. Disconnect the function generator from the circuit. Make sure the scope is in the dc input mode. Adjust scope controls to obtain a straight horizontal line display that just fills the screen from left to right. (H controls, V and H position controls.)

PROJECT 35 Basic Operation: Vertical Controls and DC V

PROCEDURE

1. Adjust the scope to obtain a straight horizontal line display that just fills the screen from left to right and is centered vertically. Use the horizontal frequency controls and vertical and horizontal position controls, as appropriate.

2. Connect a variable voltage dc source to the vertical input of the scope (– terminal to ground) (+ terminal to vertical input), Figure 35-1.

3. Set the output of the dc source to approximately 5 V.

4. "Make" and "break" the input connection from the dc source to the vertical input jack and observe the display.

OBSERVATION Does the horizontal trace (line) move when the dc voltage is applied? **Yes**. Which direction? **up**.

CONCLUSION The dc voltage applied to the vertical input causes deflection of the scope trace. If the polarity of the input were reversed, would the trace react differently? **Yes**.

Explain. **It would move it down instead of up due to polarity of voltage felt by the vertical deflection plates**

5. Reverse the polarity of the connections from the dc source to the vertical input of the scope.

OBSERVATION Does the direction of the display deflection reverse from that shown in step 4? **Yes**.

6. Make appropriate vertical control adjustments to cause the display to "jump" three vertical calibration squares on the scope face "graticule" when the input is connected and disconnected from the scope. When you have made the adjustments, demonstrate this action for your instructor.

NOTE ➤ When NOT connected, the trace should be in the middle of the screen.

OBSERVATION Demonstrate that gain controls have been adjusted to achieve specified results.

➤ Instructor initial: ____________

CONCLUSION Was more than one control adjusted to achieve the desired results? **Yes**. What were the names of the controls? **v/div switch and variable volts**

Lab Projects pages 171–172

PROJECT 36 Basic Operation: Observing Various Waveforms

PROCEDURE

1. If appropriate signal/function generator(s) are available, demonstrate to your instructor that you can obtain the waveforms shown in Figure 36-1.

   OBSERVATION Sine wave; square wave; triangular wave.

   ➤ **Instructor initial:**____________

PROJECT 37 Voltage Measurements

SPECIAL NOTE:

The following facts and considerations are important when using the oscilloscope to make voltage measurements:

1. Waveform deflection on the scope indicates the *peak-to-peak* value of the voltage under test.
2. For any **sine wave**, the peak-to-peak deflection on the screen is *directly proportional* to the peak and rms values of the ac applied voltage that produce the peak-to-peak waveform. For example, if the deflection on the CRO screen is 1″ (peak-to-peak deflection) when 1-V rms is applied, it will be 2″ of total deflection when a 2-V_{rms} ac signal is applied. Therefore, it is easy to measure ac voltage values with the scope even though the deflection is peak-to-peak in nature.
3. It is easier to read voltages on the CRO screen when the presentation is a vertical line. To achieve this type of display, set the appropriate horizontal control.

PROCEDURE

1. Connect the circuit as shown in Figure 37-1.
2. Monitor input voltage to the scope with the meter and adjust the ac input voltage to 2 volts rms (as indicated by the meter).
3. Adjust the scope positioning and V and H controls to obtain a vertical deflection of **one square** (peak-to-peak).

STOP **CAUTION: Do not move the vertical V/div or variable controls once this is done unless directed to do so. However, you can change the vertical position control for ease of viewing.**

CONCLUSION One large square of deflection equals __2__ V_{rms}, __2.8__ V_P, and __5.6__ $V_{P\text{-}P}$. Each division on the screen equals __2__ V_{rms}.

4. Remove the meter. Adjust the source voltage to obtain two squares of deflection.

OBSERVATION Screen deflection is __2__ squares.

CONCLUSION What rms value does this deflection represent? __4__ V.

5. Now measure the voltage with the meter.

OBSERVATION Voltage measures __4__ V.

CONCLUSION Does this answer agree with the scope measurement? __Yes__.

Lab Projects pages 175–176

6. Use the meter and adjust the source voltage to 6 volts rms.

OBSERVATION Amount of deflection equals ___3___ squares.

CONCLUSION Is the change in deflection essentially linear in nature? ___Yes___.

7. If a low-voltage transformer is available, **calibrate** the scope and measure an unknown transformer secondary voltage using the oscilloscope as the measuring device.

STOP **CAUTION: Do not use exposed 120 VAC connections anywhere in the circuit setup! Be careful! 120 VAC can be lethal!**

OBSERVATION Scope is calibrated so that each large square = ________ $V_{P\text{-}P}$ or ________ V_{rms}. The unknown voltage is causing ________ large squares of deflection.

CONCLUSION The value of the unknown voltage must equal approximately ________ V_{rms}.

8. Use the meter and verify the measurement taken in step 7 above.

OBSERVATION Meter measures ________ V_{rms}.

CONCLUSION The scope and meter voltage measurements (*are, are not*) ___are___ close to equal.

PROJECT 38 Phase Comparisons

SAFETY HINTS CAUTION: Do not use exposed 120 VAC connections anywhere in the circuit setup! Be careful! 120 VAC can be lethal!

SPECIAL NOTE:

In this project, you will briefly look at an approach that may be used to compare the phase of two sine-wave signals (of the same frequency) using an oscilloscope. This method is by direct comparison of signals using a dual-trace scope.

PROCEDURE

1. Connect the variable phase-shift circuit shown in Figure 38-1.
2. Connect the outputs of the network to the dual-trace scope vertical inputs 1 and 2, as appropriate. Adjust the scope H frequency, variable controls, and centering controls to achieve superimposed waveforms, centered on the face of the scope, similar to that shown in Figure 38-2.

   OBSERVATION *V* input 1 gain control set at: ________ V/div.

   V input 2 gain control set at: ________ V/div.

   CONCLUSION Are the *V* levels fed to the two vertical inputs equal? (*close, no, yes*) ________.

3. Refer to Figure 38-3. Using the technique shown, determine the phase difference between the two signals you are displaying on your scope.

   OBSERVATION Distance X = ________ calibration marks on the scope screen.

   Distance Y = ________ calibration marks on the scope screen.

   CONCLUSION The number of degrees difference between the two signals is ________ degrees.

4. Change the setting on the 100-kΩ potentiometer enough to see a noticeable change in the phase and repeat steps 2 and 3 from the previous diagram in Figure 38-1.

   OBSERVATION Distance X = ________ calibration marks on the scope screen.

   Distance Y = ________ calibration marks on the scope screen.

   CONCLUSION The number of degrees difference between the two signals is ________ degrees.

Lab Projects pages 177–178

PROJECT 39 Determining Frequency

SPECIAL NOTE:

One method used to determine frequency with an oscilloscope is the direct method. This method uses a scope having a triggered sweep with calibrated sweep times. Since the horizontal sweep is linear and the calibrated sweep provides information about how many milliseconds or microseconds are required for the sweep to travel 1 div. horizontally on the screen, we can easily determine the time for one cycle of the waveform being viewed in Figure 39-1. Once we know the time it takes for one cycle of the waveform (its period), we can then use the $f = 1/T$ formula to find the frequency of the signal causing the pattern. For example, if the calibrated sweep time is set at 2 milliseconds per division and the scope display shows that one cycle of the signal's waveform starts and finishes in 5 horizontal divisions, the signal must have a period of 5×2 ms, or 10 ms. The frequency of the signal equals $1/T$, or $1/0.01$ sec = 100 Hz.

PROCEDURE

1. Connect the output of an audio oscillator or function generator to a vertical input on a scope having a calibrated sweep system.
2. Set the sweep for a time of 1 ms per division.

OBSERVATION The number of ms for the sweep to travel all the way across the screen is **10** ms.

CONCLUSION A signal having a frequency of **100** Hz would cause a waveform display wherein one cycle would take 10 div. on the scope display.

3. Adjust the signal generator frequency to obtain a one-cycle display across 10 divisions.

OBSERVATION Generator frequency dial calibration reads **100** Hz.

CONCLUSION The measured period for this signal is **10** ms. This indicates that the frequency is **100** Hz.

Does the generator calibration approximately agree with the measured signal frequency? **Yes** What could cause any differences? **Frequency dial calibration and/or error in reading the scope display or frequency dial** ___.

4. Double the signal input frequency with the scope sweep setting still at 1 ms per division.

OBSERVATION How many cycles of the signal are now displayed? **2** cycle(s).

CONCLUSION The time for one cycle of this frequency is **5** ms; $f =$ **200** Hz.

Increasing the frequency of an input signal while keeping the sweep speed the same causes (*more, fewer*) **more** cycles to be displayed.

Lab Projects pages 179–180

5. Have the instructor set the frequency of the signal source to a completely different setting. Preferably, the frequency should be set to one that will cause you to change sweep speeds in order to determine the frequency of the signal.

OBSERVATION To get a readable display, the sweep speed had to be changed to ________________ per div.

CONCLUSION The period of one cycle is ________. The frequency determined by the scope's direct measurement system is ________ kHz.

6. Have the instructor check your results in step 5.

OBSERVATION *Instructor:* Indicate whether the measurement is correct.

____ Yes. ____ No.

7. Practice determining other frequencies, as time permits.

Lab Projects pages 179–180

PART 8 Summary

Complete the multiple-choice questions by placing a check in the box next to the best answer option. Respond to the other types of questions by filling in the blanks or responding appropriately.

1. Four important uses of the oscilloscope are
   - ☐ Waveform display, voltage measurement, current measurement, and power measurement
   - ☐ Waveform display, voltage measurement, determining frequency, and power measurement
   - ☐ Waveform display, voltage measurement, determining phase, and power measurement
   - ☑ Waveform display, voltage measurement, determining phase, and determining frequency

2. Why should the intensity or brightness control be set at the lowest point that makes the display readable?

   To prevent damaging the screen.

3. What oscilloscope terminal is used as the input terminal for the signal whose waveform is to be viewed?

   Vertical input terminals.

4. If an ac voltage of 125 V_{rms} causes 2 div. deflection on the scope screen, what is the peak-to-peak voltage of a signal that causes a deflection of 5 div. with the scope controls unchanged? Show your calculations below.

   Voltage is **883.75** $V_{P\text{-}P}$.

5. If a frequency of 150 Hz gives a display on the scope screen of three cycles, what is the frequency of a signal that gives a display of four cycles with the scope controls unchanged? Show your calculations below.

   Frequency is **200** Hz.

6. The most common way in which the scope might be used to show phase comparisons between two signals is

   Direct comparison.

7. A scope having a calibrated sweep system can be used to directly determine frequency because the time per division of the **horizontal** sweep allows us to determine the **period** of the signal being observed.

8. What is the frequency of a signal whose display indicates a period of 0.1 ms?

   10 kHz.

9. What is the period of a 25-kHz signal? **40 µs**.

Lab Projects pages 181–182

10. By which of the following two methods can you more accurately measure a given signal's frequency?
    - [x] Setting the sweep speed so that two cycles of the signal's waveform covers the entire distance of the horizontal sweep
    - [ ] Setting the sweep speed so that four cycles of the signal's waveform covers the entire x-axis display

Lab Projects pages 181–182

PART 9

INDUCTANCE

Special Notes to Students and Instructors

Note 1: Circuit Options

Beginning with *this project* and continuing through the *Series Resonance* and *Parallel Resonance* series of projects there are a number of projects in which two types of circuits are shown. These are a "lower-frequency" circuit and a "higher-frequency" circuit that can be used to perform the project. The instructor has the option of having the students perform each project using either the lower-frequency circuit or the higher-frequency circuit, depending on component availability and instructional program preferences. The instructor may also opt to have students connect and make required measurements and calculations for both types of circuits, as time permits.

Note 2: Measurement Cautions

1. For the *lower-frequency circuit* option, the inductors used in this project and a number of projects that follow are iron-core, filter-choke-type inductors. Under normal operating conditions, this inductor has dc current as well as ac signal components present. The manufacturer has rated the inductance value based on the normal operating environment for this type inductor. *Students and instructors* should be aware that this inductor will exhibit "apparent" inductance values in our projects that are quite different from the manufacturer's rating. Due to the inductor being used under operating conditions different from those specified by the manufacturer, such factors as "incremental permeability," and so forth, enter into the results in terms of how much acting inductance the inductor "appears" to have. Also, the voltage dropped by the inductor is due to the inductor's impedance—not just its reactance. Add into this scenario the tolerance of component values, variances in calibration of signal sources, test equipment, and so on, and it is obvious that results will vary from those that would result if the inductor were truly acting at the manufacturer-rated value and all the components and test equipment were perfectly calibrated.

2. For the *higher-frequency circuit* option, there are several important things to keep in mind. Be aware that the inductor called for in the circuits is a 100-mH inductor. Typically, the tolerance on these inductors is large; therefore, they may act like inductances from 80 mH to 100 mH+, even though rated at 100 mH.

   Most handheld DMMs have frequency limitations in terms of measuring ac voltages and currents. In many cases, the highest frequencies they should be used to measure are signals up to about 1 kHz. For other, more expensive, true RMS meters, the specifications may allow for voltage and current measurements up to about 10 kHz.

   Since a number of the higher-frequency circuit options for our projects ask for frequency runs as high as 5 kHz to 10 kHz, the measurements must be made with a DMM (bench-type or otherwise) rated to measure ac up to at least 10 kHz. Ideally, the meter should be a *True RMS* reading instrument; although, the more common *averaging type* reading instrument will probably work as long as its frequency rating is sufficient.

Alternatively, a scope can be used for measurements. However, great caution must be used to make sure that the signal source ground and the scope ground(s) are connected to the same end of the component across which voltage measurement is being made. If they are not, the two ground locations can short out a component or a portion of the circuit under test! In some cases, this may mean changing positions of components in the circuit for each measurement; in other cases, this may not be necessary.

3. Students should be careful to set the signal sources to the correct frequencies and voltage levels for changes in the circuit conditions they are examining. Normally, it will be necessary to check and reset the voltage level whenever source frequency settings, circuit components, or conditions are changed.

Lab Projects pages 183–184

PROJECT 40 Total Inductance in Series and Parallel

SPECIAL NOTE:

For this project, we will observe the property of inductance to oppose a change in current by applying a continuously changing ac voltage to the circuit and noting the current limiting effects. The higher the inductance, or L, the higher the opposition to ac current. By noting the ac circuit current with a single inductor, then two inductors in series, then two inductors in parallel, we will illustrate the effects on total inductance of connecting two coils in series and in parallel.

For convenience, the voltage drop across a 1-kΩ resistor will be used as a current indicator. Since $I = V/R$, the number of volts divided by 1 kΩ automatically yields I in mA (e.g., 10 volts across a 1-kΩ resistor indicates 10 mA through the resistor, and so on).

PROCEDURE

1. Measure the dc resistance of the two inductors that will be used for this project. Also, set the ac source voltage that will be used to 3 V_{rms}.

OBSERVATION

	Lower f	Higher f
$L_1 =$	**100** Ω.	**170** Ω.
$L_2 =$	**100** Ω. (approximately)	**170** Ω. (approximately)
$V =$	**3.0** V_{rms}. (no load)	**3.0** V_{rms}. (no load)

2. Connect the initial circuit as shown in Figure 40-1.

CONCLUSION If V_A were dc, what would be the current through this circuit? **Lower f = 3 mA; Higher f = 3 mA.**

3. Measure V_{R_1} (voltage drop across R_1) and calculate the ac current.

OBSERVATION

	Lower f	Higher f
$V_{R_1} =$	**1.4** V.	**2.1** V.
$I =$	**1.4** mA.	**2.1** mA.

CONCLUSION The back emf produced by the **inductor** is limiting the current to a lower value than it would be if dc were applied to the circuit.

4. Insert the second L (L_2) in series with the circuit. With V_A again set to 3 V_{rms}, measure V_{R_1} and calculate the ac current.

OBSERVATION

	Lower f			Higher f	
$V_{R_1} =$	**0.95**	V.	$V_{R_1} =$	**1.46**	V.
$I =$	**0.95**	mA.	$I =$	**1.46**	mA.

CONCLUSION Since the current was lower with the two inductors in series, the L total is obviously (*more, less*) **more** than with one inductor. We conclude that inductors in series add like resistors in (*series, parallel*) **series**.

5. Change the circuit so L_2 is in parallel with L_1. Measure V_1 and calculate the ac current.

OBSERVATION

	Lower f			Higher f	
$V_{R_1} =$	**2.4**	V.	$V_{R_1} =$	**2.6**	V.
$I =$	**2.4**	mA.	$I =$	**2.6**	mA.

CONCLUSION Since the current is higher with the two inductors in parallel, then L total must be (*more, less*) **less** than with one inductor. We conclude that inductors in parallel add like resistors in (*series, parallel*) **parallel**.

Lab Projects pages 185–186

PART 9 Story Behind the Numbers

Answers to Analysis Questions

1. Due to induced cemf in the inductor with ac, but not with dc. That is, there is an inductive reactance of the inductor to ac, which is not present with dc.

2. The opposition to ac current (inductive reactances) are additive in a series circuit.

3. Not exactly one-half. Ideally, if a perfect inductor with no resistance were possible, it would be; but due to resistance in each inductor's wire, the opposition to current is comprised of both inductive reactance and resistance (Z or impedance). Additionally, the inductances are probably not precisely the same value, due to tolerances, etc.

4. Again, the inductors are not perfect inductors; they have resistance, tolerances to inductance value, and so forth.

5. Theoretically, yes!

6. Total inductance for series inductances = $L_1 + L_2 + \ldots L_N$. Total inductance for parallel inductances = $1/(1/L_1 + 1/1/L_2 + \ldots$ etc.).

Lab Projects pages 187–189

Data Table

Component and Parameter Identifiers	Higher f (1 inductor) $V_A = 3\ V_{rms}$; $f = 5{,}000$ Hz Parameter Values
L_1 rated (H)	0.1
L_1 measured R (Ω)	180
R_1 color code (Ω)	1000
If $V =$ dc, I would be:	2.5
V_{R_1} measured (V_{rms})	2.1
I (ac) calculated (mA)	2.1
Component and Parameter Identifiers	**Higher f (2 series inductors) $V_A = 3\ V_{rms}$; $f = 5{,}000$ Hz Parameter Values**
L_1 rated (H)	0.1
L_2 rated (H)	0.1
L_1 measured R (Ω)	180
L_2 measured R (Ω)	180
R_1 color code (Ω)	1000
V_{R_1} (ac) measured (V_{rms})	1.44
I (ac) calculated (mA)	1.44
Component and Parameter Identifiers	**Higher f (2 parallel inductors) $V_A = 3\ V_{rms}$; $f = 5{,}000$ Hz Parameter Values**
L_1 rated (H)	0.1
L_2 rated (H)	0.1
L_1 measured R (Ω)	180
L_2 measured R (Ω)	180
R_1 color code (Ω)	1000
V_{R_1} (ac) measured (V_{rms})	2.65
I (ac) calculated (mA)	2.65

PART 9 Summary

Complete the following review questions, indicating the appropriate response by placing a check in the box next to the correct answer.

1. Inductance is that property in an electrical circuit that opposes
   - ☐ a change in voltage
   - ☑ a change in current
   - ☐ a change in resistance

2. In an ac circuit, if L is increased, the circuit current will
   - ☐ increase
   - ☐ remain the same
   - ☑ decrease

3. If two equal inductances are connected in series, total inductance will be
   - ☑ two times that of one
   - ☐ neither of these
   - ☐ one-half that of one

4. If two equal inductances are connected in parallel, total inductance will be
   - ☐ two times that of one
   - ☐ neither of these
   - ☑ one-half that of one

The instructor may wish to bring to the students' attention the four factors that influence the value of inductance:

1. the number of turns
2. the diameter of the coil
3. the core material's permeability
4. the length of the coil.

In summarizing, the instructor should be sure the students understand the concept of inductance opposing a *change* in current flow, and the effects on total inductance of connecting inductors in series, or parallel (assuming no mutual coupling).

Lab Projects page 191

PART 10

INDUCTIVE REACTANCE IN AC

PROJECT 41 Induced Voltage

SPECIAL NOTE:

The methods used during this project are used to illustrate the concept of back emf and are not a precise scientific method of measuring or calculating exact values of back emf. Also remember that the ac equivalent of a given dc value is the "effective" (rms) value of ac.

PROCEDURE

1. Connect the initial circuit as shown in Figure 41-1.
2. Measure the ac voltage source that will be used for this demonstration. Next, connect the VVPS to the circuit and adjust the dc input V to a value that matches the ac voltage you will be using in a later step.

   OBSERVATION ac source equals **3.0** V_{rms}.
   dc source connected to circuit set to **3.0** V.

3. Measure V_{R_1} (dc voltage across R_1) and calculate circuit dc current.

   OBSERVATION ____ (Using lower f circuit components) (Using higher f circuit components)
   I (dc) = **2.7** mA. I (dc) = **2.56** mA.

   CONCLUSION The circuit current is limited only by the dc resistance of R_1 and **the inductor**.

4. Remove the dc source, connect the 3 V_{rms} ac source and measure V_1. Now calculate the circuit ac current.

   OBSERVATION Lower f Higher f
   I (ac) = **1.33** mA. I (ac) = **1.1** mA.

   CONCLUSION The inductor developed a back emf that opposes the changing current. Discount the small dc resistance of the coil. The current that flowed with dc applied was equivalent to having a V_A of **lower = 2.77; higher = 2.55** V across a circuit resistance of 1 kΩ. When the same value of ac voltage was applied to the circuit, a current would be equivalent to applying only ___**lower = 1.33; higher = 1.55**___ V across the circuit resistance. The effect simulated was as if there must be a back emf of approximately ___**lower = 1.44; higher = 1.42**___ V.

NOTE ➤ This disregards any *R* in the circuit and assumes that the total limiting effect on the current is due to back emf.

PROJECT 42 Relationship of X_L to L and Frequency

SPECIAL REMINDER:

The inductors used in the lower-frequency circuit of this project, *and a number of projects following this one*, are iron-core, filter-choke-type inductors. Under normal operating conditions, this inductor has dc current as well as ac signal components present. The manufacturer has rated the inductance value based on the normal operating environment for this type of inductor. *Students and instructors* should be aware that this inductor will exhibit "apparent" inductance values quite different from the manufacturer's rating under the varying operating conditions in our projects. Because the inductor is used under different conditions from those specified by the manufacturer, such factors as "incremental permeability" enter into the results in terms of how much acting inductance the inductor appears to have. Also, the voltage dropped by the inductor is due to the inductor's impedance—not just its reactance. Add into this scenario the tolerances of component values, variances in calibration of signal sources, test equipment, and so on, and it is obvious that results will vary from those that would result if the inductor were acting at the rated value.

For example, in Project 42 (Relationship of X_L to L and Frequency): For the circuit conditions set up for Procedure step 1, the calculated X_L based on voltage drop and current calculation makes the inductor appear to have about 4.4 H inductance rather than the rated 1.5 H. For the circuit conditions set up for Procedure step 3, the calculated X_L would make the apparent inductance of the two inductors in series appear as about 7.3 H. For the conditions of Procedure step 4, the calculated X_L causes the inductance to appear as approximately 2.55 H.

The instructor should also be aware that the primary purpose of these projects is to help students understand the "concepts" being addressed. For example, students can see that the actual circuit measurements do not perfectly demonstrate that putting a second equally rated inductor in series with the first inductor doubles the total inductance in the circuit. Students, however, do see the general effect of putting inductors in series.

The difference in results between expectations and what happens can actually be used as a teaching tool. Students can be asked to think about what could cause the differences between the theory and what they measured. They can be asked to explain that the inductor displays an impedance (not just reactance); that permeability of the inductor varies with different operating parameters; that components are given their value ratings under specified conditions; that there is tolerance in the ratings of components; and so on.

Lab Projects pages 197–201

PROCEDURE

1. Connect the initial circuit as shown in Figure 42-1.

2. Set the function generator to a frequency of 100 Hz for the lower f circuit or to 5,000 Hz for the higher f circuit and set V_A from the source to 3 V. Measure V_{R_1} and calculate the circuit current. Next, measure V_L and calculate X_L by Ohm's Law ($X_L = V_L/I$).

OBSERVATION

	Lower f			Higher f	
$V_A =$	**3**	V.	$V_A =$	**3**	V.
$V_{R_1} =$	**1.25**	V.	$V_{R_1} =$	**1**	V.
$I =$	**1.25**	mA.	$I =$	**1**	mA.
$V_L =$	**2.46**	V.	$V_L =$	**2.7**	V.
$X_L =$	**1.97 k**	Ω.	$X_L =$	**2.7 k**	Ω.

CONCLUSION Since this is a simple series circuit, the current through the inductor is the same as the current through the **resistor**.

3. Insert a second inductor (the same type as L_1). You should now have a series circuit of L_1, L_2, and R_1. With the same frequency and V_A as step 2 above, measure V_{R_1}, calculate I, measure the voltage across the total inductance of L_1 and L_2, then calculate X_L total.

OBSERVATION

	Lower f			Higher f	
$V_A =$	**3**	V.	$V_A =$	**3**	V.
$V_{R_1} =$	**0.8**	V.	$V_{R_1} =$	**0.57**	V.
$I =$	**0.8**	mA.	$I =$	**0.57**	mA.
$V_{L_T} =$	**2.7**	V.	$V_{L_T} =$	**2.9**	V.
$X_{L_T} =$	**3.37**	kΩ.	$X_{L_T} =$	**5.08**	kΩ.

CONCLUSION Connecting a second inductor of nearly equal value as L_1 in series with L_2 caused the total inductance to approximately (*double, halve*) **double**. In analyzing the results of step 2 and comparing with this step, we conclude that doubling total inductance caused the total inductive reactance (X_L) to approximately (*double, halve*) **double**. We therefore conclude that inductive reactance (X_L) is (*directly, inversely*) **directly** proportional to inductance (L). Increasing L causes X_L to (*increase, decrease*) **increase** at any given frequency.

Lab Projects pages 197–201

4. Remove L_2 and replace with a jumper. For the lower f circuit, change the input frequency to 1,000 Hz and keep V_A at 3 volts. For the higher f circuit, change the input frequency to 10,000 Hz, keeping V_A at 3 volts. Measure V_{R_1}, calculate I, measure V_L, and calculate X_L.

OBSERVATION

	Lower f			Higher f	
V_A =	3	V.	V_A =	3	V.
V_{R_1} =	0.29	V.	V_{R_1} =	0.61	V.
I =	0.29	mA.	I =	0.61	mA.
V_{L_1} =	2.9	V.	V_{L_1} =	2.93	V.
X_L =	10	kΩ.	X_L =	4.8	kΩ.

CONCLUSION Increasing the frequency caused the inductive reactance (X_L) to (*increase, decrease*) **increase**. If this had been a perfect inductive circuit, would X_L have increased the same number of times as frequency was increased? **Yes**. We conclude that inductive reactance is (*inversely, directly*) **directly** proportional to f (frequency). If the frequency were decreased to one-half its original value, then theoretically the X_L would (*increase, decrease*) **decrease** to (*double, one-half*) **one-half** its original value.

5. Make a graphic plot of X_L versus f from 1,000 Hz to 2,000 Hz for the lower f circuit or from 5,000 to 10,000 Hz for the higher f circuit. Use graph coordinates similar to those shown in Figure 42-2. Make graphs on separate graph paper.

NOTE ➤ Calculate X_L and apparent L values for each 200 Hz or each 1,000 Hz change, as appropriate. See optional steps 9 and 10 for an alternative approach.

OBSERVATION

Apparent L values		Lower f		Higher f	
1.59	H	X_L at 1,000 Hz =	10 kΩ	@ 5 kHz =	2.5 kΩ.
1.54	H	X_L at 1,200 Hz =	11.6 kΩ	@ 6 kHz =	2.9 kΩ.
1.50	H	X_L at 1,400 Hz =	13.2 kΩ	@ 7 kHz =	3.4 kΩ.
1.46	H	X_L at 1,600 Hz =	14.7 kΩ	@ 8 kHz =	3.9 kΩ.
1.42	H	X_L at 1,800 Hz =	16 kΩ	@ 9 kHz =	4.4 kΩ.
1.39	H	X_L at 2,000 Hz =	17.5 kΩ	@ 10 kHz =	4.8 kΩ.

CONCLUSION Did X_L act like it is directly related to f? **Yes**.

Optional Steps

6. Use the appropriate version of the $X_L = 2\pi fL$ formula to find the apparent L of the inductor for the operating conditions used in step 2.

NOTE ➤ Use the X_L value found by Ohm's Law in step 2 when solving for L.

OBSERVATION

	Lower f Calculated apparent		Higher f Calculated apparent	
$L =$	**3.1**	H. $L =$	**85**	mH.

CONCLUSION Is the calculated (apparent) inductance different from the manufacturer's rated value for this inductor? **Yes**. What do you think the inductance value would appear to be if the inductor were operated under the conditions used by the manufacturer when rating the inductor? **1.5 H/100 mH.**.

7. Use the appropriate version of the $X_L = 2\pi fL$ formula to find the apparent L_T of the series inductors for the operating conditions used in step 3.

NOTE ➤ Use the X_L value found by Ohm's Law in step 3 when solving for L_T.

OBSERVATION

	Lower f Calculated apparent		Higher f Calculated apparent	
$L_T =$	**5.3**	H. $L_T =$	**161**	mH.

CONCLUSION Is the calculated (apparent) total inductance different from the manufacturer's rated value for these inductors? **Yes**. What do you think the total inductance value would appear to be if the inductors were operated under the conditions used by the manufacturer when rating these inductors? **3 H/100 mH.**.

8. Use the appropriate version of the $X_L = 2\pi fL$ formula to find the apparent L of the inductor for the operating conditions used in step 4.

NOTE ➤ Use the X_L value found by Ohm's Law in step 4 when solving for L.

OBSERVATION

	Lower f Calculated apparent		Higher f Calculated apparent	
$L =$	**1.59**	H. $L =$	**76.4**	mH.

CONCLUSION Is the calculated (apparent) inductance different from the manufacturer's rated value for this inductor, in this case? **Yes**. Is it different under step 4's operating conditions than it was for step 2? **Yes**. If so, what might account for the difference? **permeability change, different operating conditions, reading errors, etc.**.

Lab Projects pages 197–201

9. Use Excel (or another) spreadsheet and set up a chart similar to the one shown below for charting the data collected in step 5.

PROJECT: RELATIONSHIP OF X_L TO FREQUENCY *(Data derived from Project 42, step 5 data)*				
	A	**B**	**C**	**D**
1	**Frequency**	**Mfg Rated**	**Calc X_L**	**Calc X_L**
2	**Setting (Hz)**	***L* in H**	**X_L formula**	**From Measurements**
3	1st Freq (Hz)	*L* rating (H)	= 6.28*A3*B3	$= V_L/I$
4	2nd Freq (Hz)	*L* rating (H)	= 6.28*A4*B4	$= V_L/I$
5	(etc.)	(etc.)	(etc.)	(etc.)
6	(etc.)	(etc.)	(etc.)	(etc.)
7	(etc.)	(etc.)	(etc.)	(etc.)

10. Use the spreadsheet chart graphing capability to create a line graph showing frequency as the x-axis and X_L as the y-axis data. Show a line graph for both the X_L formula results (column C) and the X_L from measurements results (column D) on the same graph. Appropriately label the graph using the titling and labeling capabilities of the spreadsheet program.

PART 10 Story Behind the Numbers

Data Table 1

Component and Parameter Identifiers	Higher *f* (1 inductor) $V_A = 3\ V_{rms}$; @ *f* = 5,000 Hz Parameter Values
L_1 rated (H)	0.1
R_1 color code (Ω)	1000
V_{R_1} measured (V_{rms})	**1**
I (ac) calculated (mA)	**1**
V_L measured (V_{rms})	**2.7**
X_L calculated (kΩ)	**2.7**
Component and Parameter Identifiers	**Higher *f* (2 series inductors) $V_A = 3\ V_{rms}$; @ *f* = 5,000 Hz Parameter Values**
L_1 rated (H)	0.1
L_2 rated (H)	0.1
R_1 color code (Ω)	1000
V_{R_1} measured (V_{rms})	**0.57**
I (ac) calculated (mA)	**0.57**
V_{L_T} measured (V_{rms})	**2.9**
X_{L_T} calculated (kΩ)	**5.08**

Data Table 2

Component and Parameter Identifiers	Higher *f*	Circuit	(1 inductor)	$V_A = 3\ V_{rms}$	@ *f* = 5,000–10,000 Hz	
	5,000 Hz	6,000 Hz	7,000 Hz	8,000 Hz	9,000 Hz	10,000 Hz
V_{R_1} measured (V_{rms})	**1.03**	**0.87**	**0.76**	**0.67**	**0.61**	**0.54**
I (ac) calculated (mA)	**1.03**	**0.87**	**0.76**	**0.67**	**0.61**	**0.54**
V_L measured (V_{rms})	**2.77**	**2.84**	**2.87**	**2.9**	**2.92**	**2.94**
X_L calculated (kΩ)	**2.68**	**3.26**	**3.77**	**4.78**	**4.78**	**5.44**

Lab Projects pages 203–206

Graph

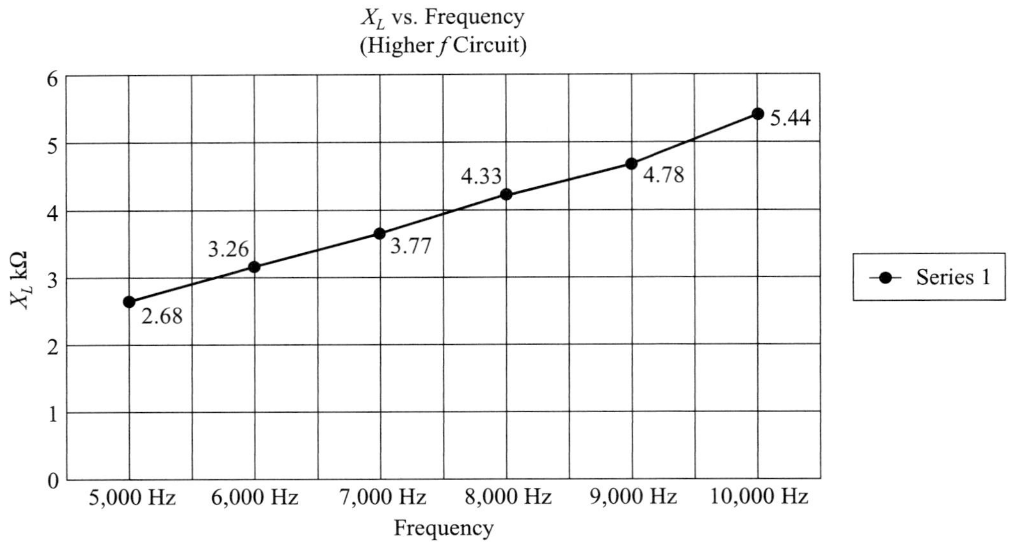

Answers to Analysis Questions

1. Their cemfs were additive. Therefore, their inductive reactances were additive, providing higher opposition to ac current flow.

2. $X_L = 2\pi fL$, thus $L = X_L/2\pi f$. $L = 2{,}800/(6.28 \times 5{,}000) = 2{,}800/31{,}400 = 89$ mH

3. $L = X_{L_T}/(6/29 \times 5{,}000) = 5{,}450/(6.28 \times 5{,}000) = 5{,}450/31{,}400 = 174$ mH

4. The inductance value of two series inductors was reasonably close to two times that of a single inductor. The inductive reactance of two series inductors was reasonably close to double that of the single inductor.

5. Inductive reactance is directly related to inductance value.

6. Inductive reactance is directly related to frequency. That is, the higher the f, the higher the inductive reactance; the lower the f, the lower the inductive reactance.

7. Single inductor at lowest frequency in run: $L = X_L/(2\pi f) = 2{,}680/(6.28 \times 5{,}000) = 85$ mH
   Single inductor at highest frequency in run:
   $L = 5{,}440/(6.28 \times 10{,}000) = 5{,}440/62{,}800 = 86.6$ mH

8. Quite close. Some factors may have been leakage flux change, measurement inaccuracies, etc.

9. Inductors were not necessarily being operated under the same circuit and frequency conditions used by the manufacturer when the inductors were rated.

PART 10 Summary

Complete the following review questions, indicating the appropriate response by placing a check in the box next to the correct answer.

1. The induced voltage and X_L of a coil are directly proportional to
   - ☐ L and R
   - ☐ L/R
   - ☑ L and f
   - ☐ R and f
   - ☐ none of these

2. If frequency is doubled and L is halved, the net resultant X_L will
   - ☐ double
   - ☐ halve
   - ☐ quadruple
   - ☑ remain the same
   - ☐ none of these

3. If two equal inductors that were in series are now parallel connected, the resulting total X_L compared to the original circuit will be
   - ☐ two times greater
   - ☐ one-half as great
   - ☐ four times greater
   - ☑ one-fourth as great

4. As f increases, the rate of change of current
   - ☑ increases
   - ☐ decreases
   - ☐ remains the same

5. The X_L formula shows that inductive reactance is
   - ☐ directly proportional to L and inversely proportional to f
   - ☐ inversely proportional to L and directly proportional to f
   - ☑ neither of these

6. To solve for L when X_L and frequency are known, use the formula
   - ☐ $2\pi f/X_L$
   - ☐ $2\pi X_L/f$
   - ☑ $X_L/2\pi f$
   - ☐ none of these

7. The unit of X_L is the
   - ☐ back emf
   - ☑ ohm
   - ☐ ampere
   - ☐ volt
   - ☐ none of these

8. The opposition that an inductor shows to ac is
   - ☐ purely inductive
   - ☐ purely resistive
   - ☑ a combination of resistance and inductive reactance
   - ☐ none of these

9. The amount of inductive reactance that a given coil exhibits is directly related to
   - ☐ the amount of current
   - ☐ the applied voltage
   - ☑ neither of these

10. If frequency is tripled and inductance halved, the resultant X_L will be
    - ☐ two-thirds of the original
    - ☑ three-halves of the original
    - ☐ six times the original
    - ☐ one-sixth of the original

Lab Projects pages 207–208

The instructor may wish to quiz the students by assigning them X_L problems with specific values of L and f given. In summarizing, the instructor should be sure the students understand the direct relationship of X_L to both L and frequency. It will also be good to clarify again that there is no such thing as a perfect inductor with zero resistance. This must be taken into account when analyzing inductive circuits. The instructor might mention the general rule that if the R of the coil is less than one-tenth the X_L, we can generally neglect it in analyzing the circuit.

Lab Projects pages 207–208

PART 11

RL CIRCUITS IN AC

PROJECT 43 *V, I, R, Z,* and θ Relationships in a Series *RL* Circuit

PROCEDURE

1. Connect the initial circuit as shown in Figure 43-1.

2. Set the frequency of the function generator to the frequency indicated for the circuit option you are using. Set the circuit input voltage to 3 volts. Measure V_A, V_R, and V_L.

OBSERVATION

Lower f		Higher f	
V_A =	**3** V.	V_A =	**3** V.
V_{R_1} =	**0.46** V.	V_{R_1} =	**1.97** V.
V_{L_1} =	**2.86** V.	V_{L_1} =	**1.97** V.

CONCLUSION Does V_A equal the arithmetic sum of V_{R_1} and V_{L_1}? **No**. We conclude that to find V_A we must vectorially (*add, subtract*) **add** V_R and V_L. We can also use the **Pythagorean** theorem.

3. Calculate I_T from V_{R_1}/R_1. Calculate X_L from V_{L_1}/I. Calculate circuit total impedance from $Z = V_T/I_T$.

OBSERVATION

Lower f		Higher f	
I_T =	**0.46** mA.	I_T =	**1.97** mA.
X_L =	**6.2 k** Ω.	X_L =	**1.0 k** Ω.
Z =	**6.5 k** Ω.	Z =	**1.5 k** Ω.

4. Determine the apparent value of L from the known frequency and X_L parameters.

OBSERVATION

Lower f		Higher f	
L =	**1.97** H.	L =	**77 m** H.

5. Determine Z using the Pythagorean theorem (using the R and X_L parameters).

OBSERVATION

Lower f		Higher f	
Z =	**6.28 k** Ω.	Z =	**1.41 k** Ω.

6. Draw an impedance diagram in the Observation section.

OBSERVATION

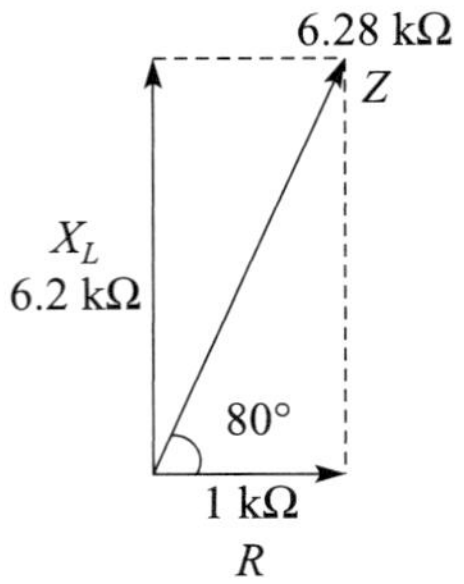

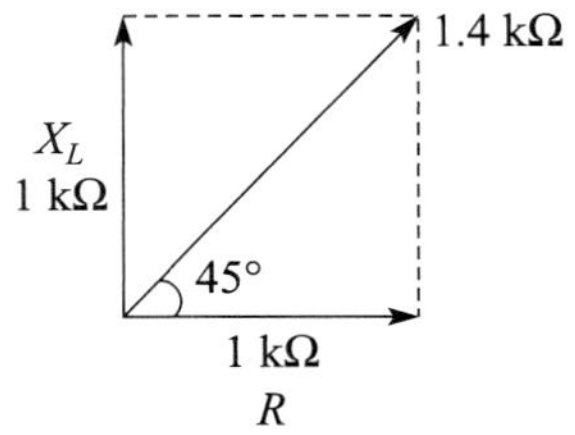

7. Use trigonometry and determine the phase angle.

OBSERVATION

Lower f: θ = **80** degrees.

Higher f: θ = **45** degrees.

CONCLUSION These conclusions are drawn for steps 3 through 7.

Does Z equal the arithmetic sum of R and X_L? **No**. We again conclude that we must use the vector sum or the Pythagorean theorem used in analysis of right **triangles**. From the observations of steps 2 and 3 we may conclude that V_R and V_L are (*in phase, not in phase*) **not in phase**. Since inductance opposes a change in current, we may assume that V_L (*leads, lags*) **leads** I_L by some angle. If L were a perfect inductor, **V_L** would lead **I_L** by 90 degrees. Since the circuit is not composed of purely resistance in which the phase angle between V and I = **0** degrees, nor purely inductance in which the phase angle between V and I equals **90** degrees, but rather a composite of both, we might expect the phase angle between V_T and I_T to be between **0** and **90** degrees. Further, the larger the X_L is compared to the circuit R, the more like a purely inductive circuit the results will be; thus the (*greater, lesser*) **greater** will be the circuit phase angle. The converse is also true.

Optional Steps

8. Use the measured and calculated data in steps 2 and 3, and draw a V-I vector diagram in the Observation section.

OBSERVATION

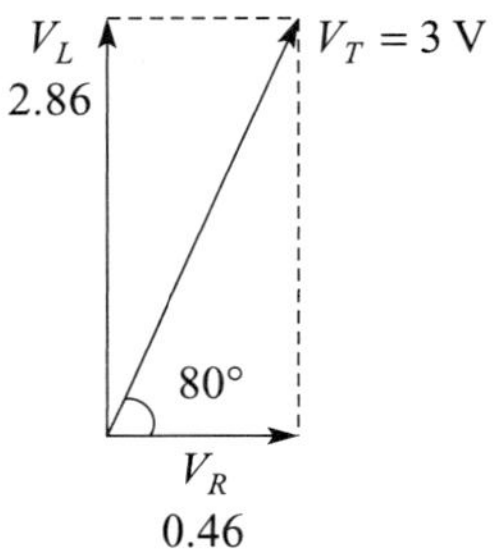

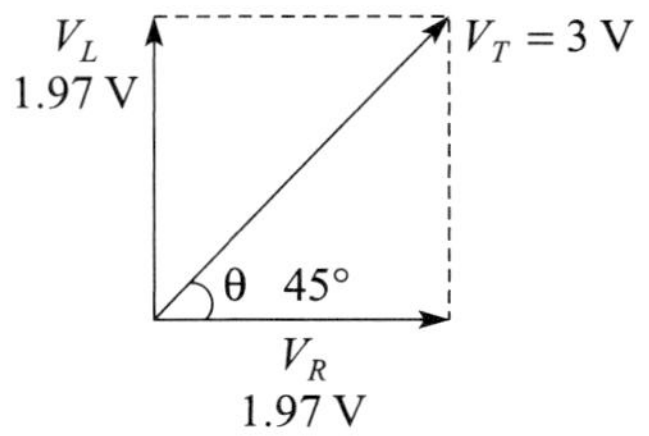

9. Use trigonometry and determine the phase angle.

OBSERVATION

	Lower *f*		Higher *f*
θ =	**80** degrees.	θ =	**45** degrees.

CONCLUSION Does the phase angle from the *V-I* vector diagram agree reasonably with the phase angle you determined from the *Z* diagram? **Yes**.

10. Use a dual-trace oscilloscope and perform a phase comparison of V_A and circuit current, (represented by the voltage across the resistor).

STOP **CAUTION: Be sure the signal source ground and the scope ground(s) are connected to the same end of the resistor when making the measurements to prevent the two grounds from shorting out a portion of the circuit!**

Determine the phase difference between the two signals. (If possible, demonstrate your scope waveforms and calculations to your instructor.)

OBSERVATION θ determined by the scope phase comparison:

	Lower *f*		Higher *f*
θ =	**80** degrees.	θ =	**45** degrees.

CONCLUSION Do the scope phase measurements and the phase angle calculations agree reasonably with your earlier findings? (Considering tolerances in components, source and scope frequency calibration tolerances, etc.) **Yes**.

The instructor may want to show the students a *V-I* vector diagram illustrating the phase relationships discussed above. The measurements above have certain measurement errors involved. Therefore, vector resultants will vary from the pure theoretical values. The instructor may wish to discuss why *I* is used as the reference vector for series circuits. Also, you may want to use scope patterns, as appropriate.

Lab Projects pages 211–214

PROJECT 44 *V, I, R, Z,* and θ Relationships in a Parallel *RL* Circuit

SPECIAL NOTE:

It should be noted that once again, we will be using the 100-Ω resistor in series with the main line as a circuit current indicator. The current equals 10 times the voltage drop.

PROCEDURE

1. Connect the initial circuit as shown in Figure 44-1.

2. Set the source for 60 Hz, 3 V_{rms} for the lower *f* circuit and for 2,000 Hz, 3 V_{rms} for the higher *f* circuit. Measure the circuit voltages.

OBSERVATION

Lower *f*			Higher *f*		
V_A =	3	V.	V_A =	3	V.
V_{R_1} =	0.35	V.	V_{R_1} =	0.42	V.
V_{R_2} =	2.75	V.	V_{R_2} =	2.68	V.
V_{R_3} =	0.157	V.	V_{R_3} =	0.263	V.
V_L =	2.73	V.	V_L =	2.63	V.

CONCLUSION Do the voltages around any given closed loop add up by addition to V_A? **No**. From this we conclude that the circuit current(s) and voltage(s) are (*in phase, out of phase*) **out of phase**.

3. Calculate the total circuit current from V_{R_1}. Calculate the current through R_2 by Ohm's Law. Calculate the current through *L* by using the voltage drop across R_3. Use Ohm's Law to solve for I_L ($I_L = V_3/R_3$) and X_L ($X_L = V_L/I_L$).

OBSERVATION

Lower *f*			Higher *f*		
I_T =	3.5	mA.	I_T =	4.2	mA.
I_{R_2} =	2.73	mA.	I_{R_2} =	2.68	mA.
I_L =	1.67	mA.	I_L =	2.63	mA.
X_L =	1.7 k	Ω.	X_L =	1.0 k	Ω.

4. Use measured and calculated values of I_{R_2} and I_L. Apply the Pythagorean theorem formula and calculate I_T.

OBSERVATION

Lower *f*			Higher *f*		
I_T calculated =	3.2	mA.	I_T calculated =	3.75	mA.

Lab Projects pages 215–217

5. Draw the appropriate *V-I* vector diagram in the Observation section.

OBSERVATION

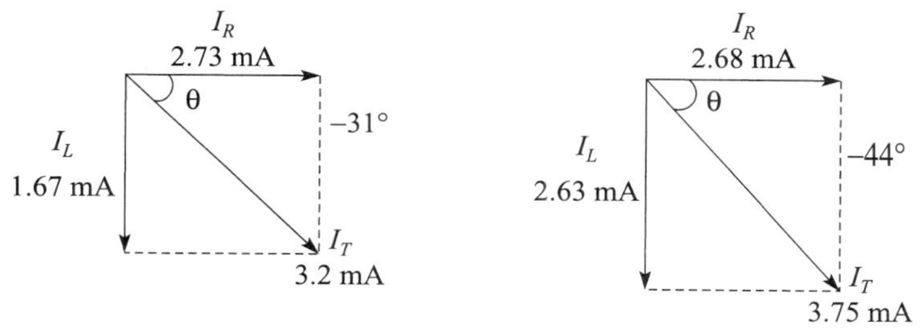

6. Use trigonometry and determine the phase angle. (Neglect R_1 parameters and use only R_2 and L parameters.)

OBSERVATION

Lower *f* — θ = **31.2** degrees. Higher *f* — θ = **44** degrees.

CONCLUSION These conclusions are drawn for steps 3 through 6.

Does total current equal the arithmetic sum of the branch currents? **No**. This is because the branch currents are **out of phase** __________. The current through R_2 is in phase with V_2 (*True* or *False*) **true**. The current through the coil (*leads, lags*) **lags** the voltage across the coil by close to **90** degrees. If the inductor were perfect, it would be exactly **90** degrees. Since the total circuit current is the vector resultant of the two branch currents, it would seem logical to assume the circuit total current would be (*leading, lagging*) **lagging** V_A by some angle between **0** and **90** degrees. Also note from our measurements and calculations that the total circuit impedance (Z) cannot be found by the product-over-the-sum method but is most easily solved by Ohm's Law, where $Z = V_T / I_T$.

Optional Step

7. Use a dual-trace oscilloscope and perform a phase comparison of the current through the R_2 branch and the current through the inductor branch. Do this by letting the voltage across R_2 represent the current through R_2 and the voltage across R_3 represent the current through the inductor branch.

STOP **CAUTION: Be sure the signal source ground and the scope ground(s) are connected to the same end (the bottom end) of the circuit network when making the measurements to prevent the grounds from shorting out a portion of the circuit!**

Determine the phase difference between the two signals. (If possible, demonstrate your scope waveforms and calculations to your instructor.)

OBSERVATION Measured θ between the resistor branch and the inductor branch as determined by the scope phase comparison:

Lower f — $\theta \cong$ **90** degrees.

Higher f — $\theta \cong$ **90** degrees.

CONCLUSION Was the phase difference between the two branches reasonably close to 90°? **Yes**. If not, what variables and factors might account for the difference? ***R* in coil, scope calibration, human reading error**.

The instructor may wish to show the students the *V-I* vector diagrams for the preceding data. Point out why V_A is used as the reference vector for parallel circuits, as contrasted to I_T being the reference vector for series circuits. Also note that in parallel circuits, the smaller the X_L, the more inductive the circuit acts, which is converse to the series R_L circuit action.

Lab Projects pages 215–217

PART 11 Story Behind the Numbers
Section 1: *RL* (Series) Circuits in AC

Data Table

Component and Parameter I.D.'s	Measured Values	Calculated Values
V_A (V)	3	—
V_R (V)	1.97	—
V_L (V)	1.97	—
I_T (mA)	—	1.97
X_L (kΩ)	—	1
Z, Ohm's Law (kΩ)	—	1.5
Z, Pythagorean theorem (kΩ)	—	1.44
Calculated apparent L (mH)	—	≈ 80

Answers to Analysis Questions (Series Circuit)

1. No. Voltages are out of phase and therefore cannot be simply added arithmetically.
2. No. They must be added vectorially.
3. Calculated = 87.5 mH; rated is 100 mH. The causes of differences could be mostly the actual tolerance % of the inductor, being operated differently from how the manufacturer rated the inductor, measurement errors in the project, and others.
4. 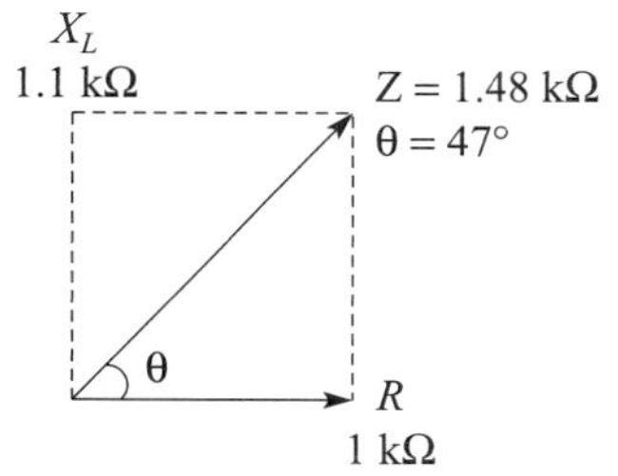

5. Arctan of X_L/R = arctan of 1,100/1,000 = arctan of 1.1 = 47 degrees

Lab Projects pages 219–222

6.

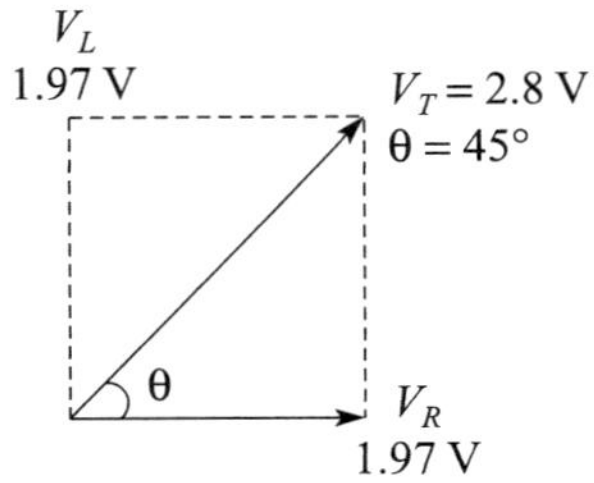

7. Arctan of V_L/V_R = arctan of 1.97/1.97 = arctan of 1 = 45 degrees

8. Phase angle = 360 x-div/y-div = approximately 47 degrees. Or, using the vertical divisions between where two signals cross the horizontal zero axis on the scope graticule, with the waveforms set up for one complete sine wave over the 10 divisions: Each division = 36 degrees and signal separations were approximately 1.3 divisions, which equals approximately 47 degrees.

9. If frequency were doubled: V_A would remain the same; V_R would decrease; V_L would increase; I_T would decrease; X_L would increase; and Z would increase.

PART 11 Story Behind the Numbers

Section 2: *RL* (Parallel) Circuits in AC

Data Table

Component and Parameter I.D.'s	Measured Values	Calculated Values
V_A (V)	3	—
V_{R_1} (V)	0.42	—
V_{R_2} (V)	2.68	—
V_{R_3} (V)	0.263	—
V_L (V)	2.63	—
I_T (mA)	—	4.2
I_{R_2} (mA)	—	2.68
X_L (kΩ)	—	1
I_T, Pythagorean theorem (mA)	—	3.75

Answers to Analysis Questions (Parallel Circuit)

1. Branch currents are out of phase with each other and must be added vectorially, not simply arithmetically.

2. Yes.

3. Current lags voltage and voltage leads current by about 90 degrees.

4.

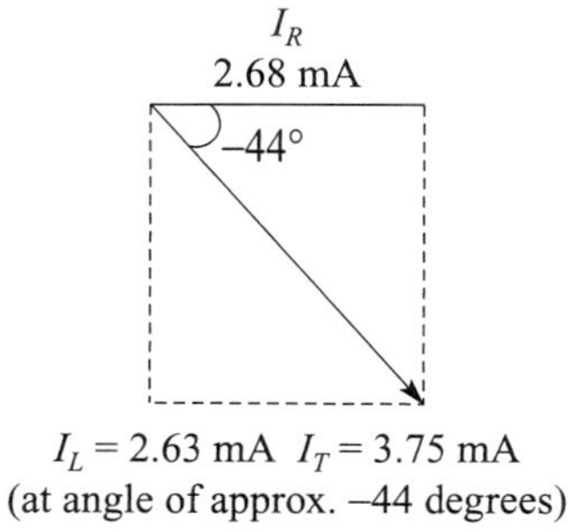

5. Phase angle = arctan of 2.63/2.68 = 44.4 degrees

6. I_T is lagging V_A by about 44 degrees. Due to the reactive component branch, branch currents are out of phase even though both branches have the same applied voltage.

7. Using a single sine-wave cycle across the scope screen, each major division across the scope screen represents 36 degrees. The scope pattern for the two signals, based on the voltage patterns used, thus indicated close to 90 degrees difference between branch currents. The resistance in the inductive branch would cause some difference, plus there are potential measurement errors, as well.

8. If frequency were doubled: V_A would remain the same; V_{R_1} would decrease; V_{R_2} would increase; V_{R_3} would decrease; V_L would decrease; I_T would decrease; I_{R_2} would increase; and X_L would increase.

Lab Projects pages 223–225

PART 11 Summary

Complete the review questions, indicating the appropriate response by placing a check in the box next to the correct answer.

1. An ac circuit whose phase angle is 45 degrees is composed of
   - ☑ an equal amount of resistance and reactance
   - ☐ an unequal amount of resistance and reactance
   - ☐ purely resistance
   - ☐ purely reactance

2. The current through an inductor
   - ☐ leads V_L
   - ☑ lags V_L
   - ☐ is in phase with V_L

3. To find the Z of a series RL circuit
   - ☐ simply add X_L total and R total
   - ☐ subtract X_L from R
   - ☑ neither of these

4. When making a V-I vector diagram of a series RL circuit, the reference vector is
   - ☑ I_T
   - ☐ V_T
   - ☐ Z_T
   - ☐ R_T
   - ☐ none of these

5. When making a V-I vector diagram of a parallel RL circuit, the reference vector is
   - ☐ I_T
   - ☑ V_T
   - ☐ Z_T
   - ☐ R_T
   - ☐ none of these

6. In a series RL circuit, if L is increased while R remains the same, the circuit phase angle (angle between V_A and I_T) will
   - ☑ increase
   - ☐ decrease
   - ☐ remain the same

7. In a series RL circuit, if R is increased while L remains the same, the circuit phase angle will
   - ☐ increase
   - ☑ decrease
   - ☐ remain the same

8. In a parallel RL circuit, if L is increased while R remains the same, the circuit phase angle will
   - ☐ increase
   - ☑ decrease
   - ☐ remain the same

9. In a parallel RL circuit, if R is increased while L remains the same, the circuit phase angle will
   - ☑ increase
   - ☐ decrease
   - ☐ remain the same

Lab Projects pages 227–228

10. The value of total circuit impedance (Z) for both series and parallel ac circuits can be solved by Ohm's Law.
    - ☑ True
    - ☐ False

The instructor may wish to further test the students' knowledge by presenting *RL* circuit problems with specific values given and requiring quantitative results. The key items to be sure the students understand are:

1. Circuit parameters cannot be solved by simple arithmetic sums but must be solved vectorially.
2. In a "perfect" inductance, V leads I by 90 degrees.
3. In a resistor, V and I are in phase.
4. In a circuit composed of both resistance and inductance, the circuit V and I are out of phase by an angle between 0 and 90 degrees.
5. The more inductive the circuit, the larger the phase angle.
6. The more resistive the circuit, the smaller the phase angle.
7. For series *RL* circuits, the larger the X_L, the more inductive the circuit (for any given R).
8. For parallel *RL* circuits, the smaller the X_L, the more inductive the circuit (for any given R).

Lab Projects pages 227–228

PART 12

BASIC TRANSFORMER CHARACTERISTICS

PROJECT 45 Turns, Voltage, and Current Ratios

SPECIAL NOTE:

Transformers may be used to "step-up" or "step-down" voltages. If it were possible to have 100% efficiency, all of the power in the primary would be transferred to the secondary. If this were true, then the product of V and I in the secondary would equal the product of V and I in the primary, and the current step-up or step-down ratio would be inverse to the voltage step-up or step-down ratio in order for the $V \times I$ products to be the same.

To illustrate the concepts of basic transformer action, we want you to assume 100% efficiency of the transformers used for any calculations you are required to perform.

STOP **Caution! When working with HIGH VOLTAGE, use all the safety procedures relative to working with high voltage.**

PROCEDURE

1. Obtain a 12.6-volt transformer and connect the circuit shown in Figure 45-1.

2. Use a voltmeter and measure the voltage present on the primary and on the secondary.

OBSERVATION Primary $V =$ ____**5**____ V.
Secondary $V =$ ____**0.52**____ V.
The primary-to-secondary voltage ratio is ____**9:1**____.

CONCLUSION The secondary-to-primary voltage ratio is ____**0.105:1**____.
What is the N_S/N_P turns ratio? ____**0.105:1**____. Are the voltage and turns ratios the same? ____**Yes**____.

3. Connect a 100-Ω, 1-W resistor between points A and B as shown in Figure 45-2. Use Ohm's Law to determine the secondary current.

OBSERVATION $V_R =$ ____**0.52**____ V.

CONCLUSION $I = V/R$

$I =$ ____**≈ 0.01**____ A.

Is the secondary current the same as the current through the load R? ____**Yes**____.

4. Use the previously determined voltage ratio from above and calculate the primary-to-secondary current ratio.

OBSERVATION $V_P{:}V_S =$ **9.5:1**.
$I_P{:}I_S =$ **1:9.6**.

CONCLUSION The current ratio is the **inverse** of the voltage ratio.
Using the secondary current previously determined in Figure 45-2, step 3, and the current ratio just calculated, what is the approximate primary current value? **0.001** A.

PROJECT 46 Turns Ratios versus Impedance Ratios

SPECIAL NOTE:

The impedance ratio of a transformer is related to the square of the turns ratio. This project will let you determine the turns ratio of a transformer, then theoretically determine the impedance ratio. Once this has been determined, you or your instructor will compare your results with the catalog specifications for the transformer you used. You can then see how closely your measurements and calculations correlate with the transformer specifications.

PROCEDURE

1. Obtain an audio output transformer and connect a circuit similar to that shown in Figure 46-1.

2. Apply a 1,000-Hz signal at a 3–5 V voltage level to the transformer primary. Measure the secondary voltage with an appropriate measuring instrument.

OBSERVATION Voltage applied to the primary is ____________________ V.

Measured voltage on secondary is ____________________ V.

CONCLUSION The N_P/N_S ratio must be about ________ :1.
Using the formula $Z_P/Z_S = N_P^2/N_S^2$, the primary-to-secondary impedance ratio is ________ :1.

3. Assume that the secondary load is going to be 4 ohms. Using the impedance ratio previously calculated, compute the nominal impedance of the circuit that should be connected to the primary of the output transformer.

OBSERVATION Nominal Z calculated = ____________________ Ω.

CONCLUSION Using the catalog parameters for the transformer, the primary-to-secondary Z ratio is ________ :1.
Does the impedance ratio you measured and calculated via voltage measurements reasonably agree with the catalog specifications for the transformer? ____________.
What might cause any differences?____
____.

Optional Step

4. Connect a 4-Ω speaker to the secondary, Figure 46-2. Measure the voltages and determine the Z ratio.

OBSERVATION Primary V = ____________________ V. Calculated turns ratio = ____________.
Secondary V = ____________________ V. Calculated Z ratio = ____________.

CONCLUSION Does this measured and calculated Z ratio more closely match the transformer specifications? ________.

Lab Projects pages 233–234

PART 12 Story Behind the Numbers

Data Table

Component and Parameter I.D.'s	Measured Values	Calculated Values
Circuit 1, primary *V*		—
Circuit 1, secondary *V*		—
P-S voltage ratio	—	
V_{R_L} @ points *A–B*		—
I secondary w/ 100 Ω	—	
P-S current ratio	—	
Approx. *I* primary current	—	
I measured, primary		—
I measured, secondary		—
Circuit 2, secondary *V*		—
P-S turns ratio	—	
P-S *Z* ratio, calculated	—	

NOTE: Values in the table will depend upon the specific transformer used at the student's laboratory site.

Answers to Analysis Questions

1. Voltage ratio is directly related to turns ratio; current ratio is inversely related to turns ratio. Therefore, voltage and current ratios are inversely related to each other.
2. In Circuit 1 we are using a step-down transformer setup. If the transformer connections were such that the source was connected to points *A* and *B* and the load connected to the other winding, it might be used as a step-up situation.
3. Assumed virtually 100% efficiency.
4. Items such as current ratings, voltage ratings, whether center-tapped or not, mounting style, dimensions, and method of connecting to circuits or circuit boards.
5. Items such as nominal primary impedance, secondary impedance, dc mA rating, primary dc resistance, secondary dc resistance, output power rating, physical dimensions, mounting method, and so on.
6. When using the parameters from your setup, use the formula: $Z_P = Z_S (N_P/N_S)^2$

Lab Projects pages 235–238

PART 12 Summary

Answer the following questions with "T" for true and "F" for false. (Put your answer in the appropriate blank.)

1. A transformer's voltage and current ratios are inverse. ___T___
2. The impedance ratio of a transformer is related to the square root of the turns ratio. ___F___
3. If a transformer has a 3:1 turns ratio (secondary-to-primary) and a primary impedance of 5,000 ohms, the load impedance on the secondary should be 45,000 ohms to affect a proper impedance match. ___T___
4. If a transformer's turns ratio is doubled, the related impedance ratio will quadruple. ___T___
5. A transformer with a secondary-to-primary turns ratio of 10:1 will have a secondary voltage of 50 volts with 0.5 volts applied to its primary. ___F___
6. If a transformer's current ratio is 1:3, then the related voltage ratio will be 1:3. ___F___
7. If a transformer's voltage ratio is doubled, its related current ratio will halve. ___T___
8. If the turns on a transformer secondary are doubled, the impedance of the secondary will be four times as great as the original impedance. ___T___
9. Transformer efficiency is typically 100%. ___F___
10. The voltage across a transformer secondary will be higher when it is "unloaded" as compared to when it is supplying current to a load. ___T___

Lab Projects page 239

PART 13

CAPACITANCE
(DC CHARACTERISTICS)

PROJECT 47 Charge and Discharge Action and *RC* Time

PROCEDURE

1. Connect the initial circuit as shown in Figure 47-1.
2. Set V_A at 20 volts. Insert a jumper wire between points *A* and *B* and observe the meter action.

OBSERVATION Current flowed for approximately **4 or 5** seconds as evidenced by the voltage measured across *R*. The rate of charge was (*linear, nonlinear*) **nonlinear**.

CONCLUSION The current that flowed was the charging current that charged the capacitor to a voltage equal to (V_R, V_A) **V_A**. Was charging current maximum at the beginning of the charge time or near the end? **at the beginning**. At the beginning of the charge time, the voltage across *R* was equal to (V_C, V_A) **V_A** (the first instant). At the end of the charge time, the voltage across the resistor is **0** volts; the voltage across the capacitor is equal to V_A.

NOTE ➤ The charged capacitor voltage is equal to V_A and series opposing the source. Hence, no current can flow once the capacitor is charged.

3. Remove the jumper from points *A* and *B*. Reverse the polarity of the voltmeter. Insert a jumper between points *C* and *D* and observe the meter action during discharge of the capacitor.

OBSERVATION Discharge time was approximately **4 or 5** seconds. The *rate* of discharge was (*linear, nonlinear*) **nonlinear**.

CONCLUSION Did the capacitor take the same time to discharge through *R* as it did to charge? **Yes**. At the end of the discharge time V_C = **0** ; V_R = **0** V.

4. Remove the jumper from points *C* and *D*. Change the DMM polarity from the original setup if it is not an "autopolarity"-type meter. Change *R* to a 10-MΩ resistor. Repeat the sequence of steps 2 and 3.

OBSERVATION Charge time was approximately **30 to 50** seconds. Discharge time was **30 to 50** seconds.

CONCLUSION Increasing *R* increased the charge time because the charging current was limited to a smaller value. Thus, it took longer to obtain a given potential difference or **V** across the *C*.

Lab Projects pages 243–244

5. Change the *C* value to a 0.1-μF capacitor. Insert a jumper between points *A* and *B* and note the charge time.

OBSERVATION Charge time was approximately **3 to 5** seconds.

CONCLUSION Changing *C* from a 1.0-μF to a 0.1-μF capacitor caused the charge time to **decrease**. We conclude that both the value of ***R*** and of ***C*** determine charge and discharge time. It should be noted that if it were not for the effect of the multimeter resistance, the time to charge the capacitor or discharge it would be directly proportional to *R* and to *C*. The formula relating to this is called the formula for the *time constant*. This formula states that one *RC* time constant = *R* in ohms times *C* in farads, and the answer is in seconds.

Also, it should be observed that it takes 5 time constants to charge or discharge the capacitor. Calculate the R_e of *R* and the meter circuit's resistance in parallel and determine if the charge time is about equal to the expected 5 TC. Is it? **Yes**.

6. Use the *RC* time-constant formula and determine how long it would take to charge a 1-μF capacitor in series with a 100-kΩ resistor.

OBSERVATION 5 *RC* time constants = **0.5** seconds.

7. Connect the circuit described in step 6 and note the charge time.

OBSERVATION Charge time measured approximately **0.5** seconds.

CONCLUSION A multimeter with a meter circuit *R* of 5 MΩ or greater does not alter the circuit resistance of 100 kΩ very much. The measured charge time was (*close, not close*) **close** to the theoretical value.

Key points to emphasize regarding capacitance in dc are:

1. the nonlinear characteristic of charge and discharge. (Perhaps introduce the exponential *TC* graph.)
2. the *RC* time-constant formula and (optional) use of epsilon approach.
3. 5 time constants are required for completion of charge and discharge action.

If time permits, you may want to note the $Q = CV$ relationships, which do complement the *RC* time concept.

Lab Projects pages 243–244

PROJECT 48 Total Capacitance in Series and Parallel

SPECIAL NOTE:

For this project, we will take advantage of the fact that the charge time of a capacitor is directly proportional to capacitance. By noting the charge time of a single capacitor, then noting the charge time for two capacitors in series and then in parallel, we should be able to conclude the effect on total capacitance of connecting capacitors in series or parallel. A DMM is highly preferred over a VOM for voltage measurements here.

PROCEDURE

1. Connect the initial circuit as shown in Figure 48-1.

2. Set V_A to 20 volts. Insert a jumper between points A and B and note the charge time by observing the voltmeter.

OBSERVATION Charge time was approximately **20 to 40** seconds.

CONCLUSION One RC time-constant is approximately **20 to 40** seconds. This means that the R_e of the meter and the 10-MΩ resistor is approximately **5 M** Ω.

3. Obtain a second 1-μF capacitor. Remove the jumper from points A and B. Carefully discharge C_1, then insert the second capacitor (C_2) in series with C_1. Insert a jumper between points A and B and note the charge time of C_1 and C_2 in series.

OBSERVATION Charge time was approximately **10 to 20** seconds.

CONCLUSION Since the charge time has decreased to **1/2** the value it was with only C_1 in the circuit, it may be concluded that the total capacitance has (*increased, decreased*) **decreased**. Since the new RC time is (*double, half*) **half** the original and the R has not been changed, we conclude that the new total capacitance is **.5** μF. Our observations tell us that capacitors in series add like resistors in (*series, parallel*) **parallel**.

4. Remove the jumper from points A and B. Discharge the capacitors. Change the circuit as required to achieve a circuit with C_1 and C_2 in parallel, and this combination in series with R. Insert the jumper again, and note the charge time of C_1 and C_2 in parallel.

OBSERVATION Charge time was approximately **40+** seconds.

CONCLUSION The charge time for this step is approximately (*2, 3, 4*) **2** times the time recorded in step 2. This indicates that the total capacitance of C_1 and C_2 in parallel is (*1/2, 2×,*

3×) ___**2×**___ the capacitance of C_1 alone. We may conclude that capacitors in parallel add like resistors in (*series, parallel*) ___________**series**__________. If C_2 were not the same value as C_1, would the statements concerning total capacitance of capacitors in series and parallel still hold true? (That is, series *C*s add like parallel *R*s, and parallel *C*s add like series *R*s.) ___**Yes**___.

PART 13 Story Behind the Numbers

Answers to Analysis Questions

1. Yes. The time would be the same. The charging current would be higher for the higher applied voltage, but charge time and discharge time are dependent upon values of *R* and *C*, not on the value of applied voltage.
2. Yes. It takes five *RC* time constants to change from one voltage to any other given voltage, again based on the *RC* time constant of the circuit.
3. Approximately equal. The *RC* time for each circuit is the same. Therefore, five time constants for either circuit will be the same if components are actually at rated values.
4. Yes, they would be different. The lower resistance of the VOM would have impacted the charge and discharge times by changing the circuit *RC* time-constant value.
5. The circuit with the larger 10-MΩ *R* value.
6. It takes five *RC* time constants.

Data Table

Component and Parameter I.D.'s	Measured Values	Calculated Values
V_A (V)	20	—
Charge time (1 μF & 1 MΩ) sec.	**4–5 seconds**	—
Discharge time (1 μF & 1 MΩ) sec.	**4–5 seconds**	—
Charge time (0.1 μF & 10 MΩ) sec.	**4–5 seconds**	—
Discharge time (0.1 μF & 10 MΩ) sec.	**4–5 seconds**	—
RC time, (1 μF & 100 kΩ calc.) sec.	**0.5 seconds**	**0.5 seconds**

Lab Projects pages 247–249

PART 13 Summary

Complete the following review questions, indicating the appropriate response by placing a check in the box next to the correct answer.

1. A charged capacitor has a difference of potential between its plates due to
   - [x] an excess of electrons on one plate and a deficiency of electrons on the other
   - [ ] an excess of electrons on both plates
   - [ ] neither of these

2. A capacitor in a given circuit will take the same amount of time to discharge as it does to charge.
   - [x] True
   - [ ] False

3. If the value of R is doubled and the value of C is halved in a given RC circuit, the time it will take to charge the capacitor will
   - [ ] increase
   - [ ] decrease
   - [x] remain the same

4. One RC time constant is equal to
   - [ ] $R + C$
   - [ ] R/C
   - [ ] $R - C$
   - [x] $R \times C$
   - [ ] none of these

5. In order for a capacitor to fully charge or discharge, it takes
   - [ ] one time constant
   - [ ] two time constants
   - [ ] four time constants
   - [x] five time constants

6. For a given RC circuit, increasing the value of V_A will cause the time needed for the capacitor to fully charge or discharge to
   - [ ] increase
   - [ ] decrease
   - [x] remain the same

7. The total capacitance of a 0.05-μF capacitor and a 0.1-μF capacitor in parallel is
   - [ ] 0.05 μF
   - [ ] 0.1 μF
   - [ ] 0.06 μF
   - [x] 0.15 μF
   - [ ] 0.033 μF
   - [ ] none of these

8. The total capacitance of a 0.05-μF capacitor and a 0.1-μF capacitor in series is
   - [ ] 0.05 μF
   - [ ] 0.1 μF
   - [ ] 0.06 μF
   - [ ] 0.15 μF
   - [x] 0.033 μF
   - [ ] none of these

9. How long would it take two parallel 0.1-μF capacitors to charge through a 1-MΩ resistance?
   - [ ] 0.2 seconds
   - [x] 1 second
   - [ ] 0.1 seconds
   - [ ] 5 seconds
   - [ ] none of these

Lab Projects pages 251–252

10. How long would it take two series 0.1-μF capacitors to charge through a 1-MΩ resistance?

- ☐ 1/4 second
- ☑ 1/2 second
- ☐ 3/4 second
- ☐ 1 second
- ☐ 2.5 seconds
- ☐ none of these

You may wish to include facts relating to the $Q = CV$ formula in your summary discussion. At any rate, you should be sure the students understand the concepts of:

1. charge and discharge action
2. V_C (charged capacitor voltage) being series-opposing to the source
3. 1 $TC = R \times C$
4. 5 time constants are needed to fully charge or discharge a capacitor through a resistance (Optional: discussion of epsilon approach to calculations).
5. series capacitors adding like parallel resistances and parallel capacitors adding like series resistances.

For example, C_T of series capacitors = the product-over-the-sum (two capacitors) and C_T for parallel capacitors = $C_1 + C_2 + \dots C_n$.

Lab Projects pages 251–252

PART 14

CAPACITIVE REACTANCE IN AC

PROJECT 49 Capacitance Opposing a Change in Voltage

PROCEDURE

1. Connect the initial circuit as shown in Figure 49-1.
2. Temporarily remove one lead from the VVPS and then set its output voltage at 10 volts. Observe how much time it takes the capacitor to stop charging after the lead is reinserted into the VVPS by watching the voltmeter measuring V_C.

OBSERVATION Approximate time to reach a steady state for V_C was **≈ 25** seconds.

CONCLUSION V_C did not take 50 seconds to change because the (*wire, meter*) **meter** resistance formed a voltage divider with the 10-MΩ resistor.

3. Quickly turn the VVPS voltage control knob to a higher voltage setting, and note how much time it takes V_C to reach its new steady-state value.

STOP **CAUTION: Do not set V_A higher than the multimeter voltage range setting.**

OBSERVATION Approximate time to reach the new steady-state value was **≈ 25** seconds.

CONCLUSION It takes the voltage on the capacitor (*the same, a different*) **the same** amount of time to change from some given value to a new value, as it does for V_C to change from zero to any steady-state value.

4. Quickly turn the VVPS voltage control knob to a lower voltage setting and observe if V_C changes instantaneously, or takes time.

OBSERVATION It (*did, didn't*) **did** take time for V_C to decrease to its new steady-state value.

CONCLUSION From our observations, we conclude that a capacitor seems to oppose a change in (*current, voltage*) **voltage** in a similar fashion to a coil or inductor opposing a change in (*current, voltage*) **current**.

The instructor may desire to enrich this project by demonstrating the effect of capacitors in waveshaping. The instructor might use a square-wave generator and scope to show the effects of different T/RC ratios in an RC coupling network on the coupled resultant waveform.

PROJECT 50 X_C Related to Capacitance and Frequency

PROCEDURE

1. Connect the initial circuit as shown in Figure 50-1.
2. Set the function generator or audio oscillator to a frequency of 100 Hz and V_A to 3 volts. Measure V_R and V_C. Calculate the circuit I by Ohm's Law, and also calculate X_C ($X_C = V_C/I$).

OBSERVATION V_A = **3.02** V. I_T = **0.096** mA.
V_R = **2.59** V. X_C = **15.9 k** Ω.
V_C = **1.53** V.

CONCLUSION Since the current is the same through all parts of a series circuit and the voltage drop across R is approximately 2 times V_C, it appears that X_C must be about (*double, half*) **half** the value of R.

3. Change C to a 1.0-μF capacitor and repeat step 2.

OBSERVATION V_A = **3.02** V. I_T = approximately **.112** mA.
V_R = **3.01** V. X_C = approximately **1.6 k** Ω.
V_C = **.18** V.

CONCLUSION Increasing the value of C while maintaining all other parameters the same caused X_C to (*increase, decrease*) **decrease**. The X_C for the 1.0-μF capacitor was approximately (*10 times, one-tenth*) **one-tenth** the X_C of the 0.1-μF C. From this we conclude that X_C is (*directly, inversely*) **inversely** proportional to capacitance. This means as C decreases, X_C (*increases, decreases*) **increases** or, if X_C has decreased, then C must have (*increased, decreased*) **increased**, all other factors being constant.

4. Keeping the 1.0-μF capacitor, change the frequency to 200 Hz and maintain 3 volts V_A. Repeat the measurements and calculations of the previous steps.

OBSERVATION V_A = **3.0** V. I_T = approximately **.112** mA.
V_R = **3.0** V. X_C = approximately **727** Ω.
V_C = **.08** V.

CONCLUSION Increasing the frequency and keeping the C the same caused X_C to (*increase, decrease*) **decrease**. If there were no voltmeter loading effects on the circuit, the indicated X_C at 200 Hz would have been (*two times, one-half*) **one-half** the X_C value at 100 Hz. From this we see that X_C is (*directly, inversely*) **inversely** proportional to frequency.

5. Keep f at 200 Hz and change C back to a 0.1-μF capacitor. Repeat the measurements and calculations of the previous steps.

OBSERVATION V_A = **3.01** V. I_T = **0.106** mA.
V_R = **2.88** V. X_C = **8 k** Ω.
V_C = **.85** V.

CONCLUSION Referring back to step 2 observations, does it appear that the X_C of the 0.1-μF C at 200 Hz (this step) is about one-half that at 100 Hz? **Yes**. We conclude from the preceding that X_C is inversely proportional to ***C*** and ***f***.

6. Plot a graph of X_C versus f from 200 Hz to 1,000 Hz using the 0.1-μF capacitor.

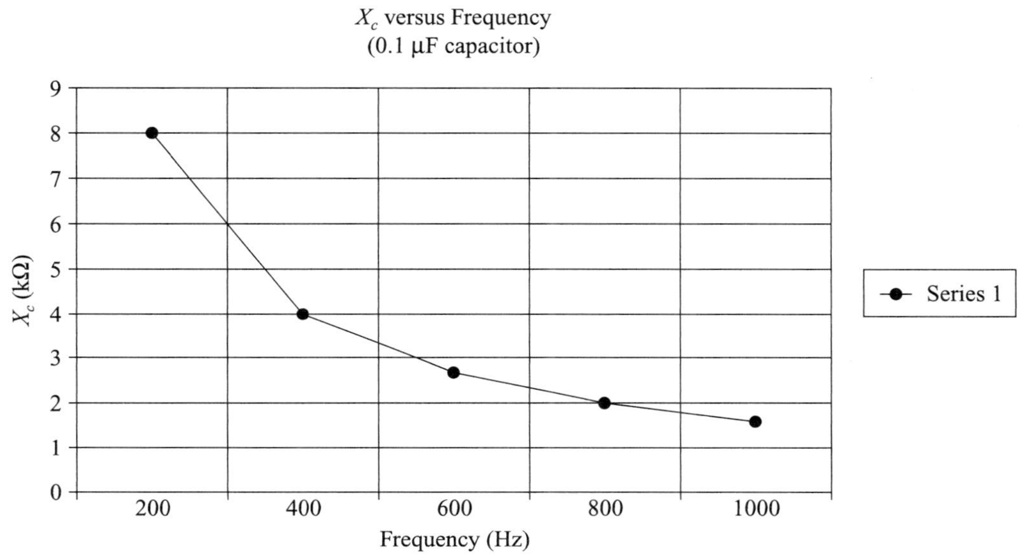

NOTE ➤ Calculate X_C for each 200-Hz change in frequency and plot with coordinates similar to those shown in Figure 50-2. Plot the graph on a separate sheet of graph paper.

OBSERVATION X_C at 200 Hz = **7.96 kΩ** X_C at 800 Hz = **1.99 kΩ**
X_C at 400 Hz = **3.98 kΩ** X_C at 1,000 Hz = **1.59 kΩ**
X_C at 600 Hz = **2.65 kΩ**

CONCLUSION Did X_C act inversely proportional to f? **Yes**.

The instructor should be aware that meter loading has quite an effect on the above results if using a VOM.

PROJECT 51 The X_C Formula

SPECIAL NOTE:

As shown in the previous project, the X_C of a capacitor in ohms is inversely proportional to the frequency and also to the capacitance value. This relationship is shown by the X_C formula,

$$X_C = \frac{1}{2\pi fC}$$

The rationale we might use in understanding this formula is as follows. A larger C requires more charge (stored electrons) to arrive at a given difference of potential between its plates. This means that more charging (and discharging) current must flow in the circuit in order for the capacitor to "follow" the ac voltage applied to it at a given frequency than a smaller C would require under the same conditions. More current flowing for a given applied voltage indicates a lower opposition. A higher frequency also causes more charge and discharge current to flow per unit time in order for the voltage across the capacitor to follow the applied voltage.

PROCEDURE

1. Connect the initial circuit as shown in Figure 51-1.
2. Set the function generator at a frequency of 150 Hz and V_A at 3 volts. Measure V_R and then calculate I_T. Measure V_C and use the calculated I to determine X_C by Ohm's Law. Calculate X_C by the X_C formula.

OBSERVATION $V_A =$ **3.0** V $I =$ **.206** mA.
$V_R =$ **2.06** V. $V_C =$ **2.17** V.
X_C by Ohm's Law approximately = **10.5 k** Ω.
X_C by X_C formula = **10.6 k** Ω.

CONCLUSION Since this is a series circuit and V_C is virtually the same as V_R, it is apparent that the X_C must essentially be equal to ***R***. Was the X_C calculated by the capacitative reactance formula close to the value determined by Ohm's Law? **Yes**.

3. Change the frequency to 75 Hz and keep V_A at 3 volts. Make the measurements and calculations described in step 2 above.

OBSERVATION $V_A =$ **3.0** V. $I =$ **0.13** mA.
$V_R =$ **1.30** V. $V_C =$ **2.71** V.
X_C by Ohm's Law approximately = **20.8 k** Ω.
X_C by X_C formula = **21.2 k** Ω.

CONCLUSION Lowering the frequency to one-half its previous value caused the X_C to (*increase, decrease*) **increase** to a value nearly (*double, half*) **double** the original. Two reasons for the X_C values calculated by Ohm's Law method and X_C

Lab Projects pages 261–262

formula not being the same might be the resistor and capacitor **tolerance** and meter **loading** effects.

4. Keep the frequency at 75 Hz, but change C to 1 μF. Measure and calculate as before.

OBSERVATION V_A = **3.0** V. I = **0.293** mA.
V_R = **2.93** V. V_C = **0.62** V.

X_C by Ohm's Law approximately = **2.1 k** Ω.
X_C by X_C formula = **2.12 k** Ω.

CONCLUSION Increasing C by a factor of 10 caused the X_C to (*increase, decrease*) **decrease**. Did the X_C approximately change by a factor of 10? **Yes**. From the observations we have made, does it appear that the X_C formula is functional for predicting parameters in practical circuits? **Yes**.

PART 14 Story Behind the Numbers

Data Table

Component and Parameter I.D.'s	0.1 μF @ 100 Hz	1.0 μF @ 100 Hz	f_1 0.1 μF @ 200 Hz	f_2 0.1 μF @ 400 Hz	f_3 0.1 μF @ 600 Hz	f_4 0.1 μF @ 800 Hz	f_5 0.1 μF @ 1,000 Hz
V_A (V)	3	3	3	3	3	3	3
V_R measured (V_{rms})	**2.59**	**3**	**3**	**2.98**	**2.99**	**3**	**3**
V_C measured (V_{rms})	**1.53**	**0.18**	**0.08**	**0.43**	**0.29**	**0.22**	**0.17**
I_T (ac) calculated (mA)	**0.096**	**0.112**	**0.112**	**0.116**	**0.11**	**0.11**	**0.11**
X_C calculated (kΩ)	**15.9**	**1.6**	**7.96**	**3.98**	**2.65**	**1.99**	**1.59**

Graph

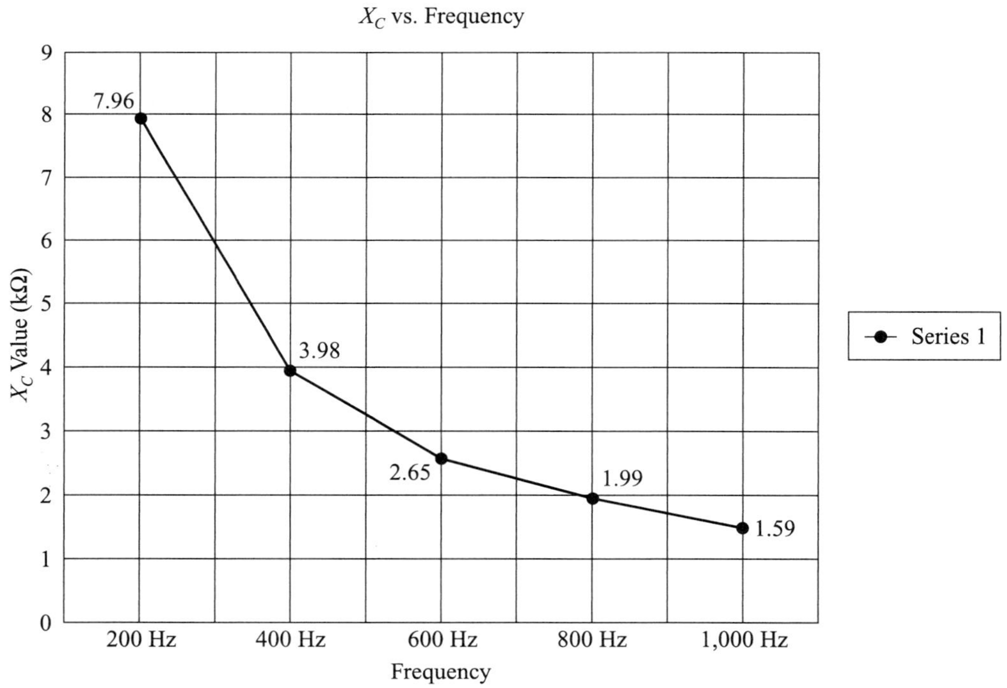

Lab Projects pages 263–265

Answers to Analysis Questions

1. Capacitance value and the value of capacitive reactance are inversely proportional. At 100 Hz the capacitive reactance of the 0.1-μF capacitor was about 15.9 kΩ and at this same frequency, the 1.0-μF capacitor had a capacitive reactance of about 1.6 kΩ, or roughly one-tenth that of the smaller capacitance. If components and measurements were perfect, it would show the 10 to 1 relationship precisely.

2. Capacitive reactance is inversely proportional to frequency. For example, at 200 Hz the capacitive reactance of the 0.1-μF capacitor was 7.96 kΩ; whereas, at 400 Hz the capacitive reactance of this same 0.1-μF capacitor was 3.98 kΩ.

3. It changes nonlinearly. The graph shows a curved line, not a straight line.

4. It has an inverse relationship to frequency. When frequency doubled from 200 to 400 Hz, the capacitive reactance approximately halved from 7.96 kΩ to about 3.98 kΩ.

5. $X_C = 1/2\pi fC$

6. $C = 1/2\pi fX_C$

Lab Projects pages 263–265

PART 14 Summary

Complete the following review questions, indicating the appropriate response by placing a check in the box next to the correct answer.

1. X_C is measured in ohms because it limits ac current to a value of
   - ☐ $0.159 \times f \times C$
   - ☐ $0.159/fC$
   - ☑ V/X_C
   - ☐ none of these

2. The formula for X_C is
   - ☐ $X_C = 2\pi/fC$
   - ☐ $X_C = fC/0.159$
   - ☑ $X_C = 1/2\pi fC$
   - ☐ none of these

3. For a given value of C, if f is increased, then X_C will
   - ☐ increase
   - ☐ remain the same
   - ☑ decrease

4. For a given value of f, if C is decreased, then X_C will
   - ☑ increase
   - ☐ remain the same
   - ☐ decrease

5. A capacitor appears to oppose a change in voltage because current must flow before a difference of potential can be established across a capacitor.
   - ☑ True
   - ☐ False

6. Since capacitors in series add like resistors in parallel, the X_C of two series capacitors will be
   - ☑ greater than either one alone
   - ☐ less than either one alone

7. Capacitive reactances in parallel add like resistances in
   - ☐ series
   - ☐ neither of these
   - ☑ parallel

8. If C and f are both doubled in a given circuit, the X_C will
   - ☐ increase two times
   - ☐ increase four times
   - ☐ decrease two times
   - ☑ decrease four times

9. If C is doubled and f is halved in a given circuit, the X_C will
   - ☐ increase
   - ☑ remain the same
   - ☐ decrease

10. What is the total capacitive reactance of two series 1-μF capacitors at a frequency of 200 Hz?
    - ☐ 3,180 Ω
    - ☑ 1.59 kΩ
    - ☐ 15.9 Ω
    - ☐ 31.8 kΩ
    - ☐ 318 Ω
    - ☐ none of these

Lab Projects pages 267–268

The instructor may wish to teach the students how to use the X_C formula to solve for an unknown C, when X_C and f are known, or how to solve for an unknown f, when C and X_C are known. ($C = 0.159/fX_C$; and $f = 0.159/CX_C$)
Key points to cover in summarizing are:

1. the inverse relationship of X_C to both C and f.

2. X_Cs in series add like Rs in series.

3. X_Cs in parallel add like parallel Rs.

It is important to emphasize that capacitance does the opposite of the X_C in this regard.

Lab Projects pages 267–268

PART 15

RC CIRCUITS IN AC

PROJECT 52 *V, I, R, Z,* and θ Relationships in a Series *RC* Circuit

SPECIAL NOTE:

By way of review, recall that in a purely resistive series ac circuit, the circuit current was in phase with the applied voltage. Also, the voltage drops across the individual resistors were in phase with current and with each other. The total opposition to current flow in the purely resistive circuit of this type was the arithmetic sum of the individual resistances. If a vector diagram of voltage and current were drawn, we would use the current as the reference vector, since it is the common factor in a series circuit. Summarizing: $Z = R_T$, $\theta = 0$ degrees, V_T = simple sum of $V_1 + V_2 +\ldots$ (etc.) for purely resistive series ac circuits. In this project, you will examine parameters for a series *RC* circuit.

PROCEDURE

1. Connect the initial circuit as shown in Figure 52-1.
2. Set the frequency of the function generator to 100 Hz and V_A to 3 volts. Measure V_A, V_R, and V_C.

OBSERVATION V_A = **3.0** V. V_C = **1.49** V.
V_R = **2.60** V.

CONCLUSION Does the sum of V_R and V_C equal V_A? **No**. This is because V_R and V_C are (*in phase, out of phase*) **out of phase**.

3. Calculate I_T from V_R/R; X_C from V_C/I, and Z from V_T/I_T.

OBSERVATION I_T = **0.096** mA. Z = **31.25 k** Ω.
X_C = **15.5 k** Ω.

CONCLUSION Does Z equal the arithmetic sum of R and X_C? **No**. We may conclude that since V_A is not equal to $V_R + V_C$ and Z is not equal to $R + X_C$, these values are the resultant of two out-of-phase vectors. We can solve for the resultant vectors by means of the Pythagorean theorem or trigonometry. Is the voltage across the resistor in phase with the current through it? **Yes**. Since a capacitor opposes a change in voltage, we might assume that the I_C (*leads, lags*) **leads** V_C. In a perfect capacitor, I_C (*leads, lags*) **leads** V_C by 90 degrees. We should also conclude that since this is a series circuit, the larger the amount of X_C (the smaller the

Lab Projects pages 271–274

C), the more like a purely (*resistive, capacitive*) ____**capacitive**____ circuit the circuit will act. This means the larger the X_C compared to the R, the (*greater, smaller*) ____**greater**____ will be the resultant phase angle between V_A and I_T.

4. Calculate Z using the Pythagorean approach.

OBSERVATION Z calculated = ____**31 k**____ Ω.

5. Draw an impedance diagram for the circuit in the Observation section.

OBSERVATION

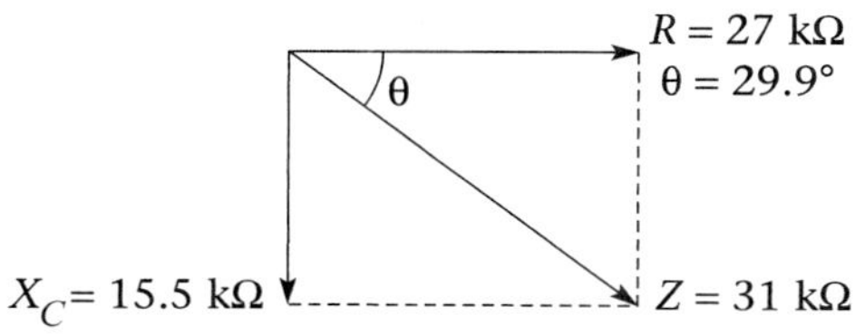

6. Use the Z diagram and trigonometry to determine the phase angle.

OBSERVATION θ = ____**30**____ degrees.

7. Repeat the previous steps 2 through 6. This time set f = 200 Hz and keep V_A at 3 volts.

OBSERVATION

V_A = **3.0** V.		X_C = **7.66 k** Ω.	
V_R = **2.88** V.		Z = **28 k** Ω.	
V_C = **.82** V.		θ = **15.8** degrees.	
I_T = **.107** mA.			

CONCLUSION Increasing frequency while keeping all other factors the same caused Z to ____**decrease**____, X_C to ____**decrease**____, θ to ____**decrease**____.

Optional Steps

8. Use the measured and calculated data in the previous steps 2 and 3, and draw a V-I vector diagram in the Observation section.

OBSERVATION

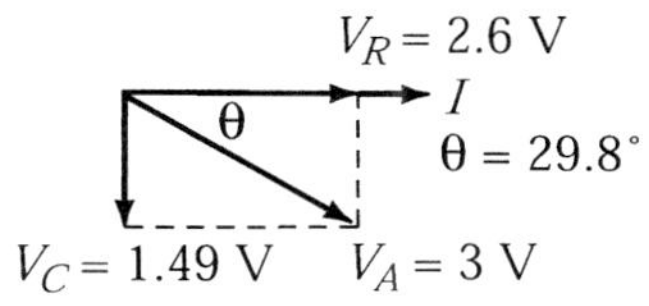

9. Use trigonometry to determine the phase angle.

OBSERVATION $\theta =$ **29.8** degrees.

CONCLUSION Does the phase angle from the *V-I* vector diagram agree reasonably with the angle you determined from the *Z* diagram in step 5? **Yes**.

10. Again, set the source for 100 Hz and a V_A of 3 volts. Use a dual-trace oscilloscope and perform a phase comparison of V_A and circuit current (represented by the voltage across the resistor).

STOP **CAUTION: Be sure the signal source ground and the scope ground(s) are connected to the same end (the bottom end in the diagram) of the resistor when making the measurements to prevent the grounds from shorting out a portion of the circuit!**

Determine the phase difference between the two signals.

OBSERVATION θ determined by the scope phase comparison:

$\theta \cong$ **30** degrees.

CONCLUSION Do the scope phase measurements and the phase angle calculations agree reasonably with your earlier findings? (Consider tolerances in components, source and scope frequency calibration tolerances, etc.) **Yes**.

Lab Projects pages 271–274

PROJECT 53 *V, I, R, Z,* and θ Relationships in a Parallel *RC* Circuit

SPECIAL NOTE:

By way of review, recall that in a purely resistive parallel ac circuit, the following conditions exist:

> $Z = R_T$; $\theta = 0$ degrees; I_T = arithmetic sum of branch currents; and branch voltage is in phase with branch current for a purely resistive circuit.

In this project, you will examine parameters in a parallel RC circuit. It should be noted again that for this project we will be using the 100-Ω resistor in series with the main line as a circuit current indicator. (Current through R_1 equals 10 times its voltage drop.)

PROCEDURE

1. Connect the initial circuit as shown in Figure 53-1.
2. Set the audio oscillator to a frequency of 500 Hz and V_A to 3 volts. Measure the circuit voltages.

OBSERVATION V_A = **3.0** V. V_{R_2} = **2.63** V.
V_{R_1} = **0.084** V. V_C = **2.63** V.

CONCLUSION Does the addition of the voltages around any closed loop equal V_A? **No**. We conclude from this that V_R and V_C are (*in phase, out of phase*) **out of phase**.

3. Calculate I_T from V_{R_1}. Calculate I_{R_2} by Ohm's Law. Calculate X_C by using the X_C formula, then calculate I_C using the formula:

$$I_C = V_C/X_C$$

Also, calculate θ using the arctan of I_C/I_R.

OBSERVATION I_T = **0.84** mA. I_C = **0.827** mA.
I_{R_2} = **.263** mA. θ = **72.3** degrees.
X_C = **3.18 k** Ω.

CONCLUSION Does the total current equal the arithmetic sum of the branch currents? **No**.

This is because the branch currents are **out of phase**.
Is the current through R_2 in phase with V_2? **Yes**. The current through the capacitor branch (*leads, lags*) **leads** the voltage across the capacitor by close to **90** degrees. Since I_T is the vector resultant of I_R and I_C, we would logically conclude that I_T (*leads, lags*) **leads** V_A by an angle that is between **0** and **90** degrees.

4. Calculate the circuit Z_T from V_T/I_T.

OBSERVATION $Z_T =$ **3.57 k** Ω.

CONCLUSION It is interesting to note that the impedance of a circuit consisting of two parallel 10-kΩ branches, whose currents are 90 degrees out of phase, would ≅ 7.07 kΩ. Does our demonstration circuit result compare closely to this? **No**. Why not? **X_C branch is not 10 kΩ in value**

5. Use I_{R_2} and I_C values and determine I_T by means of the Pythagorean theorem.

OBSERVATION I_T calculated = **0.868** mA.

6. Repeat the previous steps 2 through 5. This time set $f = 250$ Hz and keep V_A at 3 volts.

OBSERVATION

$V_A =$ **3.0** V. $I_{R_2} =$ **0.29** mA.
$V_{R_1} =$ **0.063** V. $X_C =$ **6.37 k** Ω.
$V_{R_2} =$ **2.95** V. $I_C =$ **0.46** mA.
$V_C =$ **2.95** V. $Z_T =$ **4.76 k** Ω.
$I_T =$ **0.63** mA. $\theta =$ **57.7** degrees.

CONCLUSION Decreasing frequency caused Z to **increase**, X_C to **increase**, and θ to **decrease**.

PART 15 Story Behind the Numbers

Data Table

SERIES *RC*, CIRCUIT 1				
Component and Parameter I.D.'s	**Measured Values @ 100 Hz**	**Calculated Values @ 100Hz**	**Measured Values @ 200 Hz**	**Calculated Values @ 200 Hz**
V_A (V)	3	—	3	—
V_R (V)	2.6	—	2.88	—
V_C (V)	1.49	—	0.82	—
I_T (mA)	—	0.096	—	0.107
X_C (kΩ)	—	15.5	—	7.66
Z, Ohm's Law (kΩ)	—	31.25	—	28
Z, Pythagorean (kΩ)	—	31	—	28
Phase angle (trig)	—	29.8°	15.9°	15.9°

PARALLEL *RC*, CIRCUIT 2				
Component and Parameter I.D.'s	**Measured Values 500 Hz**	**Calculated Values 500 Hz**	**Measured Values @ 250 Hz**	**Calculated Values @ 250 Hz**
V_A (V)	3	—	3	—
V_{R_1} (V)	0.084	—	0.063	—
V_{R_2} (V)	2.63	—	2.95	—
V_C (V)	2.63	—	2.95	—
I_T, Ohm's Law (mA)	—	8.4	—	0.63
I_2 (mA)	—	0.263	—	0.29
X_C (kΩ)	—	3.18	—	6.37
I_C (mA)	—	0.827	—	0.46
Z_T, Ohm's Law (kΩ)	—	3.57	—	4.76
Phase angle (trig)	—	72.3°	—	57.7°
I_T, Pythagorean (mA)	—	0.87		0.543

Lab Projects pages 279–282

Answers to Analysis Questions

1. No. Since voltages are out of phase, they must be added vectorially.
2. No, Z does not equal the arithmetic sum of R and X_C. Z is greater than R, but less than the sum of R and X_C. This value is less than the sum because it must be added vectorially.
3. Voltage across R is in phase with current through R. Voltage across C is approximately 90 degrees behind current, or about 90 degrees out of phase.
4. Phase differential is called the circuit's "phase angle." At 100 Hz the phase angle was about 30 degrees. At 200 Hz the phase angle was about 16 degrees. The difference in phase angle between 100-Hz and 200-Hz situations is due to the capacitive reactance decreasing in value as frequency is increased. Thus, the phase angle changed.
5. When frequency was increased from 100 Hz to 200 Hz, the circuit impedance decreased. It did not double or halve. Again, this was due to vectorial addition of resistive and reactive values.
6. Phase angle changed when frequency changed. The phase angle was less at a higher frequency because the circuit acted less capacitively with a lower value of capacitive reactance.
7. Yes, within reason.
8. Not exactly. Due to branch currents being out of phase, you must use vectorial math to find total current and related voltage drop(s).
9. No. Again, out-of-phase branch currents must be added vectorially to find total current.
10. No. Current through the capacitor leads voltage across the capacitor by about 90 degrees.
11. Circuit current leads the circuit applied voltage by an angle of between 0 and 90 degrees. The higher the capacitive reactance compared to the R value, the smaller the angle. If the circuit were purely resistive, the phase angle would be 0 degrees. If the circuit were acting purely capacitively, the phase angle would be 90 degrees.
12. The circuit phase angle was greater at 500 Hz due to higher capacitor branch current. At 500 Hz, capacitive reactance (opposition to ac current) is about half the value of what it is at 250 Hz.
13. In series *RC* circuits, the higher the capacitive reactance, the more capacitively the circuit acts as the capacitive reactance becomes a larger portion of the circuit impedance and therefore has a larger effect on circuit current, etc. Thus, in series *RC* circuits, the lower the frequency, the higher the capacitive reactance, and the greater the circuit phase angle. In parallel *RC* circuits, the lower the capacitive reactance, the more influence the capacitor has on the circuit because the capacitive branch has higher current through it. Thus, the higher the frequency, the lower the capacitive reactance, the greater the capacitive branch current, and the greater the phase angle.
14. No. You cannot use the resistive technique of the product-over-the-sum to find impedance. It turns out that circuit impedance for a simple parallel *RC* circuit, which contains equal R branch and X_C branch values, would have an impedance of 70.7% of one branch. It would not be 50%, as would be true in a purely resistive parallel R circuit situation. You must either find circuit total current through vectorial means and then use Ohm's Law (V_T/I_T), or use the special formula:

$$Z = RX_C \Big/ \sqrt{R^2 + X_C^2} \text{ to find } Z.$$

Lab Projects pages 279–282

15. For the series RC circuit: A decrease in frequency would cause Z to ___**I**___, X_C to ___**I**___, and θ to ___**I**___.

    For the parallel RC circuit: A decrease in frequency would cause Z to ___**I**___, X_C to ___**I**___, and θ to ___**D**___.

    For a series RC circuit: As X_C increases, circuit phase angle will ___**I**___.

    For a parallel RC circuit: As X_C increases, circuit phase angle will ___**D**___.

Lab Projects pages 279–282

PART 15 Summary

Complete the following review questions, indicating the appropriate response by placing a check in the box next to the correct answer.

1. The higher the frequency of V_A applied to any *RC* circuit, either series or parallel, the ________ the circuit impedance will be.
   - ☐ higher
   - ☑ lower

2. In a purely capacitive series circuit, the arithmetic sum of the individual voltage drops equals V_A.
   - ☑ True
   - ☐ False

3. In a series *RC* circuit, the arithmetic sum of the individual voltage drops equals V_A.
   - ☐ True
   - ☑ False

4. In a parallel *RC* circuit, the arithmetic sum of the branch currents equals I_T.
   - ☐ True
   - ☑ False

5. In a series *RC* circuit, if frequency, resistance, or capacitance increases, the circuit phase angle will
   - ☐ increase
   - ☑ decrease
   - ☐ remain the same

6. In a parallel *RC* circuit, if frequency, resistance, or capacitance increases, the circuit phase angle will
   - ☑ increase
   - ☐ decrease
   - ☐ remain the same

7. As the resistance in a parallel *RC* circuit is increased, the circuit will become more
   - ☐ resistive
   - ☑ capacitive

8. As the capacitance in a series *RC* circuit is increased, the circuit will become more
   - ☑ resistive
   - ☐ capacitive

9. A circuit that is capacitive is one in which the circuit voltage is lagging the circuit current, and I_C and V_C are 90 degrees out of phase.
   - ☑ True
   - ☐ False

10. The impedance of a parallel *RC* circuit will increase as frequency is decreased.
    - ☑ True
    - ☐ False

The instructor may wish to quiz the students by giving problems with specific values and having them draw *V-I* vector diagrams for the problems.

Lab Projects pages 285–286

In summarizing, the instructor should be sure to stress the following key points:

1. I leads V by 90 degrees for a capacitor.

2. For series RC circuits, the higher the X_C compared to the R, the more capacitive the circuit acts.

3. For parallel RC circuits, the higher the X_C, the less capacitive the circuit.

4. For either series or parallel RC circuits, the higher the X_C, the higher the circuit impedance (Z).

5. Total voltage in series RC circuits equals the vector resultant of V_C and V_R.

6. Total current in parallel RC circuits equals the vector resultant of branch currents I_C and I_R.

Finally, the students should clearly understand the inverse relationship of X_C to BOTH capacitance (C) and frequency (f).

Lab Projects pages 285–286

PART 16

SERIES RESONANCE

PROJECT 54 X_L and X_C Relationships to Frequency

PROCEDURE

1. Connect the initial circuit as shown in Figure 54-1.
2. Set the function generator to the frequency indicated for the circuit option you are using. Set V_A to 3 volts. Measure V_R and calculate I_T. Measure V_L and V_C and calculate X_L and X_C by Ohm's Law.

▲ OBSERVATION

	Lower f			Higher f	
$V_A =$	3	V.	$V_A =$	3	V.
$V_R =$	0.036	V.	$V_R =$	0.25	V.
$I_T =$	0.36	mA.	$I_T =$	2.5	mA.
$V_L =$	1.15	V.	$V_L =$	1.35	V.
$V_C =$	4.2	V.	$V_C =$	4.25	V.
$X_L =$	3.19 k	Ω.	$X_L =$	0.54 k	Ω.
$X_C =$	11.6 k	Ω.	$X_C =$	1.7 k	Ω.

▲ CONCLUSION If the X_C formula were used, the X_C of a 0.1-μF capacitor at 150 Hz would be calculated as **10.6 k** ohms for the lower f circuit. For the higher f circuit, the X_C at 1 kHz would be calculated as **1,592** ohms. Can the difference between the Ohm's Law results and the X_C formula be attributed to meter loading effects, capacitor tolerance, and audio oscillator dial calibration tolerance? **Yes**.

3. Change the function generator frequency to 300 Hz for the lower-frequency circuit or to 2 kHz for the higher-frequency circuit. Keep V_A at 3 volts. Measure V_R and calculate I_T. Measure V_L and V_C and calculate X_L and X_C by Ohm's Law.

▲ OBSERVATION

	Lower f			Higher f	
$V_A =$	3	V.	$V_A =$	3	V.
$V_R =$	0.12	V.	$V_R =$	0.95	V.
$I_T =$	1.2	mA.	$I_T =$	9.5	mA.
$V_L =$	8.8	V.	$V_L =$	9.5	V.

V_C = **6.4** V. V_C = **7.9** V.

X_L = **7.3 k** Ω. X_L = **1.0 k** Ω.

X_C = **5.3 k** Ω. X_C = **0.83 k** Ω.

▲ CONCLUSION When frequency was doubled, the X_L (*increased, decreased*) **increased** by approximately **2** times, and the X_C (*increased, decreased*) **decreased** by about **2** times. We conclude from these results that X_L is (*directly, inversely*) **directly** proportional to frequency, and X_C is (*directly, inversely*) **inversely** proportional to frequency. In a circuit that has both inductance and capacitance, there must be some frequency where X_L and X_C would be **equal**. In our circuit, would the frequency where this would occur have to be higher or lower than the present settings? **lower**.

The instructor may wish to further illustrate the X_L and X_C relationships to frequency by changing the frequency several times and going through the above-described procedure. Also, the instructor may wish to show the students the graphical representation of X_L and X_C versus frequency. The students must understand that if it were possible to have perfect reactances (no meter loading effect), and perfect frequency calibration, then the above results would have coincided even more closely with the theoretical values.

PROJECT 55 V, I, R, Z, and θ Relationships When $X_L = X_C$

PROCEDURE

1. Connect the initial circuit as shown in Figure 55-1.
2. Set V_A at 3 volts and adjust the function generator frequency while monitoring V_R until V_R is maximum. Measure V_R and calculate I_T. Measure V_L and V_C. Calculate X_L and X_C by Ohm's Law.

OBSERVATION

	Lower f		Higher f
V_A =	3 V.	V_A =	3 V.
Frequency =	246 Hz. (approximately)	Frequency =	1,810 Hz. (approximately)

	Lower f		Higher f
V_R =	0.187 V.	V_R =	1.0 V.
I_T =	1.87 mA.	I_T =	10 mA.
V_L =	12.3 V.	V_L =	9.8 V.
V_C =	12.4 V.	V_C =	9.8 V.
X_L =	6.5 k Ω.	X_L =	0.98 k Ω.
X_C =	6.6 k Ω.	X_C =	0.98 k Ω.

CONCLUSION Are X_L and X_C close to being equal at the frequency of maximum V_R? **Yes**. Would they be equal if circuit and measurement conditions were perfect? **Yes**. The voltage across the coil (*leads, lags*) **leads** the current by about **90** degrees; whereas, the voltage across the capacitor (*leads, lags*) **lags** the current by close to **90** degrees. This means that V_L and V_C are close to **180** degrees out of phase with each other. Since I is the same throughout the series circuit and the $I \times X_L$ drop is essentially **180** degrees out of phase with the $I \times X_C$ drop, we may conclude that X_L and X_C reactances are opposite (vectorially) and at resonance are equal. They therefore cancel each other's effects on Z. Notice that V_L and V_C both are greater than V_A. This is due to the cancelling effect of the reactances. If X_L and X_C perfectly cancelled (as would be true with perfectly equal and opposite reactances), the circuit Z would equal ***R***. The resultant circuit current would be (*in phase, out of phase*) **in phase** with V_A, and the circuit would be acting purely **resistive**.

3. Change the frequency to a new frequency that is well above resonance. Measure V_L and V_C and determine whether the circuit is now acting inductively or capacitively.

OBSERVATION V_L is now **greater** than V_C.

Lab Projects pages 291–293

CONCLUSION The circuit is now acting like an (*RL*, *RC*) **RL** circuit, since X **$_L$** is more than cancelling out X **$_C$**. Thus, the circuit is acting like R in series with a resultant X **$_L$**. The value of this resultant X is equal to a value of X **$_L$** total minus X **$_C$**. The result is that I_T (*leads, lags*) **lags** V_A by an angle between **0** and **90** degrees; thus the circuit is (*inductive, capacitive*) **inductive**.

4. Keep the same circuit as shown in Figure 55-1.
5. Change the input frequency to a new frequency that is well below the resonant frequency determined in the earlier steps. Measure V_L and V_C and determine whether the circuit is now acting inductively or capacitively.

OBSERVATION New frequency is lower than **f_r**.

V_L is now **less** than V_C.

CONCLUSION Since V **$_C$** is now greater than V **$_L$** and the current is the same through both reactances, it indicates that X **$_C$** is greater than X **$_L$**. This means that X **$_C$** is more than cancelling X **$_L$** and the circuit is equivalent to an (*RL*, *RC*) **RC** circuit. The circuit impedance therefore would equal the vector resultant of R and a series X **$_C$** whose value equals the difference between **X_C** and **X_L**. We may conclude from the preceding that at a frequency where $X_L = X_C$ (resonance), the circuit acts essentially like a purely **resistive** circuit. At frequencies below resonance, a series *RLC* circuit will act equivalent to a simple R **C** circuit. At frequencies above resonance, a *series RLC* circuit will act equivalent to a simple R **L** circuit. We may also conclude that the larger the value of C or L, the (*lower, higher*) **lower** will be the resonant frequency and vice versa. Summarizing our observations for a series resonant circuit, we conclude that at resonance Z is (*minimum, maximum*) **minimum** since X_L and X_C cancel.

In addition to the fact that Z is minimum for a series *RLC* circuit at resonance, the instructor may wish to emphasize that current is maximum, phase angle is zero, and V_C and V_L are greater than V_A. Further clarification of these facts may be illustrated by drawing *V-I* vector diagrams.

PROJECT 56 *Q* and Voltage in a Series Resonant Circuit

SPECIAL NOTE:

The figure of merit for a resonant circuit is called "Q." In general terms, the higher the ratio of reactance at resonance to series resistance, the higher the Q. If we had a perfect inductor and capacitor tuned to resonance and there were zero resistance in the circuit, the circuit impedance would be zero. Thus, current would be infinite. However, this is impossible in practical circuits. Therefore, Q is considered to be the relationship X_L/R, or X_C/R for series resonant circuits. Also, because the cancelling effect of X_L and X_C may allow high circuit current at resonance, there is a voltage magnification across the reactive components at resonance. During this project we will illustrate some of these facts.

PROCEDURE

1. Connect the initial circuit as shown in Figure 56-1.

2. Set V_A at 3 volts and adjust the frequency, while monitoring V_L for maximum V_L. Measure V_A and V_L and calculate the circuit Q from the ratio of V_L to V_A.

OBSERVATION

	Lower f	Higher f
$V_A =$	3 V.	3 V.
$V_L =$	13 V. (approximately)	15 V. (approximately)
$Q =$	4.3 (approximately)	5 (approximately)
$f =$	250 Hz	1,850 Hz

CONCLUSION The ratio of X_L/R for this circuit must be approximately 4–5.

3. With the same V_A and frequency as the previous step, measure V_C and calculate the circuit Q from V_C/V_A.

OBSERVATION

	Lower f	Higher f
$V_A =$	3 V.	3 V.
$V_C =$	12 V. (approximately)	14.6 V. (approximately)
$Q =$	4 (approximately)	4.9 (approximately)
$f =$	250 Hz	1,850 Hz

CONCLUSION The ratio of X_C/R for this circuit must be approximately **4–5**. What is the X_C of the capacitor according to the X_C formula? Approximately **lower *f* 6,359; higher *f* 860**. From the preceding, we may conclude that V_L or V_C equals ***Q*** times V_A; that Q equals the ratio of **X_L** or **X_C** to R; and the higher the Q, the (*higher, lower*) **higher** the circuit current will be at resonance.

Optional Steps

4. Assuming that the 0.1-μF capacitor is very close to its rated value, determine what the resonant frequency of the circuit would be if the initial circuit inductor were acting at its rated value.

OBSERVATION If the inductor really were acting as rated, the resonant frequency would be approximately

411 Hz. (Lower f)　　**1,592** Hz. (Higher f)

5. Assuming a C of 0.1 μF and based on the measured resonant frequency in step 2, how much apparent inductance does this inductor have under the operating conditions used?

OBSERVATION "Apparent" inductance is approximately

4 H. (Lower f)　　**74** mH. (Higher f)

CONCLUSION It may be concluded that for a given capacitance value, the higher the inductance value used in conjunction with the capacitor, the (*higher, lower*) **lower** the resonant frequency will be.

6. Perform appropriate calculations and compare the Q calculated from data in step 2 of Project 55 to the Q determined in step 2 of this project (Project 56).

OBSERVATION Q calculated from step 2 of Project 55 =

4. (Lower f)　　**3.26**. (Higher f)

Q calculated from step 2 of this project =

4.3. (Lower f)　　**5**. (Higher f)

CONCLUSION Since the same values of L, C, and V_A are used in both projects, what accounts for the difference in Q values? **No 1-kΩ resistance present in the higher *Q* project.**

The instructor also may wish to demonstrate what happens to the circuit voltage, current, and Q if a resistor is added to the circuit in series with L and C. The instructor may also wish to relate the general rule that the higher the L/C ratio, the higher the Q will generally be. Also, for an *rf* coil (not an iron-core inductor, as in this project) maximum Q generally is obtained when the X_L of the *rf* coil is approximately 1,000 ohms at resonance.

Lab Projects pages 295–297

PROJECT 57 Bandwidth Related to *Q*

SPECIAL NOTE:

Bandwidth for a series resonant circuit may be defined as the difference in frequency between the two frequencies (one below resonance and one above) at which the circuit current is 0.707 (70.7%) of the maximum current (which occurs at resonance). It is interesting to note that the higher the Q of a circuit, the higher the maximum current will be at resonance for any given V_A. If a graph is made of current versus frequency, it will be shown that the higher the Q (and thus the I), the steeper will be the slope of the resonance curve and the smaller the bandwidth. Since Q and bandwidth are related, one formula for bandwidth is: Bandwidth $= f_r$ (resonant frequency)$/Q$ and therefore $Q = f_r/\text{bandwidth}$.

PROCEDURE

1. Connect the initial circuit as shown in Figure 57-1.
2. Set V_A to 3 volts. While monitoring V_R, set the frequency of the audio oscillator for resonance (maximum V_R). Measure V_R, V_L, and V_C. Calculate I at resonance.

OBSERVATION

	Lower f			Higher f	
$V_A =$	3	V.	$V_A =$	3	V.
$V_R =$	0.18 (approximately)	V.	$V_R =$	1.13 (approximately)	V.
$V_L =$	12.6 (approximately)	V.	$V_L =$	10.3 (approximately)	V.
$V_C =$	11.7 (approximately)	V.	$V_C =$	10.1 (approximately)	V.
$I =$	1.8 (approximately)	mA.	$I =$	11 (approximately)	mA.

CONCLUSION The approximate Q of this circuit (V_L/V_A): Lower f = **4.2**; Higher f = **3.4**.

According to the formula, Bandwidth $= f/Q$, the bandwidth of this circuit should be approximately: Lower f circuit = **59** Hz; Higher f circuit = **529** Hz.

If circuit current decreased to 70.7% of I_{max}, the current would be approximately: Lower f circuit = **1.27** mA; Higher f circuit = **7.7** mA. Then, V_R would equal: Lower f circuit = **0.127** V; Higher f circuit = **0.77** V.

3. Keeping V_A at 3 volts at each frequency setting, adjust frequency to a frequency below resonance where V_R equals 0.707 of $V_{R_{max}}$. Note this frequency, then change the frequency above resonance until V_R equals 0.707 of $V_{R_{max}}$ and note this frequency. From these two frequencies determine the measured bandwidth.

OBSERVATION

	(Lower f)		(Higher f)	
Bandwidth =	**85**	Hz. Bandwidth =	**800**	Hz.

f below resonance where V_R is at 70%:

(Lower f)		(Higher f)	
219	Hz.	**1,500**	Hz.

f above resonance where V_R is at 70%:

(Lower f)		(Higher f)	
304	Hz.	**2,300**	Hz.

CONCLUSION Does the measured bandwidth approximate the value calculated from the bandwidth formula in step 2? **not very close**.

4. If time permits, change R to a 1-kΩ resistor and repeat the steps above. Note whether the Q and bandwidth increased or decreased with the higher R value.

OBSERVATION

	(Lower f)		(Higher f)
$Q =$	**decreased**.	$Q =$	**decreased**.
Bandwidth =	**increased**	Bandwidth =	**increased**

CONCLUSION The higher the Q, the **narrower** the bandwidth. The higher the circuit R, the **broader** the bandwidth.

5. Make graphic plots of V_R versus f that illustrate bandpass and bandwidth characteristics. Use the data collected in steps 2 through 4 and make your plots on coordinates similar to those shown in Figure 57-2. Use separate graph paper for these plots.

PART 16 Story Behind the Numbers

Data Table

CIRCUIT 1: STEPS1–8					
Component Parameter I.D.'s	**Measured Values @ 1,000 Hz**	**Calculated Values @ 1,000Hz**	**Measured Values @ 2,000 Hz**	**Calculated Values @ 2,000Hz**	
V_A (V)	3	—	3	—	
V_R (V)	0.24	—	0.7	—	
Calculated I_T (mA)	—	2.4	—	7	
Calculated Z (kΩ)	—	1.25	—	0.43	
V_L (V)	1.47	—	8.32	—	
V_C (V)	4.38	—	6.2	—	
Calculated X_L (Ω)	—	612	—	1,190	
Calculated X_C (Ω)	—	1,825	—	890	
V_R (V) @ resonance	0.96	—	—	—	
I_T (mA) @ resonance	—	9.6	—	—	
Calculated Z (Ω) @ resonance	—	312	—	—	
V_L (V) @ resonance	9.7	—	—	—	
V_C (V) @ resonance	9.5	—	—	—	
X_L (Ω) @ resonance	—	1,010	—	—	
X_C (Ω) @ resonance	—	990	—	—	
V_L (V) @ above resonance	4.47	—	—	—	
V_C (V) @ above resonance	1.46	—	—	—	
V_L (V) @ below resonance	1.51	—	—	—	
V_C (V) @ below resonance	4.34	—	—	—	
CIRCUIT 1: FREQUENCY RUN STEPS 9–10					
Component Parameter I.D.'s	**1,400 Hz**	**1,600 Hz**	**1,800 Hz**	**2,000 Hz**	**2,200 Hz**
Value of I_T (V_R/R) = (mA)	5.6	8.8	9.3	7	5.1

Lab Projects pages 303–308

Data Table (*continued*)

Parameter I.D.'s	Observed Frequencies 100 Ω	Observed Frequencies 1 kΩ			
f @ 0.707 × I at resonance (below resonance)	1,480 Hz	970 Hz			
f @ 0.707 × I at resonance (above resonance)	2,040 Hz	3,100 Hz			
Approx. Bandwidth = ______ (Hz)	560 Hz	2,130 Hz			
Approx. Bandpass = from ______ to ______ Hz	1,480–2,040 Hz	970–3,100 Hz			
Circuit 2: *Q* Steps 12–14					
Component and Parameter I.D.'s	**Measured Values with V_L max**	**Calculated Values with V_L max**			
V_A (V)	3	—			
V_L (max) (V)	14.4	—			
Circuit Q (V_L / V_A)	—	4.8			
V_C (max) (V)	13.2	—			
Circuit Q (V_C/ V_A)	—	4.4			

Graph

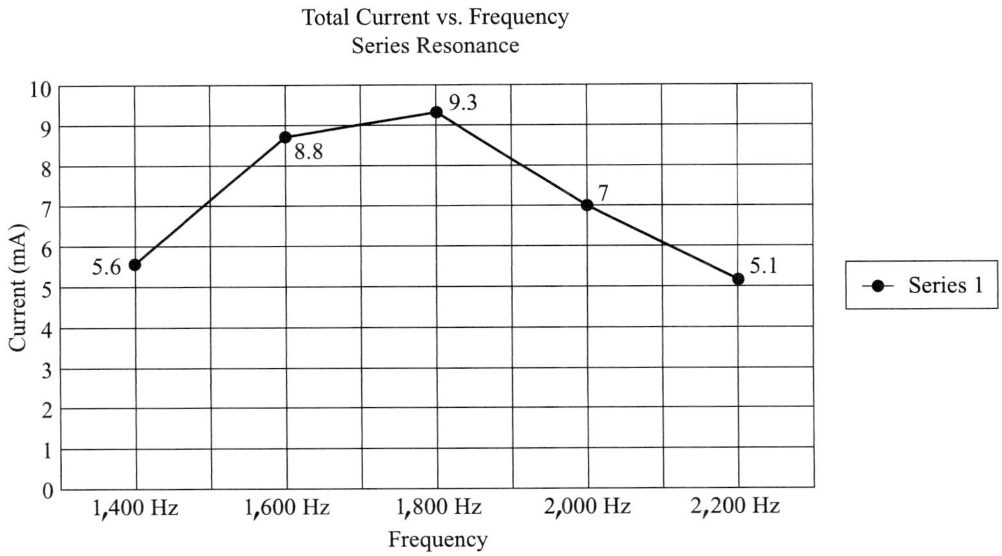

Answers to Analysis Questions

1. X_L doubled; X_C halved when frequency was doubled. X_L is directly proportional to frequency, and X_C is inversely proportional to frequency.

2. *R*. If all components were perfect, the circuit would act purely resistively. This is because, at the resonant frequency, X_L and X_C are equal and cancel one another as far as the source is concerned (assuming perfect components).

3. Above resonance, the circuit acts inductively. The circuit acts like an *RL* circuit, since X_L is greater than X_C and there is some "net" inductive reactance left in the circuit.

4. Below resonance, the circuit acts capacitively. The circuit is acting like a series *RC* circuit. Here, the capacitive reactance is greater than the inductive reactance, leaving a net capacitive reactance in series with the *R*.

5. The approximate phase relationship of the voltage across the inductor and the voltage across the capacitor is 180 degrees difference. Since V_L leads the current through it by 90 degrees and V_C lags the current by 90 degrees, the difference between the two voltages is about 180 degrees.

6. Yes. The circuit phase angle changes in both the series and the parallel circuit configurations. This is due to reactances changing when frequency is changed.

7. *Q* was higher for Circuit 2 since it did not have a discrete series resistance present, as did Circuit 1. At resonance, *Q* for Circuit 1 was about equal to $V_L/V_A = 4.4/3$ or about 1.46. *Q* for Circuit 2 = $V_L/V_A = 14.6/3 =$ about 4.8.

8. Yes. Differences might be caused by component resistance, inaccuracies in measurements, and so on.

9. Yes. The resonant frequency would be the same for both circuits when using the same reactive components and $X_C = 1/2\pi fC$ in both cases. When using Ohm's Law to find X_C in Circuit 1, X_C was calculated at about 990 ohms. For Circuit 2, if we calculate X_C using the X_C formula, the value calculates as: $1/(6.28 \times 1{,}750 \times 0.1\ \mu F)$, or about 910 ohms (assuming that f_r was about 1,750 Hz).

Lab Projects pages 303–308

PART 16 Summary

Complete the following review questions, indicating the appropriate response by placing a check in the box next to the correct answer.

1. Resonance is sometimes defined as the circuit condition when
   - ☐ $R = X_L$
   - ☐ $R = X_C$
   - ☐ $X_L = Z$
   - ☐ $X_C = Z$
   - ☑ $X_C = X_L$

2. Three circuit factors that determine whether an *RLC* circuit will be equivalent to an *RC*, an *RL*, or a resistive circuit are
   - ☐ *V*, *I*, and *Z*
   - ☐ *f*, *R*, and V_A
   - ☑ *f*, *L*, and *C*
   - ☐ none of these

3. If a series *RLC* circuit is at resonance and *f* is then increased, the circuit will begin to act
   - ☐ capacitively
   - ☑ inductively
   - ☐ resistively

4. What is the phase relationship between V_C and V_L in a series *RLC* circuit?
   - ☐ 0 degrees
   - ☐ 45 degrees
   - ☐ 90 degrees
   - ☑ 180 degrees
   - ☐ none of these

5. What is the phase relationship between I_C and I_L in a series *RLC* circuit?
   - ☑ 0 degrees
   - ☐ 45 degrees
   - ☐ 90 degrees
   - ☐ 180 degrees
   - ☐ none of these

6. As the resonant frequency is approached for any *RLC* circuit, the phase angle will
   - ☐ increase
   - ☑ decrease
   - ☐ remain the same

7. Is it possible for an *RLC* circuit to change from capacitive to inductive simply by changing the frequency of the applied voltage?
   - ☑ Yes
   - ☐ No

8. Impedance in a series *RLC* circuit at resonance is
   - ☐ maximum
   - ☑ minimum
   - ☐ neither of these

9. The voltage across either reactive component in a series resonant *RLC* circuit is equal to
   - ☐ 0 volts
   - ☐ $R \times C$
   - ☑ $Q \times V_T$
   - ☐ $V_T - V_R$
   - ☐ none of these

10. If the series resistance in a series resonant *RLC* circuit is decreased, the *Q* of the circuit will
    - ☑ increase
    - ☐ decrease
    - ☐ remain the same

Lab Projects pages 309–310

The instructor should be sure in summarizing that the students understand the following:

1. For a series resonant circuit, I is maximum, Z is minimum, phase angle is zero, voltage across the reactive components = $Q \times V_A$.
2. Bandwidth = f_r/Q.
3. The circuit acts "capacitive" below resonance and "inductive" above resonance.
4. A high Q means sharp tuning and narrow bandwidth, while a low Q means broad tuning and wide bandwidth.

Lab Projects pages 309–310

PART 17

PARALLEL RESONANCE

PROJECT 58 *V, I, R, Z*, and θ Relationships When $X_L = X_C$

SPECIAL NOTE:

Quickly reviewing, recall that X_L is directly proportional to frequency, and X_C is inversely proportional to frequency. For any given circuit containing L and C, there is a frequency where $X_L = X_C$, often referred to as the *resonant frequency*. Some interesting phenomena occur at resonance, and we will observe some of these for the parallel LC circuit. The Rs used as current indicators do have an effect on the operation of the circuit; however, not to the extent that we cannot observe some of the resonance phenomena.

PROCEDURE

1. Connect the initial circuit as shown in Figure 58-1.

2. Adjust V_A for 3 volts V_{A-B} *between points A and B.* Monitor V_{R_1} with a voltmeter and adjust the frequency for *minimum* V_{R_1}. Check that V_{A-B} is still 3 volts. Use the measured V_{R_1} and calculate I_T. Measure V_{R_2} and calculate I_C. Measure V_{R_3} and calculate I_L.

OBSERVATION

	Lower f			Higher f	
$V_{A-B} =$	3	V.	$V_{A-B} =$	3	V.
$V_{R_1} =$	0.125	V.	$V_{R_1} =$	1.35	V.
$I_T =$	0.125	mA.	$I_T =$	1.35	mA.
$V_{R_2} =$	0.05	V.	$V_{R_2} =$	0.33	V.
$I_C =$	0.5	mA.	$I_C =$	3.3	mA.
$V_{R_3} =$	0.05	V.	$V_{R_3} =$	0.33	V.
$I_L =$	0.5	mA.	$I_L =$	3.3	mA.
$f \cong$	280	Hz.	$f \cong$	≈ 1,750–1,800	Hz.

CONCLUSION Are I_C and I_L close to being equal at the frequency of minimum V_{R_1}? **Yes**.
Would they be precisely equal if we had a pure L and C and ideal measurement

Lab Projects pages 313–315

conditions? ___Yes___. Does I_T equal the sum of the branch currents? ___No___. This is because I_C and I_L are ___out of phase___. Since I_C leads V_C by 90 degrees, and I_L lags V_L by about 90 degrees, then I_C and I_L must be about ___180___ degrees out of phase with each other. Theoretically, if I_C and I_L were exactly equal and 180 degrees out of phase, the resultant I_T would be ___0___. This means that Z would be ___∞___. In practical circuits, however, Z is maximum at resonance, and its value depends on the Q of the circuit. $Z = Q \times X_L$ (or X_C).

3. Calculate the Z of the parallel resonant circuit using the resultant I_T of the branches as determined from V_{R_1} and the 3 volts V_{A-B} ($Z = V/I$).

OBSERVATION

Lower f	Higher f
$Z =$ ___24 k___ Ω. (approximately)	$Z =$ ___2.2 k___ Ω. (approximately)

CONCLUSION The X_C of the capacitor at the resonant frequency, according to the X_C formula, is about: Lower $f =$ ___5.7 k___ Ω; Higher $f =$ ___884___ Ω. If $X_L = X_C$, then X_L is about: Lower $f =$ ___5.7 k___ Ω; Higher $f =$ ___884___ Ω. Notice that Z is greater than either branch's opposition to current. This is in contrast to a purely resistive parallel circuit, where Z is less than the least R branch.

4. Keep the same circuit as shown in Figure 58-1.

5. Change the frequency to a new frequency that is well above f_r. Measure V_{R_2} and V_{R_3}. Calculate I_C and I_L and determine whether the circuit is now acting inductively or capacitively.

OBSERVATION I_C is now ___greater___ than I_L.

CONCLUSION Since (I_C, I_L) ___I_C___ is now greater than (I_C, I_L) ___I_L___, it will cancel it out as far as I_T is concerned. Therefore, the resultant I_T is (*leading, lagging*) ___leading___. The circuit at this frequency is acting equivalent to a simple R ___C___ circuit. This means a parallel LC circuit at a frequency that is above resonance will act (*inductively, capacitively*) ___capacitively___. This is in contrast to a series LC circuit, which above resonance acts ___inductively___.

6. Change the input frequency to a new frequency that is well below the resonant frequency of the circuit. Determine I_C and I_L and note whether the circuit is now acting inductively or capacitively.

OBSERVATION I_C is now ___lower___ than I_L.

CONCLUSION The circuit is now acting ___inductively___, since the predominant current is ___I_L___. Thus, we conclude that a parallel LC circuit acts ___inductively___ below resonance. This is in contrast to a series LC circuit, which acts ___capacitively___ below resonance. Summarizing our observations for a parallel resonant circuit (at resonance), we conclude that Z is (*minimum, maximum*) ___maximum___, since the branch currents cancel, causing I_T to be (*minimum, maximum*) ___minimum___ at resonance. Since the reactive currents cancel, then the resultant circuit current is (*resistive, capacitive, inductive*)

Lab Projects pages 313–315

resistive and (*in phase, out of phase*) **in phase** with V_A. This means the circuit phase angle is **0** degrees. In order for the branch currents to completely cancel, they must be **180** degrees out of phase with each other. At resonance, the parallel LC circuit acts (*resistively, capacitively, inductively*) **resistively**; above resonance acts (*resistively, capacitively, inductively*) **capacitively**; and below resonance acts (*resistively, capacitively, inductively*) **inductively**.

PROJECT 59 *Q* and Impedance in a Parallel Resonant Circuit

SPECIAL NOTE:

It should be noted that the Q of a parallel resonant circuit is often considered as the ratio of Z/X_L (or Z/X_C), and that the ratio of branch current to I_T is approximately this same number. For convenience, we will use the ratio of I_C to I_T for determining Q in this project.

PROCEDURE

1. Connect the initial circuit as shown in Figure 59-1.
2. Adjust V_A for 3 volts *between points A and B*. Monitor V_{R_1} and adjust the frequency for minimum V_{R_1} (resonance). Check that V_{A-B} is still 3 volts. Determine I_T and I_C from V_{R_1} and V_{R_2}, respectively.

OBSERVATION

	Lower f		Higher f
V_{A-B} =	**3** V.	V_{A-B} =	**3** V.
V_{R_1} =	**0.11** V.	V_{R_1} =	**1** V.
I_T =	**0.11** mA.	I_T =	**1** mA.
V_{R_2} =	**0.05** V.	V_{R_2} =	**0.32** V.
I_C =	**0.5** mA.	I_C =	**3.2** mA.

CONCLUSION I_C is approximately: Lower f circuit = **4.5**; Higher f circuit = **3.2** times greater than I_T. This indicates a Q of about: Lower f circuit = **4.5**; Higher *f circuit* = **3.2**.
The X_C of the capacitor at the resonant frequency, if calculated by the X_C formula, is about: Lower f = circuit **5.7 k** Ω; Higher f circuit = **0.88 k** Ω.
If the circuit Z is calculated from the formula $Z = Q \times X_C$, the value of Z is: Lower f circuit = **25 k** Ω; Higher f circuit = **2.8 k** Ω. If we calculate Z by Ohm's Law using the measured V_A and calculated I_T, the result is Z equals: Lower f circuit = **27 k** Ω; Higher f circuit = **3 k** Ω.
Are the two results reasonably comparable considering the measurement errors, and so on? **Yes**.

3. If time permits, connect a 10-kΩ resistor in parallel with points A and B and repeat the procedures of step 2 to determine the new Q and Z. Note whether Q and Z increased or decreased with the shunt 10-kΩ present in the circuit.

OBSERVATION

	Lower f		Higher f
Q =	**decreased**.	Q =	**decreased**.
Z =	**decreased** Ω.	Z =	**decreased** Ω.

CONCLUSION We may conclude that the lower the *R* in shunt with a parallel *LC* circuit, the (*higher, lower*) ___**lower**___ the *Q* and *Z* will be.

The instructor may wish to have the students change the shunt *R* to several different values to validate the aforementioned conclusions (as time permits). The instructor should point out that the parallel resistance across the tuned circuit not only determines *Q*, but also the series resistance internal to the branches is involved in determining *Q*. For example, the coil resistance has an effect on *Q*. The students should have grasped the concepts that the reactive branch currents equal *Q* times the total current, and *Z* is *Q* times as great as either reactance.

Lab Projects pages 317–318

PROJECT 60 Bandwidth Related to Q

SPECIAL NOTE:

Bandwidth for a parallel resonant circuit might be defined as the difference in frequency between two frequencies (one below resonance and one above) at which the circuit impedance is 70.7% of the maximum impedance, which occurs at resonance. When the circuit Z is 0.707 of maximum, the circuit current will be 1.414 times I minimum.

PROCEDURE

1. Connect the initial circuit as shown in Figure 60-1.
2. Adjust V_A for 3 volts *between points A and B*. Monitor V_{R_1} and adjust the frequency for minimum V_{R_1} (resonance). Check that V_{A-B} is still 3 volts. Determine I_T and I_C from V_{R_1} and V_{R_2}, respectively. Also determine Q from the formula: $Q = I_C/I_T$.

OBSERVATION

	Lower f		Higher f	
V_{A-B} =	3	V.	3	V.
V_{R_1} =	0.11	V.	1	V.
I_T =	0.11	mA.	1	mA.
V_{R_2} =	0.046	V.	0.32	V.
I_C =	0.46	mA.	3.2	mA.
Q =	4.1 (approximately)	.	3.2 (approximately)	.

CONCLUSION Since I_T is the resultant of the two reactive branch currents, we may determine the Z at resonance of the parallel resonant circuit by dividing the 3.0 volts V_{A-B} by I_T. Z for the parallel resonant circuit thus equals approximately **27 k** Ω for the lower f circuit and **3 k** Ω for the higher f circuit.

3. Calculate what the current will be at the two frequencies when $Z = 0.707 \times Z_{max}$ by multiplying the I_T at resonance by 1.4.

OBSERVATION I at 70.7% Z points: Lower f = **0.154** mA; Higher f = **1.4** mA.

CONCLUSION If the impedance is 70.7% of Z at resonance, then the current will be 1.4 times I_{min} because 1 divided by 0.707 = **1.414**.

4. Keeping V_{A-B} at 3 volts, find the two frequencies at which Z is 0.707 of Z_{max} by monitoring I_T (V_{R_1}). One frequency should be above the resonant frequency, one below. Determine the bandwidth from $f_{hi} - f_{lo}$.

OBSERVATION Bandwidth equals: Lower f = **62** Hz; Higher f = **630** Hz.

Lab Projects pages 319–320

CONCLUSION If bandwidth is determined from the bandwidth = f/Q formula, using the value of Q from step 2, the value would be approximately: Lower f = **68** Hz; Higher f = **562** Hz. Does the bandwidth in this step reasonably correlate to the f/Q formula results, all things considered? **roughly, yes**.

5. If time permits, connect a 27-kΩ resistor in shunt with the parallel LC circuit (between points A and B) and repeat the procedures of step 4 to determine bandwidth.

OBSERVATION The bandwidth with a shunt 27-kΩ is (*more, less*): Lower f = **more**; Higher f = **more** than without it.

CONCLUSION If the bandwidth is greater, the Q of the circuit must have (*increased, decreased*) **decreased** according to the Bandwidth = f_r/Q formula.

PART 17 Story Behind the Numbers

Data Table

CIRCUIT 1: STEPS 1–9						
Component and Parameter I.D.'s	**Measured Values**	**Calculated Values**				
$V_{A\text{-}B}$ (V) @ resonance	3	—				
V_{R_1} (V) @ resonance	1.26	—				
Calculated I_T (mA) @ resonance	—	1.26				
V_{R_2} (V) @ resonance	0.31	—				
Calculated I_C (mA) @ resonance	—	3.1				
V_{R_3} (V) @ resonance	0.3	—				
Calculated I_L (mA) @ resonance	—	3				
Calculated Z (kΩ) @ resonance	—	2.38				
$V_{A\text{-}B}$ (V) above resonance	3	—				
V_{R_2} (V) above resonance	0.5	—				
Calculated I_C (mA) above resonance	—	5				
V_{R_3} (V) above resonance	0.14	—				
Calculated I_L (mA) above resonance	—	1.4				
$V_{A\text{-}B}$ (V) below resonance	3	—				
V_{R_2} (V) below resonance	0.16	—				
Calculated I_C (mA) below resonance	—	1.6				
V_{R_3} (V) below resonance	0.45	—				
Calculated I_L (mA) below resonance	—	4.5				

Lab Projects pages 321–326

Data Table *(continued)*

Component and Parameter I.D.'s	1,200 Hz	1,400 Hz	1,600 Hz	1,800 Hz	2,000 Hz	2,200 Hz
Calculated Z (kΩ) at each frequency	1.22	1.7	2.29	2.42	2	1.55
CIRCUIT 2: STEPS 10–13						
Component and Parameter I.D.'s	**Measured Values**	**Calculated Values**	**Observed Frequency Values**			
$V_{A\text{-}B}$ (V) @ resonance	3	—	—			
Frequency @ resonance	—	—	1.74 kHz			
Calculated I_T (mA) @ resonance	—	0.94	—			
Calculated Z (kΩ) @ resonance	—	3.19	—			
Calculated 1.4 × I_T (mA) = ?	—	1.3	—			
Frequency above resonance for 1.4 × I_T	—	—	2.02 kHz			
Calculated Z (kΩ) @ this frequency	—	2.3	—			
Frequency below resonance for 1.4 × I_T	—	—	1.5 kHz			
Calculated Z (kΩ) @ this frequency	—	2.3	—			

Graph

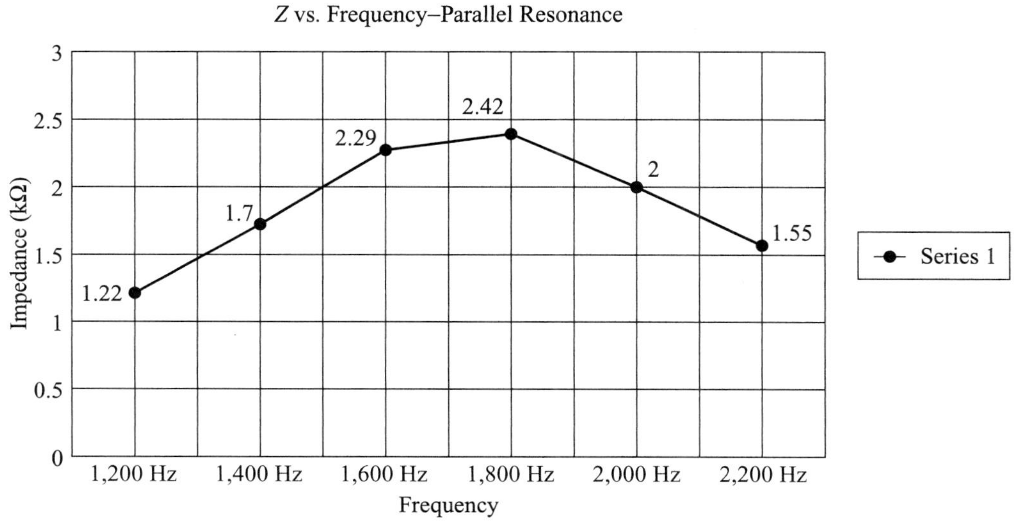

Lab Projects pages 321–326

Answers to Analysis Questions

1. Reactive branch currents are virtually canceling each other due to 180-degree phase difference between them.
2. They are approximately equal. Any differences might be caused by the fact that the inductor has some inherent R in its wire, thus the inductor branch is not as purely inductive as the capacitive branch is purely capacitive.
3. No, due to out-of-phase branch currents having to be added vectorially, not arithmetically.
4. Above resonance the circuit acts capacitively due to higher capacitive branch current. That is, above resonance X_C is lower than X_L.
5. Below resonance the circuit acts inductively due to higher inductive branch current. That is, below resonance X_L is lower than X_C.
6. With ideal components the circuit would act resistively at resonance. That is, phase angle would be 0 degrees, etc. With ideal components the X_L and X_C values would precisely cancel, leaving only resistance affecting the circuit parameters.
7. $X_C = 3\text{ V}/3.1\text{ mA} = 967\ \Omega$; $Q = 2.38\text{ k}\Omega/0.967\text{ k}\Omega = 2.46$.
8. Z indicated was maximum at resonance, and lower both above and below the resonant frequency.
9. Bandwidth was approximately 520 Hz. Bandpass was from about 1.5 kHz to 2.02 kHz. Values were determined by finding frequencies at which Z was about $0.707 \times Z$ maximum, which was determined by finding currents that were 1.4 times greater than current at resonance. *Note*: Resonant frequency was about 1,740 Hz.
10. Q would decrease. Bandwidth would increase.

Lab Projects pages 321–326

PART 17 Summary

Complete the following review questions, indicating the appropriate response by placing a check in the box next to the correct answer.

1. The impedance of a parallel LC circuit at resonance is
   - ☑ maximum
   - ☐ minimum
   - ☐ neither of these

2. The resultant total current of a parallel LC circuit at resonance is
   - ☐ maximum
   - ☑ minimum
   - ☐ neither of these

3. The current "through" either reactive branch of a parallel LC circuit at resonance is equal to
   - ☐ V_A/Z_T
   - ☐ $Q \times X_L$
   - ☑ $Q \times I_T$
   - ☐ f_r/Q
   - ☐ none of these

4. The impedance of a parallel LC circuit at resonance is equal to
   - ☐ V_A/I_C
   - ☑ $Q \times X_L$
   - ☐ X_C
   - ☐ X_L
   - ☐ $X_L + X_C$
   - ☐ none of these

5. At a frequency higher than the resonant frequency of a parallel LC circuit, the circuit acts somewhat
   - ☐ inductive
   - ☑ capacitive

6. At resonance, the phase angle between V_A and I_T for a parallel LC circuit is
   - ☑ 0 degrees
   - ☐ 45 degrees
   - ☐ 90 degrees
   - ☐ 180 degrees
   - ☐ none of these

7. If the Q of a parallel resonant circuit is increased, the bandwidth will
   - ☐ increase
   - ☑ decrease
   - ☐ remain the same

8. If the Q of a parallel resonant circuit is decreased, the Z will
   - ☐ increase
   - ☑ decrease
   - ☐ remain the same

9. At a frequency lower than the resonant frequency, a parallel LC circuit acts somewhat
   - ☑ inductive
   - ☐ capacitive

10. If the value of shunt resistance in parallel with a parallel LC circuit is changed, will the resonant frequency change?
    - ☐ Yes
    - ☑ No

Lab Projects pages 327–328

The instructor may give additional summary problems with specific values, demanding quantitative answers if more practice is desirable. The instructor may also mention some key parameters generally associated with parallel resonance: zero phase angle; $P.F. = 1$; maximum total impedance and minimum line current; $X_L = X_C$. Point out that all these events DO NOT occur *precisely* at the same frequency in practical LC circuits. The higher the circuit Q, the closer to the same frequency these events occur. The term *antiresonance* is sometimes used for the frequency at which circuit Z is maximum.

In summarizing, be sure the students understand the following key points:

1. At resonance I_T is minimum.
2. Z_T is maximum.
3. Phase angle = zero degrees.
4. Current through the reactive branches = $Q \times I_T$.
5. The parallel LC circuit acts capacitively above resonance and inductively below resonance.
6. $B.W. = f_r/Q$
7. High Q means sharp tuning and narrow bandwidth, while low Q means broad tuning and wide bandwidth.

Lab Projects pages 327–328

PART 18

THE SEMICONDUCTOR DIODE

PROJECT 61 Forward and Reverse Bias and *I* vs. *V*

SPECIAL NOTE:

For this project you will first test the diode using a DMM that has a diode testing function. Next, you will use the voltage drop across a 1-kΩ resistor in series with the diode as a means for monitoring the current. Since this is a 1-kΩ resistor, the amount of current in milliamperes is the same as the values you read for the voltage drop across the resistor. Example: If you read 22 V across the 1-kΩ resistor, the current is equal to 22 mA.

Ideally, you will use two meters for this experiment: one for monitoring the value of V_A and the second for measuring the voltages across the resistor and diode.

PROCEDURE

1. Set the DMM to the diode test position. Connect the diode to the DMM with the negative lead to the cathode and the positive lead to the anode. Record the DMM reading.

OBSERVATION Forward voltage for a good diode: $V_{forward}$ = **0.7** V.

CONCLUSION For a good diode, the forward DMM reading is relatively (*high, low*) **low**.

2. Switch the leads to the diode, so that the negative lead of the DMM is connected to the anode and the positive lead to the cathode. Record the DMM reading.

OBSERVATION Reverse voltage for a good diode: $V_{reverse}$ = **O L** V.

CONCLUSION For a good diode, the reverse DMM reading is relatively (*high, low*) **high**.

3. Connect the initial circuit as shown in Figure 61-2.

4. Apply 5 volts to the circuit with the anode connected to the positive side of the source. Determine the circuit current from V_R for the forward-biased diode.

OBSERVATION I = **4.3** mA. (approximately)

CONCLUSION Is the diode conducting with the anode positive with respect to the cathode? **Yes**. The voltage drop across the forward-biased diode is approximately **0.7** V.

5. Turn the diode around so that the positive side of the source will be connected to the cathode of the diode as shown in Figure 61-3.

6. Apply 5 volts to the circuit. Determine the circuit current for V_R for the reverse-biased diode.

OBSERVATION $I =$ **0** mA.

CONCLUSION Is the diode conducting with the cathode positive with respect to the anode? **No**. For the reverse-biased condition, the diode is acting like (*an open, a short*) **an open** circuit. The resistance of the reverse-biased diode is virtually **infinite**. Since $V_R = 0$ V, the voltage across the diode must equal **5 V**.

7. Turn the diode around once more in order to forward bias it, then increase V_A in small steps, as indicated in the Observation section, and note the current at each step.

OBSERVATION

For $V_A = 0.5$ V: $I =$ **0.06** mA.
For $V_A = 1$ V: $I =$ **0.3** mA.
$V_A = 2$ V: $I =$ **1.3** mA.
$V_A = 3$ V: $I =$ **2.3** mA.
$V_A = 4$ V: $I =$ **3.3** mA.
$V_A = 5$ V: $I =$ **4.3** mA.
$V_A = 6$ V: $I =$ **5.3** mA.
$V_A = 7$ V: $I =$ **6.3** mA.
$V_A = 8$ V: $I =$ **7.3** mA.
$V_A = 9$ V: $I =$ **8.3** mA.
$V_A = 10$ V: $I =$ **9.3** mA.

CONCLUSION As V_A was increased, the current (*increased, decreased*) **increased**. We can deduce, therefore, that diode conduction is (*directly, inversely*) **directly** proportional to V_A when forward biased. What is the approximate dc resistance of the diode when V_A is 7 volts? **111** Ω. We conclude from the preceding that a diode will conduct when the anode is (*positive, negative*) **positive** with respect to the cathode.

8. With the diode forward biased, measure the voltage drop across the diode for each increase in V_A as shown in the Observation section.

OBSERVATION

$V_A = 0.5$ V: $V_D =$ **0.44** V.
$V_A = 1$ V: $V_D =$ **0.7** V.
$V_A = 2$ V: $V_D =$ **0.7** V.
$V_A = 4$ V: $V_D =$ **0.7** V.
$V_A = 10$ V: $V_D =$ **0.7** V.

CONCLUSION As V_A reached a certain level of forward biasing, the voltage drop across the diode (*decreased, increased, stayed about the same*) **stayed about the same**. This implies that the resistance of the diode changed (*linearly, nonlinearly*) **linearly** as V_A was increased.

PROJECT 62 Diode Clipper Circuits

PROCEDURE

1. Connect the circuit exactly as shown in Figure 62-1.

2. Set the function generator (sine-wave mode) or audio oscillator to a frequency of 1 kHz and an amplitude (V_{in}) of 6 $V_{P\text{-}P}$. Draw the waveform for V_{in} (2 cycles), clearly indicating the positive and negative peak voltage values.

OBSERVATION Waveform:

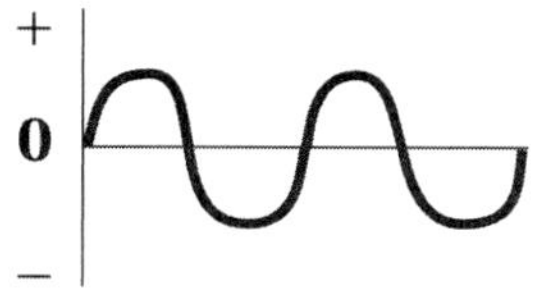

V_{in} positive peak voltage = **3** V.

V_{in} negative peak voltage = **3** V.

3. Draw the waveform observed at V_{out}. Again, clearly indicate the positive and negative peak voltage values.

OBSERVATION Waveform:

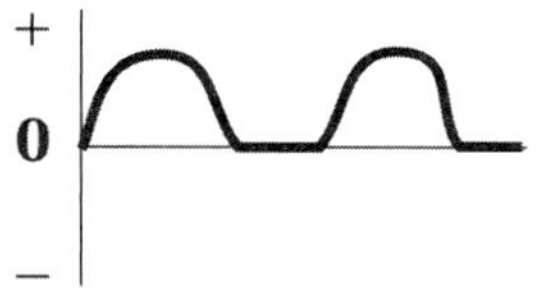

V_{out} positive peak voltage = **2.3** V.

V_{out} negative peak voltage = **0** V.

CONCLUSION This circuit is a (*series, parallel*) **series** diode clipper. The output waveform indicates this circuit is a (*positive, negative*) **negative** clipper. During the positive half-cycle of V_{in}, the diode is (*forward, reverse*) **forward** biased and current (*is, is not*) **is** flowing through the resistor.

4. Connect the circuit exactly as shown in Figure 62-2.

5. Set the function generator (sine-wave mode) or audio oscillator to a frequency of 1 kHz and an amplitude (V_{in}) of 6 $V_{P\text{-}P}$. Sketch the waveform for V_{in} (2 cycles), clearly indicating the positive and negative peak voltage values.

OBSERVATION Waveform:

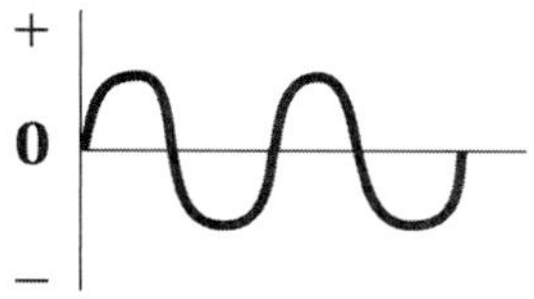

Lab Projects pages 335–339

V_{in} positive peak voltage = 3 V.

V_{in} negative peak voltage = 3 V.

6. Draw the waveform observed at V_{out}, and show the positive and negative peak voltage values.

OBSERVATION Waveform:

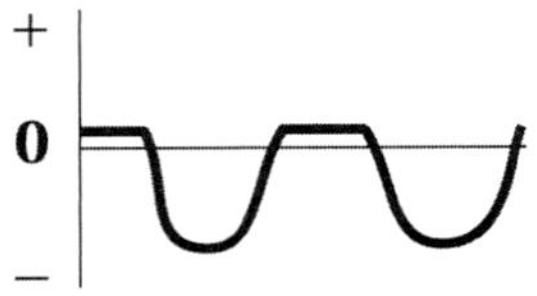

V_{out} positive peak voltage = 0 V.

V_{out} negative peak voltage = 2.3 V.

CONCLUSION This circuit is a (*series, parallel*) **series** diode clipper. The output waveform indicates this circuit is a (*positive, negative*) **positive** clipper. During the positive half-cycle of V_{in}, the diode is (*forward, reverse*) **reverse** biased and current (*is, is not*) **is not** flowing through the resistor.

7. Connect the circuit exactly as shown in Figure 62-3.

8. Set the input waveform (V_{in}) as a 6-$V_{P\text{-}P}$, 1-kHz sine wave. Sketch the waveform (2 cycles), clearly indicating the positive and negative peak voltage values.

OBSERVATION Waveform:

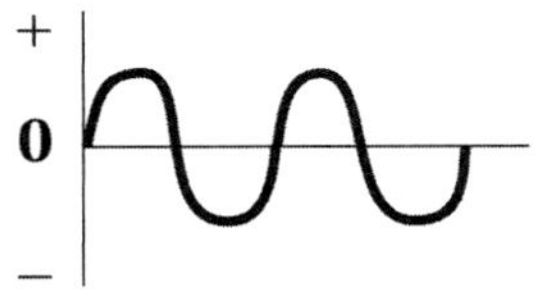

V_{out} positive peak voltage = 3 V.

V_{out} negative peak voltage = 3 V.

9. Sketch the waveform observed at V_{out}, clearly indicating the positive and negative peak voltage values.

OBSERVATION Waveform:

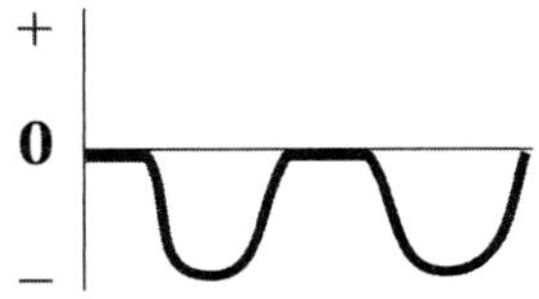

V_{out} positive peak voltage = 0.7 V.

V_{out} negative peak voltage = 3 V.

CONCLUSION This circuit is a (*series, parallel*) **parallel** diode clipper. The output waveform indicates this circuit is a (*positive, negative*) **positive** clipper. During the positive half-cycle of V_{in}, the diode is (*forward, reverse*) **forward** biased and current (*is, is not*) **is** flowing through the resistor.

10. Connect the circuit exactly as shown in Figure 62-4.

11. Set the input waveform (V_{in}) as a 6-$V_{P\text{-}P}$, 1-kHz sine wave. Sketch the waveform (2 cycles), clearly indicating the positive and negative peak voltage values.

OBSERVATION Waveform:

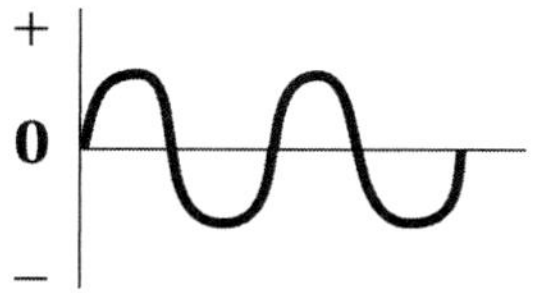

V_{out} positive peak voltage = ___3___ V.

V_{out} negative peak voltage = ___3___ V.

12. Sketch the waveform observed at V_{out}, clearly indicating the positive and negative peak voltage values.

OBSERVATION Waveform:

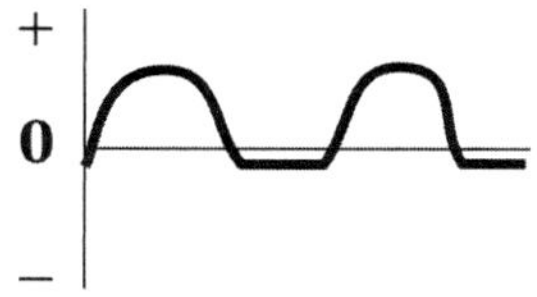

V_{out} positive peak voltage = ___3___ V.

V_{out} negative peak voltage = ___0.7___ V.

CONCLUSION This circuit is a (*series, parallel*) ___**parallel**___ diode clipper. The output waveform indicates this circuit is a (*positive, negative*) ___**negative**___ clipper. During the positive half-cycle of V_{in}, the diode is (*forward, reverse*) ___**reverse**___ biased and current (*is, is not*) ___**is not**___ flowing through the resistor.

PART 18 Summary

Complete the following review questions, indicating the appropriate response by placing a check in the box next to the correct answer.

1. In analyzing the symbol for the semiconductor diode,
   - ☐ the arrow is the cathode and the "flat bar" is the anode
   - ☑ the arrow is the anode and the "flat bar" is the cathode
   - ☐ neither of these

2. When a diode is forward biased,
   - ☐ the anode is negative with respect to the cathode
   - ☐ the cathode is positive with respect to the anode
   - ☑ the anode is positive with respect to the cathode
   - ☐ none of these

3. A diode in which the cathode is more negative than the anode is
   - ☑ forward biased
   - ☐ reverse biased
   - ☐ neither of these

4. As the forward bias on a diode is increased, the current will
   - ☑ increase
   - ☐ decrease
   - ☐ remain the same

5. A reverse-biased rectifier diode acts like
   - ☐ a short
   - ☑ an open
   - ☐ a 10-kΩ resistor

6. The voltage drop across a forward-conducting silicon diode is approximately
   - ☐ 0.25 V
   - ☐ 0.5 V
   - ☑ 0.7 V
   - ☐ 1.0 V
   - ☐ 2.0 V

7. When testing a good rectifier diode with a DMM in diode test mode and with the positive lead connected to the anode,
   - ☐ the reading is 0
   - ☑ the reading is 0.7
   - ☐ the reading is 0 L
   - ☐ none of these

8. When testing a shorted rectifier diode with a DMM in diode test mode and with the positive lead to the anode,
   - ☑ the reading is 0
   - ☐ the reading is 0.7
   - ☐ the reading is 0 L
   - ☐ none of these

9. In a series diode clipper circuit, the voltage output is taken across the
   - ☐ power source
   - ☑ resistor
   - ☐ diode

Lab Projects pages 341–342

10. Reversing the polarity of the diode in a parallel diode clipper circuit
    - [x] reverses the clipping polarity
    - [ ] changes it to a series clipper circuit
    - [ ] shorts out the power source
    - [ ] has no effect on the circuit

Lab Projects pages 341–342

PART 19

SPECIAL-PURPOSE DIODES

PROJECT 63 Zener Diodes

SPECIAL NOTE:

You can closely approximate the amount of current flowing through the circuit (in mA) by measuring the voltage across the 100-Ω resistor and multiplying the value by 10. Example: If you read 2.2 V across the resistor, the current through the circuit will be close to 22 mA. Ideally, you will use two meters for this experiment: one for monitoring V_A, and the second for taking measurements across the resistor (V_R) and across the diode (V_Z).

PROCEDURE

1. Connect the circuit shown in Figure 63-1.

CONCLUSION The zener diode in this circuit is properly (*forward, reverse*) **reverse** biased. What is the rated zener voltage (V_{ZT}) for this zener diode? **5.1** V.

2. Set V_A to 3 volts. Measure the voltage across the zener diode (V_Z) and the voltage across the resistor (V_R), and determine the current through the circuit (I_R). Increase V_A in small steps, as indicated in the Observation section, and determine the values of V_Z, V_R, and I_R for each step.

OBSERVATION

For V_A = 3 V: V_Z = **3** V, V_R = **0** V, I_R = **0** mA.

For V_A = 4 V: V_Z = **4** V, V_R = **0** V, I_R = **0** mA.

For V_A = 5 V: V_Z = **5** V, V_R = **0** V, I_R = **0** mA.

For V_A = 6 V: V_Z = **5.1** V, V_R = **0.9** V, I_R = **9** mA.

For V_A = 7 V: V_Z = **5.1** V, V_R = **1.9** V, I_R = **19** mA.

CONCLUSION As long as V_A remains below V_{ZT}, the voltage across the zener diode (*equals V_A, remains close to the diode's V_{ZT} value, remains very close to zero volts*) **equals V_A** ________. While V_A is above V_{ZT}, the voltage across the zener diode (*equals V_A, remains close to the diode's V_{ZT} value, remains very close to zero volts*) **remains close to the diode's V_Z value**.

While V_A remains below V_{ZT}, the current through the zener diode (*changes with V_A, remains at a regulated level, remains very close to zero*) **remains very close to zero**

____________. Once V_A rises above V_{ZT}, the current through the zener diode (*changes with V_A, remains at a regulated level, remains very close to zero*) **changes with V_A**.

3. Return V_A to 0 volts. Determine the exact zener breakdown voltage (V_{ZT}) by gradually increasing the value of V_A while observing the voltage across the zener diode. Note the values where further increasing V_A no longer causes a significant increase in the zener voltage.

OBSERVATION V_{ZT} = **5.1** V.

CONCLUSION The voltage across a zener diode remains at the fixed V_{ZT} level as long as V_A is (*below, above*) **above** V_{ZT}.

4. Replace the VVPS for V_A with the function generator or audio generator (Figure 63-2). Set the function generator (sine-wave mode) to 1 kHz at 9 V_{PK}.

5. Sketch the waveform you find at V_A, indicating the positive and negative peak values.

OBSERVATION Waveform:

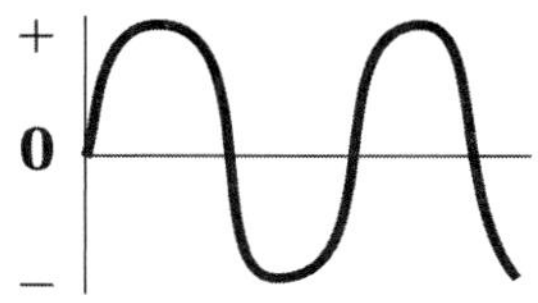

Positive peak V_A = **9** V.

Negative peak V_A = **9** V.

6. Sketch the waveform you find across the zener diode. Indicate the peak positive and peak negative voltages.

OBSERVATION Waveform:

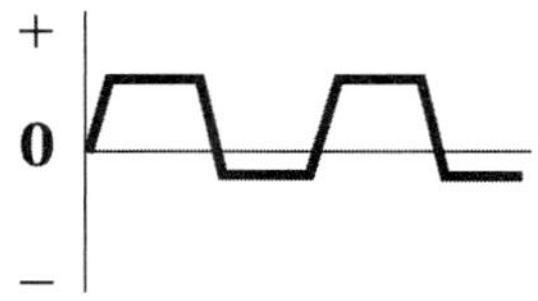

Positive peak V_Z = **5.1** V.

Negative peak V_Z = **0.7** V.

CONCLUSION Explain the shape of the waveform that occurs across the zener diode during the positive half-cycle of the input waveform. Account for the shape of the waveform that occurs across the zener diode during the negative half-cycle of the input waveform. Describe how this circuit could be considered a clipping circuit.

NOTE ➤ Use space below for explanations.

Voltage on the positive half-cycle rises to 5.1 V (V_Z), where zener conduction begins. On the negative half-cycle, the zener acts as a normal forward-biased diode that clips the voltage at –0.7 V.

PROJECT 64 Light-Emitting Diodes

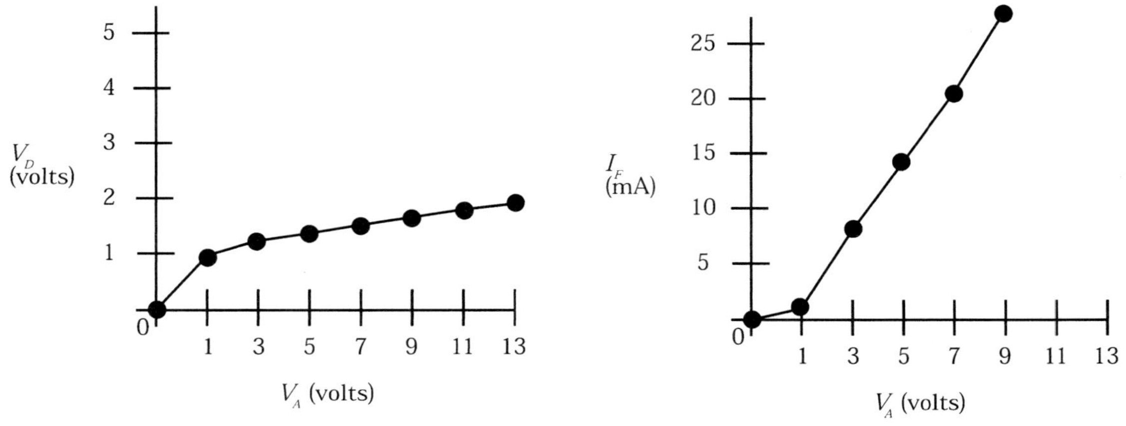

FIGURE 64-2

SPECIAL NOTE:

You will need the following formula to calculate the amount of forward current through the LED.

$$I_D = (V_A - V_D)/R$$

where:

- I_D is the forward current through the diode
- V_A is the voltage applied to the circuit
- V_D is the forward voltage drop across the LED
- R is the value of the resistor connected in series with the LED

PROCEDURE

1. Connect the circuit as shown in Figure 64-1.
2. Set V_A to 1 Vdc. Measure the voltage across the LED, calculate the amount of current flowing through the circuit, and note the intensity of the light (none, dim, moderately bright, bright, very bright). Increase V_A in steps indicated in the Observation section. Determine V_D and I_D, and note the brightness level in each case.

OBSERVATION

V_A = 1 V; V_D = **1** V, I_D = **0** mA, brightness = **none**.
V_A = 3 V; V_D = **1.1** V, I_D = **7** mA, brightness = **moderately bright**.
V_A = 5 V; V_D = **1.2** V, I_D = **14** mA, brightness = **bright**.
V_A = 7 V; V_D = **1.2** V, I_D = **21** mA, brightness = **bright**.
V_A = 9 V; V_D = **1.3** V, I_D = **28** mA, brightness = **very bright**.
V_A = 11 V; V_D = **1.4** V, I_D = **36** mA, brightness = **very bright**.
V_A = 13 V; V_D = **1.5** V, I_D = **43** mA, brightness = **very bright**.

CONCLUSION Complete the graphs in Figure 64-2 which show how V_D increases with V_A, and how I_D increases with V_A. The brightness of the LED seems to correlate better with the amount of (V_D, I_D) **I_D** than with the amount of (V_D, I_D) **V_D**. Based on the data you observed, the forward junction potential for this LED is about **1.2** V.

3. Set V_A to 3 Vdc, replace R with a 47-Ω resistor, and complete the blanks in the Observation section. Repeat the steps for the values of V_A and R indicated.

OBSERVATION
$V_A = 3$ V; $R = 47\ \Omega$, $I_D =$ **38** mA, brightness = **very bright**.
$V_A = 6$ V; $R = 220\ \Omega$, $I_D =$ **22** mA, brightness = **bright**.
$V_A = 9$ V; $R = 330\ \Omega$, $I_D =$ **24** mA, brightness = **bright**.

CONCLUSION It is possible to use a given LED with just about any amount of applied dc as long as you select the correct value of series resistor for approximately 20 mA of forward current. (*True, False*) **True**.

PART 19 Summary

Complete the following review questions, indicating the appropriate response by placing a check in the box next to the correct answer.

1. A zener diode is normally connected into a dc circuit in a direction that causes it to be
   - ☑ reverse biased
   - ☐ forward biased

2. When a zener diode is properly connected into a variable-voltage dc circuit, the zener conducts whenever the dc source
   - ☐ is greater than zero volts
   - ☐ exceeds the zener's forward junction potential
   - ☑ exceeds the zener's reverse breakdown voltage

3. A forward-biased zener diode behaves much the same as a forward-biased silicon rectifier diode.
   - ☑ True
   - ☐ False

4. When a zener diode rated at 12 V is operating from an 18-V source and the series resistor has a value of 100 Ω, the current through the zener diode is approximately
   - ☐ 12 mA
   - ☑ 60 mA
   - ☐ 18 mA
   - ☐ 120 mA

5. Used as a voltage clipper, a zener diode clips one polarity at the forward junction potential and the opposite polarity at the reverse breakdown potential.
   - ☑ True
   - ☐ False

6. The schematic symbols for an LED and a photodiode
   - ☑ are the same except the arrows point in different directions
   - ☐ are exactly the same
   - ☐ bear no resemblance to one another

7. An LED is connected into a dc circuit in a direction that causes it to
   - ☐ be reverse biased
   - ☐ gather light energy
   - ☑ be forward biased
   - ☐ break down and conduct backward

8. The forward junction potential for an LED tends to be ________ that of a rectifier diode.
   - ☐ less than
   - ☑ greater than
   - ☐ about the same as

9. The reverse breakdown voltage for an LED tends to be ________ that of a rectifier diode.
   - ☑ less than
   - ☐ greater than
   - ☐ about the same as

Lab Projects pages 351–352

10. Suppose you want to operate an LED from a 12-Vdc source, and the known specifications for the LED are $V_D = 1.7$ V when $I_D = 15$ mA. Which of the following practical resistors is closest to the correct value for the series resistor you should use?

- ☐ 270 Ω
- ☐ 330 Ω
- ☐ 470 Ω
- ☑ 680 Ω
- ☐ 1 kΩ

Lab Projects pages 351–352

PART 20

POWER SUPPLIES

PROJECT 65 Half-Wave Rectifier

SPECIAL NOTE:

The following formulas will be helpful for drawing proper conclusions about the results of your work:

Input $V_{rms} = 0.707 \times V_{pk}$ Effective ac

Output $V_{dc} = 0.318 \times V_{pk}$ Average dc out

PROCEDURE

1. Connect the initial circuit as shown in Figure 65-1.
2. Set the function generator (sine-wave mode) or audio oscillator to a frequency of 100 Hz and V_A to 3 V_{rms}. Sketch two complete cycles of the input waveform in the Observation section.

OBSERVATION Waveform:

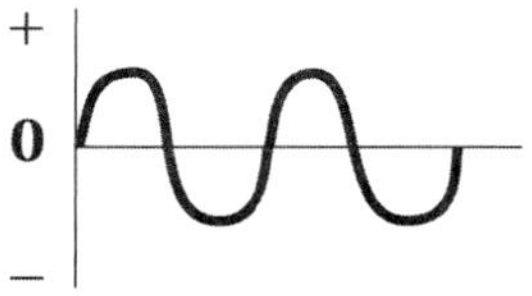

V_p input = ____4.2____ V.

3. Measure the dc voltage output across R_L (dc_{out}).

OBSERVATION dc_{out} = ____1.13____ V.

CONCLUSION The dc output voltage should be approximately what percentage of the rms input voltage? Approximately ____45____ %. Since the average voltage value over one ac alternation is $0.637 \times V_p$ and the effective value is $0.707 \times V_{pk}$, then V_{dc} must be about ____9____ tenths of V_{rms}, because 0.637 is about ____9____ tenths of 0.707. Since the diode can conduct only half the time with ac applied, the average dc_{out} for the half-cycle the diode *does not* conduct is ____0____ V.

Lab Projects pages 355–357

4. Connect the oscilloscope across R_L. Sketch the waveform for two complete cycles in the Observation section. Indicate the maximum and minimum voltage levels.

OBSERVATION Waveform:

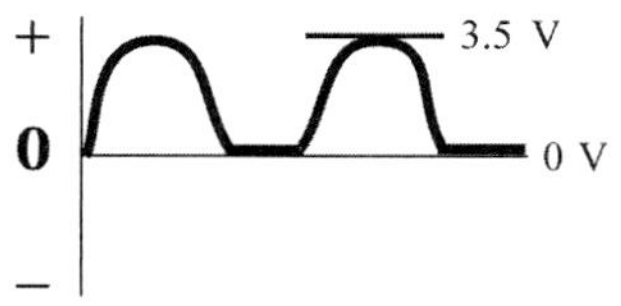

CONCLUSION For the half-cycle the diode does conduct, the peak V_{out} should be about **1.4** times V_{rms} (neglecting the small diode voltage drop). The end result is that the average dc_{out} over the *entire cycle* of ac input is about **0.45** times V_{rms} (again, neglecting the diode voltage drop).

5. Connect the 0.1-µF filter capacitor in parallel R_L as shown in Figure 65-2.

6. Set the function generator (sine-wave mode) to a frequency of 100 Hz and an amplitude of 3 V_{rms}. Sketch two complete cycles of the input waveform. Measure and record the ac input voltage (ac_{in}).

OBSERVATION ac_{in} = **3** V.

Waveform:

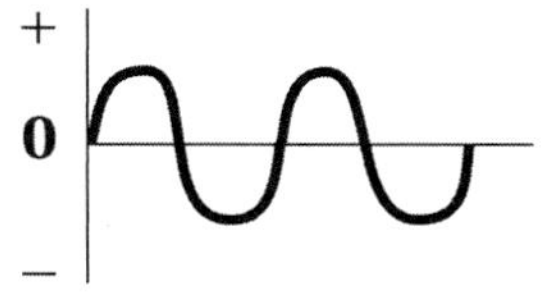

7. Measure and record the dc voltage output across R_L (dc_{out}). Sketch two complete ac cycles of the oscilloscope waveform found across R_L.

OBSERVATION dc_{out} = **3.4** V.

Waveform:

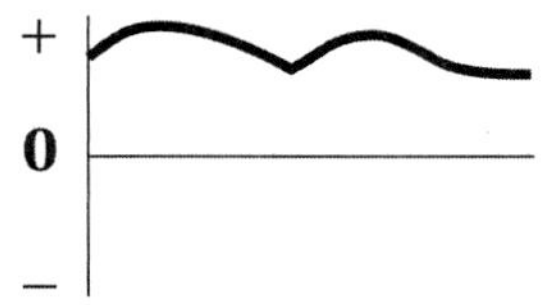

CONCLUSION Is the dc output voltage higher or lower than the effective ac input voltage? **higher**. This can be explained by the fact that the filter capacitor charges to the **peak** value of the input voltage, rather than the average or effective values. Since the charge path for the capacitor is through the low resistance of the forward-conducting diode, the capacitor has time to charge to the **peak** value of the input voltage. But when the input voltage begins to decrease from its **peak** value, the capacitor begins to **discharge** slowly through the load resistor, R_L. Before the capacitor can discharge completely, the next cycle of ac will reach a point where it is higher than the voltage remaining on the capacitor. The diode thus begins to conduct again and charge the capacitor to the **peak** voltage value again.

Lab Projects pages 355–357

8. Replace the 0.1-μF filter capacitor with a 1.0-μF capacitor.

9. Record the dc_{out} and sketch two complete ac cycles of the oscilloscope waveform found across R_L.

OBSERVATION dc_{out} = **3.5** V.

Waveform:

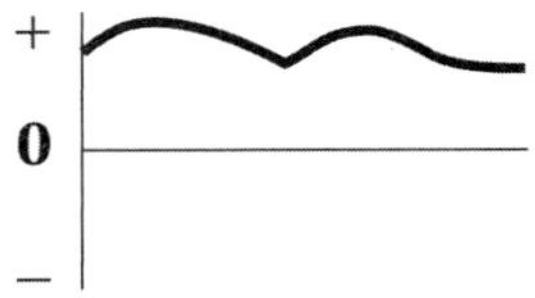

CONCLUSION Is dc_{out} higher or lower with the 1.0-μF filter capacitor? **higher**. This is because the *RC* discharge time with the 1.0-μF filter capacitor is much **longer** than with the 0.1-μF capacitor. The larger the value of filter capacitor, the (*higher, lower*) **lower** the amount of discharge during each nonconducting half-cycle for the diode. This means the average dc output voltage (*increases, decreases*) **increases** with increasing values of filter capacitance.

Lab Projects pages 355–357

PROJECT 66 Bridge Rectifier

PROCEDURE

1. Connect the initial circuit as shown in Figure 66-1.
2. Apply 120 V, 60 Hz to the primary. Measure the dc voltage output across R_L (dc_{out}).
3. Sketch the secondary waveform. Indicate the maximum and minimum voltage levels. Measure the secondary voltage.

OBSERVATION Waveform:

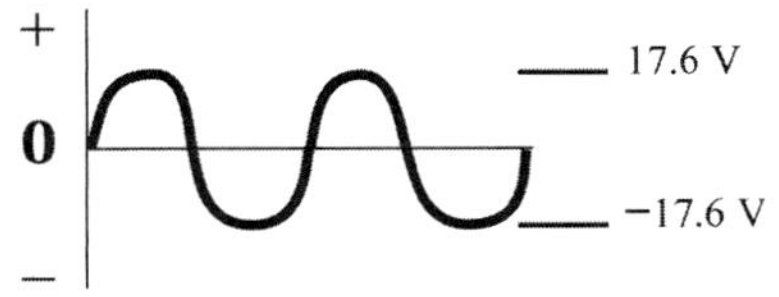

ac_{sec} = **12.6** V_{rms}.

4. Measure the dc voltage output across R_L (dc_{out}).

OBSERVATION dc_{out} = **10.3** V.

CONCLUSION Current (*flows, does not flow*) **flows** through R_L on both alternations of each ac input cycle. This means the bridge rectifier is a (*half-wave, full-wave*) **full-wave** rectifier. According to theory, if there were no diode voltage drops, the average dc output of the bridge circuit without filtering should be **0.9** × V_{rms} of the applied ac. Does your measured value of dc_{out} for this step agree reasonably with theory? **Yes**. How do you account for most of the difference, if any? **forward drop of diodes**

5. Connect the oscilloscope across R_L. Sketch the waveform for two complete cycles in the Observation section. Indicate the maximum and minimum voltage levels.

OBSERVATION Waveform:

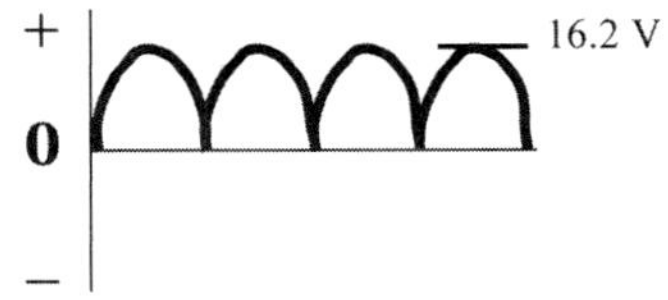

CONCLUSION The frequency of the waveform across the output resistor is (*one-half, equal to, double*) **double** the secondary frequency.

6. Connect the 1.0-μF filter capacitor across R_L as shown in Figure 66-2.

7. Measure and record the ac voltage at the secondary (ac_{sec}) and the dc voltage output across R_L (dc_{out}). Sketch two complete ac cycles of the oscilloscope waveform found across R_L.

OBSERVATION ac_{sec} = **12.6** V_{rms}.
dc_{out} = **11.76** V_{rms}.

Waveform:

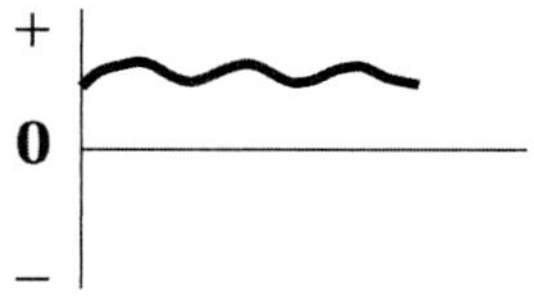

CONCLUSION What is the highest voltage output we could get from this circuit with a secondary voltage of 7 V_{rms} and no load current? About **17.6** V. What is the value of load current according to our measured value of V_{out}? **1.18** mA. The filter capacitor causes the dc level of V_{out} to be (*higher, lower*) **higher** than if it were not in the circuit. This filter capacitor has (*more, less*) **less** time to discharge between charging pulses in a bridge circuit than in a half-wave rectifier circuit. The reason for this answer is that the bridge rectifier has **two** charging pulses for each complete cycle of the ac input waveform. The capacitor in a bridge rectifier will not discharge to as low a value of voltage between charging pulses as in a half-wave rectifier. When compared to a half-wave rectifier, the average voltage output of a bridge rectifier will be (*higher, lower*) **higher** and remain closer to (*peak, effective, average*) **peak** of the secondary voltage than for a half-wave rectifier having the same R_L and filter capacitor.

Lab Projects pages 359–361

PROJECT 67 Voltage Multiplier

PROCEDURE

1. Connect the circuit shown in Figure 67-1.

2. Set the function generator (sine-wave mode) or audio oscillator to 100 Hz and V_A to 6 V_{rms}. Connect the oscilloscope across V_A and measure the peak voltage value.

⚠ OBSERVATION V_A (peak) = ____8.5____ V.

⚠ CONCLUSION If you have used a meter to set V_A to 6 V_{rms}, the peak value as measured with the oscilloscope should be ____1.414____ times the V_{rms} value. Does your measured value for V_A (peak) match up reasonably well with that ratio? ____Yes____.

3. Connect the oscilloscope across R_L. Sketch the waveform for two complete cycles in the Observation section. Indicate the maximum and minimum voltage values and note the peak-to-peak ripple of the output voltage:

$$\text{Peak-to-peak ripple of } V_{out} = V_{out}\,(\text{max}) - V_{out}\,(\text{min})$$

⚠ OBSERVATION Waveform:

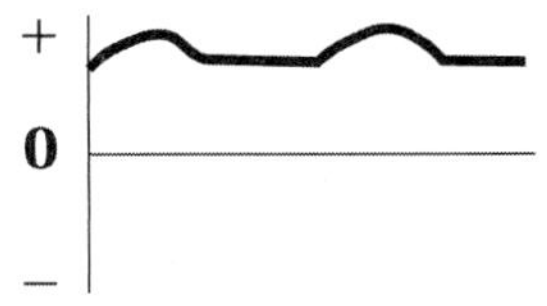

V_{out} (max) = ____17____ V.
V_{out} (min) = ____15.5____ V.
Peak-to-peak ripple of V_{out} = ____1.5____ V.

⚠ CONCLUSION The peak value of the output voltage should be about ____2____ times the peak value of the ac input voltage. Do your measurements confirm this? ____Yes____. The frequency of the ripple voltage at the output of this circuit is (*one-half, equal to, twice*) ____equal to____ the frequency of the input waveform.

4. Replace capacitor C_2 with the 10-μF capacitor. If this is an electrolytic capacitor, make sure you connect the positive terminal of the capacitor to the positive output terminal (at the cathode of diode D_2).

5. Connect the oscilloscope across R_L. Measure and record the maximum and minimum voltage values. Calculate the peak-to-peak ripple of the output voltage.

⚠ OBSERVATION V_{out} (max) = ____17.2____ V.
V_{out} (min) = ____16.8____ V.
Peak-to-peak ripple of V_{out} = ____0.4____ V.

Lab Projects pages 363–364

CONCLUSION Increasing the value of the output capacitor (*increased, decreased*) **decreased** the output ripple voltage. Increasing the value of the output capacitor increased the average dc output voltage. How can you account for this? **minimum ripple closer to peak output**

Lab Projects pages 363–364

PART 20 Story Behind the Numbers

Table 1

Measurements			Calculations					
V_{in}	V_{out}	R_L	I_L	P_L	V_{REG}	P_{REG}	P_{in}	η
V	V	Ω	A	W	V	W	W	%
7	5	∞	0	0	2	29.4 m	29.4 m	0
25	5	∞	0	0	20	150 m	150 m	0
7	5	20	250 m	1.25	2	529.4 m	1.78	70
7	5	10	500 m	2.5	2	1.03	3.53	71

Table 2

Input regulation	Output regulation	% load regulation	Output source resistance
V	V	%	Ω
0.003*	0.005*	0	0.017*

* Typical values from the data sheet

Answers to Analysis Questions

1. When the input voltage of the 7805 is between 2 to 20 V greater than the output voltage, the output voltage is ___d___.
   a. constant
   b. 5 V
   c. regulated
   d. a, b, and c
   e. not in this list

2. According to the 7805 specifications, the minimum input voltage to provide the regulated 5-V output is ___7 V___.

Lab Projects pages 365–368

3. According to the 7805 specifications, the maximum input voltage to provide the regulated 5-V output is **35 V**.

4. For a 5-Ω load resistor connected to the 7805 with an input voltage of 7 V, calculate the
   a. load current = **1 A**
   b. load power = **5 W**
   c. regulator power = **2.03 W**
   d. input power = **7.03 W**
   e. efficiency = **71%**

5. In question 4, will the voltage regulator require a heat sink? (Explain your answer.)
   Yes, the power dissipated by the regulator exceeds the 7805's. There is no heat sink power rating of 2 W at 25°C free-air temperature.

Lab Projects pages 365–368

PART 20 Summary

Complete the following review questions, indicating the appropriate response by placing a check in the box next to the correct answer.

1. The dc output voltage of a half-wave unfiltered rectifier is ________ if you neglect the forward voltage drop across the diode.
   - ☐ $0.707 \times V_{rms}$
   - ☑ $0.45 \times V_{rms}$
   - ☐ $0.318 \times V_{rms}$
   - ☐ none of these

2. The average dc output voltage of a half-wave rectifier with capacitor filtering is higher than the dc output of a half-wave rectifier without a filter.
   - ☑ True
   - ☐ False

3. A typical voltage drop across a conducting silicon diode is approximately
   - ☐ 0.0 V
   - ☑ 0.7 V
   - ☐ 0.25 V
   - ☐ 0.9 V
   - ☐ 0.5 V

4. The ripple frequency at the output of a half-wave rectifier is
   - ☐ half the ac input frequency
   - ☑ the same as the ac input frequency
   - ☐ twice the ac input frequency
   - ☐ none of these

5. The bridge rectifier is one form of
   - ☐ half-wave rectification
   - ☐ voltage multiplication
   - ☑ full-wave rectification
   - ☐ none of these

6. For one cycle of ac input, a bridge rectifier yields
   - ☐ one-half pulse of output
   - ☑ two pulses of output
   - ☐ one pulse of output
   - ☐ four pulses of output

7. Assuming a half-wave rectifier and a full-wave rectifier have identical values of output load resistance and filter capacitance, which will yield the lesser amount of output ripple voltage?
   - ☐ The half-wave rectifier
   - ☑ The full-wave rectifier

8. In a bridge rectifier circuit, how many diodes are conducting during a given alternation of the input ac waveform?
   - ☐ One
   - ☐ Three
   - ☑ Two
   - ☐ All four

9. Neglecting the forward voltage drop across the diodes, the peak output voltage of a half-wave cascade voltage doubler is ________ the peak voltage of the ac input waveform.
   - ☐ close to one-half
   - ☑ about twice
   - ☐ very nearly equal to
   - ☐ somewhat more than twice

Lab Projects pages 371–372

10. In a half-wave cascade voltage doubler, the frequency of the output ripple is ________ times the frequency of the ac input waveform.
    - [ ] one-half
    - [x] one
    - [ ] two
    - [ ] more than two

Lab Projects pages 371–372

PART 21

BJT CHARACTERISTICS

PROJECT 68 BJT Biasing

SPECIAL NOTE:

A normally biased BJT has the emitter-base junction forward biased and the collector-base junction reverse biased. Forward biasing, in this case, means that the N-type material is connected to the negative side of a dc source and the P-type material is connected to the positive side of the same dc source. So, with an NPN transistor, this means that the base should be positive with respect to the emitter (emitter-base bias) and the collector should be positive with respect to the base (collector-base bias).

The typical emitter-base junction voltage of a normally operating silicon transistor is about 0.7 volts. The collector-base junction voltage for the same transistor depends on the source voltage and values of the external components.

The following formulas can be helpful for completing the work in this project:

Formula 1 $V_{BE} = V_B - V_E$

where:

V_{BE} is the base-emitter voltage
V_B is the voltage measured from base to common
V_E is the voltage measured from emitter to common

Formula 2 $I_E = V_E / R_E$

where:

I_E is the emitter current
V_E is the voltage from emitter to common
R_E is the value of the emitter resistor

Formula 3 $V_{RC} = V_{CC} - V_C$

where:

V_{RC} is the voltage across the collector resistor
V_{CC} is the amount of supply voltage
V_C is the voltage measured from the collector to common

Formula 4 $I_C = V_{RC} / R_C$

where:

I_C is the collector current
V_{RC} is the voltage across the collector resistor
R_C is the value of the collector resistor

Lab Projects pages 375–378

Formula 5 $V_{CE} = V_C - V_E$

where:

V_{CE} is the collector-emitter voltage
V_C is the voltage measured from the collector to common
V_E is the voltage from the emitter to common

PROCEDURE

1. Connect the circuit shown in Figure 68-1.

CONCLUSION The total resistance between the base and $+V_{CC}$ is __5__ kΩ. The total collector resistance (combination of R_5 and R_6) is __5__ kΩ.

2. Set the dc source voltage to 9 V. Consider the negative side of the source as common, and measure the circuit voltages cited in the Observation section.

OBSERVATION V_{CC} = __9__ V. Collector to common = __5__ V.
Emitter to common = __0.8__ V. Base to emitter = __0.7__ V.
Base to common = __1.5__ V. Collector to emitter = __4.2__ V.

CONCLUSION The base is (*positive, negative*) __positive__ with respect to the emitter. Is the base-to-emitter voltage close to 0.7 V? __Yes__. Is the collector-to-emitter voltage less than V_{CC}? __Yes__. Is the collector-to-emitter voltage greater than 0.3 V? __Yes__. How much current must be passing through the emitter resistor? __0.8__ mA. What is the voltage drop across the collector load resistance (V_{RC})? __4__ V. This means the collector current must be approximately __0.8__ mA. The collector is more (*positive, negative*) __positive__ than the base. This means the collector-base junction is (*forward, reverse*) __reverse__ biased. The transistor is operating in (*cutoff, linear, saturation*) __linear__ region.

3. Remove resistor R_4 from the base circuit.

CONCLUSION The total resistance between the base and $+V_{CC}$ is __10__ kΩ.

4. Make sure the dc source voltage is set at 9 V, and measure the circuit voltages cited in the Observation section.

OBSERVATION Emitter to common = __0.1__ V. Base to emitter = __0.7__ V.
Base to common = __0.8__ V. Collector to emitter = __8.4__ V.
Collector to common = __8.5__ V.

CONCLUSION The total resistance between the base and $+V_{CC}$ is __10__ kΩ. Is the base-to-emitter voltage close to 0.7 V? __Yes__. Is the collector-to-emitter voltage less than V_{CC}? __Yes__. Is the collector-to-emitter voltage greater than 0.3 V? __Yes__. How much current must be passing through the emitter resistor? __0.1__ mA. What is

Lab Projects pages 375–378

the voltage drop across the collector load resistance (V_{RC})? ___0.5___ V. This means the collector current must be approximately ___0.1___ mA. Has changing the value of the base-to-V_{CC} resistance changed:

a. The collector-to-emitter voltage? ___Yes___ If so, how? ___greater___
______________________________.

b. The current? ___Yes___ If so, how? ___less___
______________________________.

c. The base-to-emitter voltage? ___No___. If so, how much? ___0___ V.

d. The emitter current? ___Yes___. If so, how? ___less___
______________________________.

The transistor is operating in (*cutoff, linear, saturation*) ___linear___ region.

5. Replace resistor R_4 as shown in Figure 68-1, and remove resistor R_6 from the collector circuit.

CONCLUSION The collector resistance is now ___10___ kΩ.

6. After making sure the dc source voltage is set at 9 V, measure the circuit voltages cited in the Observation section.

OBSERVATION

Emitter to common = ___0.8___ V. Base to emitter = ___0.7___ V.
Base to common = ___1.5___ V. Collector to emitter = ___0.2___ V.
Collector to common = ___1___ V.

CONCLUSION The total resistance between the base and $+V_{CC}$ is ___5___ kΩ. Is the base-to-emitter voltage close to 0.7 V? ___Yes___. Is the collector-to-emitter voltage less than V_{CC}? ___Yes___. Is the collector-to-emitter voltage greater than 0.3 V? ___No___. How much current must be passing through the emitter resistor? ___0.8___ mA. What is the voltage drop across the collector load resistance (V_{RC})? ___8___ V. This means the collector current must be approximately ___0.8___ mA. Has changing the value of the collector resistance (compared with your results in step 2) changed:

a. The collector-to-emitter voltage? ___Yes___ If so, how? ___less___
______________________________.

b. The current? ___No___ If so, how? ___—___
______________________________.

c. The base-to-emitter voltage? ___No___. If so, how much? ___—___ V.

d. The emitter current? ___No___. If so, how? ___—___
______________________________.

The transistor is operating in (*cutoff, linear, saturation*) ___saturation___ region.

Lab Projects pages 375–378

PART 21 Story Behind the Numbers

Results may be different. The current gain (β) should be between 100 and 400. The current ratio (α) should be between 0.9 and 1.0.

Table 1

V_{BB} = 1.7 V data

Measurements				Calculations				
V_{BB}	V_{BE}	V_{CC}	V_{CE}	I_B	I_C	I_E	β	α
V	V	V	V	mA	mA	mA		
1.7	0.7	0	0	0.01	0	0.01	0	0
1.7	0.7	1	0.15	0.01	0.85	0.86	85	0.98837
1.7	0.7	2	0.3	0.01	1.7	1.71	170	0.99415
1.7	0.7	4	2	0.01	2	2.01	200	0.99502
1.7	0.7	6	4	0.01	2	2.01	200	0.99502
1.7	0.7	8	6	0.01	2	2.01	200	0.99502
1.7	0.7	10	8	0.01	2	2.01	200	0.99502
1.7	0.7	12	10	0.01	2	2.01	200	0.99502
1.7	0.7	14	12	0.01	2	2.01	200	0.99502
1.7	0.7	16	14	0.01	2	2.01	200	0.99502
1.7	0.7	18	16	0.01	2	2.01	200	0.99502
1.7	0.7	20	18	0.01	2	2.01	200	0.99502

Table 2

V_{BB} = 2.7 V data

Measurements				Calculations				
V_{BB}	V_{BE}	V_{CC}	V_{CE}	I_B	I_C	I_E	β	α
V	V	V	V	mA	mA	mA		
2.7	0.7	0	0	0.02	0	0.02	0	0
2.7	0.7	1	0.1	0.02	0.9	0.92	45	0.97826
2.7	0.7	2	0.2	0.02	1.8	1.82	90	0.98901
2.7	0.7	4	0.3	0.02	3.7	3.72	185	0.99462
2.7	0.7	6	2	0.02	4	4.02	200	0.99502
2.7	0.7	8	4	0.02	4	4.02	200	0.99502
2.7	0.7	10	6	0.02	4	4.02	200	0.99502
2.7	0.7	12	8	0.02	4	4.02	200	0.99502
2.7	0.7	14	10	0.02	4	4.02	200	0.99502
2.7	0.7	16	12	0.02	4	4.02	200	0.99502
2.7	0.7	18	14	0.02	4	4.02	200	0.99502
2.7	0.7	20	16	0.02	4	4.02	200	0.99502

Table 3

V_{BB} = 3.7 V data

Measurements				Calculations				
V_{BB}	V_{BE}	V_{CC}	V_{CE}	I_B	I_C	I_E	β	α
V	V	V	V	mA	mA	mA		
3.7	0.7	0	0	0.03	0	0.03	0	0
3.7	0.7	1	0.05	0.03	0.95	0.98	31.6667	0.96939
3.7	0.7	2	0.1	0.03	1.9	1.93	63.3333	0.98446
3.7	0.7	4	0.25	0.03	3.75	3.78	125	0.99206
3.7	0.7	6	0.3	0.03	5.7	5.73	190	0.99476
3.7	0.7	8	2	0.03	6	6.03	200	0.99502
3.7	0.7	10	4	0.03	6	6.03	200	0.99502
3.7	0.7	12	6	0.03	6	6.03	200	0.99502
3.7	0.7	14	8	0.03	6	6.03	200	0.99502
3.7	0.7	16	10	0.03	6	6.03	200	0.99502
3.7	0.7	18	12	0.03	6	6.03	200	0.99502
3.7	0.7	20	14	0.03	6	6.03	200	0.99502

Table 4

V_{BB} = 4.7 V data

Measurements				Calculations				
V_{BB}	V_{BE}	V_{CC}	V_{CE}	I_B	I_C	I_E	β	α
V	V	V	V	A	A	A		
4.7	0.7	0	0	0.04	0	0.04	0	0
4.7	0.7	1	0.03	0.04	0.97	1.01	24.25	0.96040
4.7	0.7	2	0.06	0.04	1.94	1.98	48.5	0.97980
4.7	0.7	4	0.2	0.04	3.8	3.84	95	0.98958
4.7	0.7	6	0.25	0.04	5.75	5.79	143.75	0.99308
4.7	0.7	8	0.3	0.04	7.7	7.74	192.5	0.99483
4.7	0.7	10	2	0.04	8	8.04	200	0.99502
4.7	0.7	12	4	0.04	8	8.04	200	0.99502
4.7	0.7	14	6	0.04	8	8.04	200	0.99502
4.7	0.7	16	8	0.04	8	8.04	200	0.99502
4.7	0.7	18	10	0.04	8	8.04	200	0.99502
4.7	0.7	20	12	0.04	8	8.04	200	0.99502

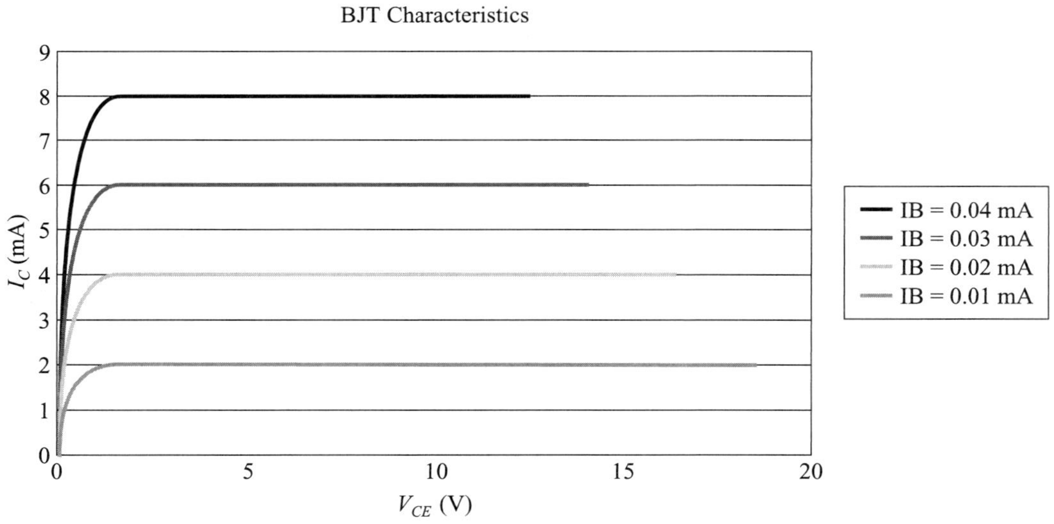

Lab Projects pages 379–386

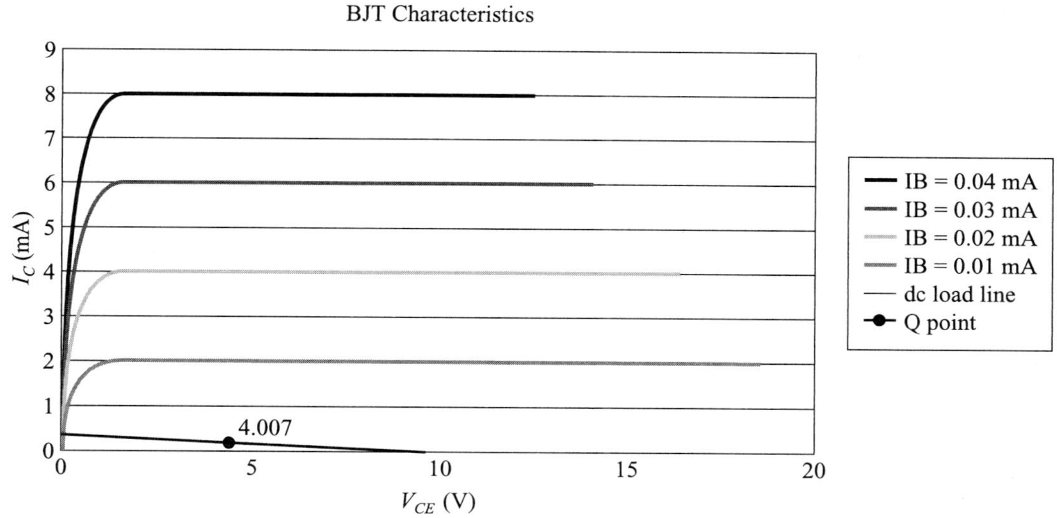

Answers to Analysis Questions

1. Evaluate the Table 1 data and answer the questions.
   a. Does V_{BE} change as V_{CE} increases? **No**
   b. Does I_B change as V_{CE} increases? **No**
   c. For $V_{CE} > 0.3$ V, does I_C change as V_{CE} increases? **Very small increase**
   d. For $V_{CE} > 0.3$ V, does the current gain β change as V_{CE} increases? **Very small increase**
   e. Is the current gain $\alpha < 1$? **Yes**

2. Evaluate the Table 2 data and answer the questions.
   a. Does V_{BE} change as V_{CE} increases? **No**
   b. Does I_B change as V_{CE} increases? **No**
   c. For $V_{CE} > 0.3$ V, does I_C change as V_{CE} increases? **Very small increase**
   d. For $V_{CE} > 0.3$ V, does the current gain β change as V_{CE} increases? **Very small increase**
   e. Is the current gain $\alpha < 1$? **Yes**

3. Evaluate the Table 3 data and answer the questions.
   a. Does V_{BE} change as V_{CE} increases? **No**
   b. Does I_B change as V_{CE} increases? **No**
   c. For $V_{CE} > 0.3$ V, does I_C change as V_{CE} increases? **Very small increase**
   d. For $V_{CE} > 0.3$ V, does the current gain β change as V_{CE} increases? **Very small increase**
   e. Is the current gain $\alpha < 1$? **Yes**

4. Evaluate the Table 4 data and answer the questions.
   a. Does V_{BE} change as V_{CE} increases? **No**
   b. Does I_B change as V_{CE} increases? **No**
   c. For $V_{CE} > 0.3$ V, does I_C change as V_{CE} increases? **Very small increase**
   d. For $V_{CE} > 0.3$ V, does the current gain β change as V_{CE} increases? **Very small increase**
   e. Is the current gain $\alpha < 1$? **Yes**

Lab Projects pages 379–386

5. Evaluate the data in Tables 1 through 4 and answer the questions.
   a. Does V_{BE} change as V_{BB} increases? **Very small increase**
   b. Does I_B change as V_{BB} increases? **Yes**
   c. For $V_{CE} > 0.3$ V, does I_C change as V_{BB} increases? **Yes**
   d. For $V_{CE} > 0.3$ V, does the current gain β change as V_{BB} increases? **Very small increase**

Lab Projects pages 379–386

PART 21 Summary

Complete the following review questions, indicating the appropriate response by placing a check in the box next to the correct answer.

1. The emitter-base junction of an NPN transistor is said to be forward biased when
   - ☑ the base is more positive than the emitter
   - ☐ the emitter is more positive than the base
   - ☐ the emitter and base have the same voltage

2. The base-collector junction of an NPN transistor is properly biased for normal transistor operation when
   - ☐ the base is more positive than the collector
   - ☑ the collector is more positive than the base
   - ☐ the base and collector have the same applied voltage

3. For the NPN BJT transistor to operate in the linear region, the emitter-base and base-collector junctions must be forward biased at the same time.
   - ☐ True
   - ☑ False

4. When a BJT circuit operating in the linear region has a resistor of known value connected between the emitter and common, you can determine the amount of emitter current by
   - ☑ measuring the emitter-to-common voltage and dividing by the amount of emitter resistance
   - ☐ measuring the base-to-common voltage, measuring the emitter-to-common voltage, and dividing the difference by the amount of emitter resistance
   - ☐ measuring the positive supply voltage, measuring the emitter-to-common voltage, and dividing the difference by the value of the emitter resistance

5. V_{CE} for a properly operating BJT circuit can be determined by taking the difference between the voltage measured between the collector and common and the voltage measured between the emitter and common.
   - ☑ True
   - ☐ False

6. The dc beta of a good BJT is
   - ☑ always much greater than 1
   - ☐ always a bit less than 1

7. The alpha of a good BJT is
   - ☐ always much greater than 1
   - ☑ always a bit less than 1

8. When the emitter-base junction of an NPN transistor is shorted, a DMM diode test will show
   - ☐ 0
   - ☑ 0.7
   - ☐ 0 L

Lab Projects pages 387–388

9. When the base-collector junction of an NPN transistor is in good working order, a DMM diode test will show
   - ☐ 0
   - ☐ 0.7
   - ☑ 0 L

10. The DMM diode test between the emitter and collector of a good BJT will be 0 L in one direction and 0.7 when the DMM leads are reversed.
    - ☐ True
    - ☑ False

Lab Projects pages 387–388

PART 22

BJT AMPLIFIER CONFIGURATIONS

PROJECT 69 Common-Emitter Amplifier

PROCEDURE

DC Bias Circuit

1. Connect the circuit as shown in Figure 69-1. DO NOT connect the signal source (function generator) at this time.

2. Set the dc source for a V_{CC} of +9 V. Consider the negative side of the source as common (ground), and measure the dc voltages required in the Observation section.

OBSERVATION V_B (voltage from base to common) = **1.58** V_{dc}

V_C (voltage from collector to common) = **5.26** V_{dc}.

V_E (voltage from emitter to common) = **0.88** V_{dc}.

V_{CE} (voltage from collector to emitter) = **4.38** V_{dc}.

V_{BE} (voltage from base to emitter) = **0.7** V_{dc}.

CONCLUSION Using the circuit values and the dc operating point formulas, calculate:

V_B = **1.58 V**; V_E = **0.88 V**; V_{CE} = **4.38 V**; I_E = **1.870 mA**; r_e' = **13.4 Ω**.

Are the voltages calculated in this column close to the observed voltages (Yes/No) for: V_B **Yes**; V_E **Yes**; V_{CE} **Yes**.

AC Operation

3. With the dc source still connected to V_{CC}, connect the signal source (function generator or audio oscillator) to V_{in}. Adjust the signal source for an input signal of a 0.5-V peak-to-peak, 1-kHz sine wave.

4. Connect channel 1 of the oscilloscope to V_{in}. Connect channel 2 of the oscilloscope to V_{out}. The scope ground is connected to the circuit ground. If the signal you find at V_{out} is clipped (flattened) on either or both half cycles, **reduce the input signal level until the clipping action is no longer observed**. Sketch the waveforms and determine the readings specified in the Observation section.

OBSERVATION V_{in} Waveform:

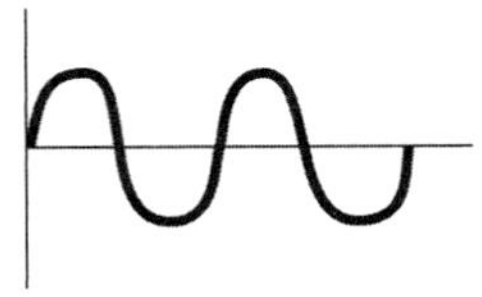

V_{out} Waveform:

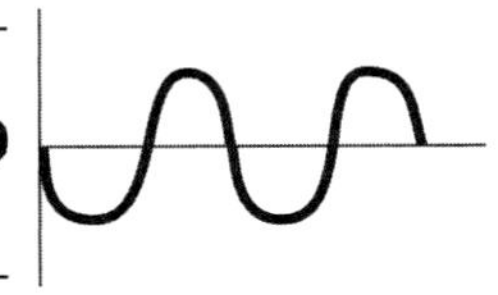

V_{in} = **0.5** $V_{P\text{-}P}$; V_{out} = **2.07** $V_{P\text{-}P}$.

CONCLUSION The oscilloscope traces indicate that the output voltage from a common-emitter amplifier is (*in phase, 180° out of phase*) **180° out of phase** with its input waveform. Use the readings you found for V_{in} and V_{out} to determine the voltage gain (A_V) of this circuit. A_V = **–4.14**. You have seen that the voltage gain of a common-emitter amplifier can be greater than 1. (*True, False*) **True**.
Using the values in this column and the ac operation formula, calculate the voltage gain. A_V = **–4.14**.

DC Bias Circuit

5. Remove the 2 Ω collector resistor (R_4) and replace it with a 1.5 kΩ resistor.

6. Remove the signal source from the circuit. Set the dc source for a V_{CC} of +9 V. Measure the dc voltages required in the Observation section.

OBSERVATION

V_B = **1.58** V_{dc}. V_{CE} = **5.32** V_{dc}.
V_C = **6.20** V_{dc}. V_{BE} = **0.7** V_{dc}.

V_E = **0.88** V_{dc}.

CONCLUSION Using the circuit values and the dc operating point formulas, calculate:
V_B = **1.58 V**; V_E = **0.88 V**; V_{CE} = **5.32**; I_E = **1.87 mA**; r_e' = **13.4 Ω**.

Use a check mark in the following list to indicate the parameter(s) that are different from step 2:
V_B ✔; V_E ✔; V_{CE} ✔.

AC Operation

7. With the dc source still connected, connect the signal source to V_{in}. Adjust the signal source for an input signal of a 0.5 $V_{P\text{-}P}$, 1-kHz sine wave.

Lab Projects pages 391–394

8. Connect channels 1 and 2 of the oscilloscope to V_{in} and V_{out}. If either peak of the waveform at V_{out} is clipped, reduce the input signal level until the clipping effect is no longer apparent. Gather the information requested in the Observation section.

OBSERVATION V_{in} Waveform:

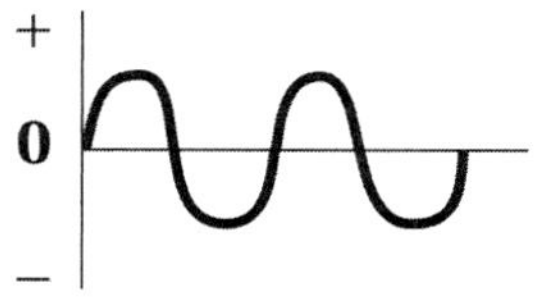

V_{out} Waveform:

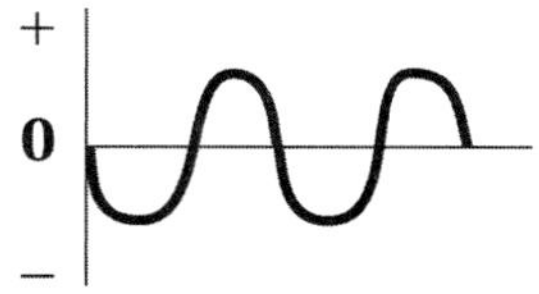

V_{in} = **0.5** $V_{P\text{-}P}$; V_{out} = **1.55** $V_{P\text{-}P}$. (Gain = 157)

CONCLUSION Use the readings you found for V_{in} and V_{out} to determine the voltage gain (A_V) of this circuit. A_V = **−3.1**. When you increased the value of the load resistor (R_4), did you notice a significant change in the amount of voltage gain? **Yes**. If so, describe the amount and direction of change.

0.75 times the gain using the 2-kΩ collector resistor

Using the values in this column and the ac operation formula, calculate the voltage gain: A_V = **−3.1**.

Lab Projects pages 391–394

PROJECT 70 Common-Collector Amplifier

PROCEDURE

DC Bias Circuit

1. Connect the circuit as shown in Figure 70-1. DO NOT connect the signal source (function generator) at this time.

2. Set the dc source for a V_{CC} of +6 V. Consider the negative side of the source as common (ground), and measure the dc voltages required in the Observation section.

OBSERVATION V_B (voltage from base to common) = **3.3** V_{dc}.
V_C (voltage from collector to common) = **6** V_{dc}.
V_E (voltage from emitter to common) = **2.6** V_{dc}.
V_{CE} (voltage from collector to emitter) = **3.4** V_{dc}.
V_{BE} (voltage from base to emitter) = **0.7** V_{dc}.

CONCLUSION Using the circuit values and the dc operating point formulas, calculate:
V_B = **3.3 V**; V_E = **2.6 V**; V_{CE} = **3.4 V**; I_E = **0.55 mA**; r_e' = **46 Ω**.
Are the voltages calculated in this column close to the observed voltages (Yes/No) for: V_B **Yes**; V_E **Yes**; V_{CE} **Yes**.

AC Operation

3. With the dc source still connected to V_{CC}, connect the signal source (function generator or audio oscillator) to V_{in}. Adjust the signal source for an input signal of a 0.5-V peak-to-peak, 1-kHz sine wave.

4. Connect channel 1 of the oscilloscope to V_{in}. Connect channel 2 of the oscilloscope to V_{out}. The scope ground is connected to the circuit ground. If the signal you find at V_{out} is clipped (flattened) on either or both half cycles, **reduce the input signal level until the clipping action is no longer observed**. Sketch the waveforms and determine the readings specified in the Observation section.

OBSERVATION V_{in} Waveform:

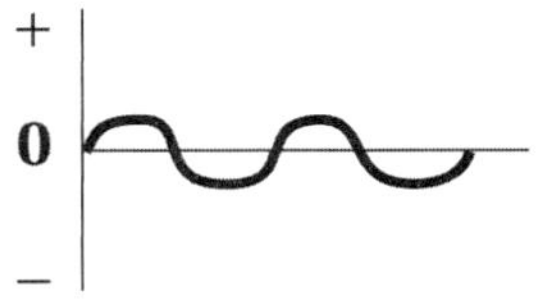

V_{out} Waveform:

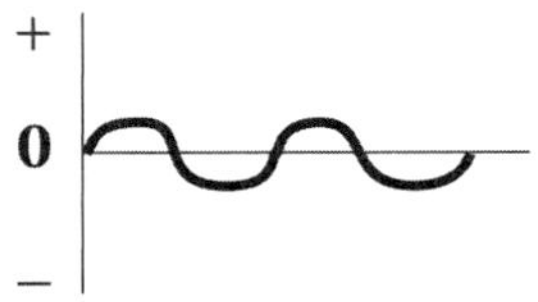

V_{in} = **0.5** V_{P-P}; V_{out} = **0.5** V_{P-P}.

Lab Projects pages 395–397

▲ CONCLUSION The oscilloscope traces indicate that the output voltage from a common-collector amplifier is (*in phase, 180° out of phase*) ____**in phase**____ with its input waveform. Use the readings you found for V_{in} and V_{out} to determine the voltage gain (A_V) of this circuit. A_V = ____**1**____. (Recall that $A_V = V_{out}/V_{in}$.) You have seen in this circuit that the voltage gain of a common-collector amplifier can be greater than 1. (*True, False*) ____**False**____.

DC Bias Circuit

5. Remove the signal source from the circuit. Increase the setting of the dc source for a V_{CC} of +9 V. Measure the dc voltages required in the Observation section.

▲ OBSERVATION V_B = ____**4.9**____ V_{dc}. V_{CE} = ____**4.8**____ V_{dc}.
V_C = ____**9**____ V_{dc}. V_{BE} = ____**0.7**____ V_{dc}.
V_E = ____**4.2**____ V_{dc}.

▲ CONCLUSION Using the circuit values and the dc operating point formulas, calculate:

V_B = **4.9 V**; V_E = **4.2 V**; V_{CE} = **4.8 V**; I_E = **0.9 mA**; r_e' = **27.7 Ω**.
Use a check mark in the following list to indicate the parameter(s) that are different from step 2: V_B ✓; V_E ✓; V_{CE} ✓.

AC Operation

6. With the dc source still connected, connect the signal source to V_{in}. Adjust the signal source for an input signal of a 0.5-V_{P-P}, 1 kHz sine wave.

7. Connect channels 1 and 2 of the oscilloscope to V_{in} and V_{out}. If either peak of the waveform at V_{out} is clipped, reduce the input signal level until the clipping effect is no longer apparent. Gather the information requested in the Observation section.

▲ OBSERVATION V_{in} Waveform:

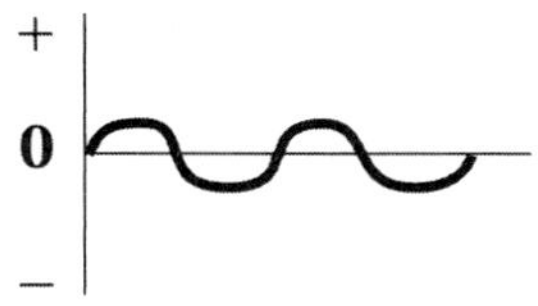

V_{out} Waveform:

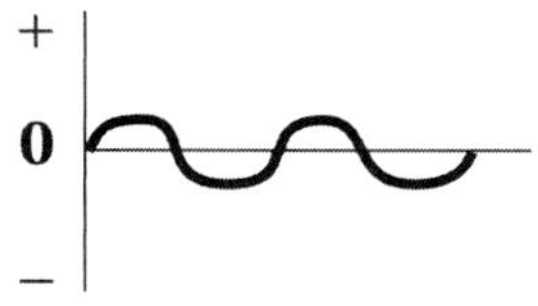

V_{in} = ____**0.5**____ V_{P-P}; V_{out} = ____**0.5**____ V_{P-P}.

▲ CONCLUSION Use the readings you found for V_{in} and V_{out} to determine the voltage gain (A_V) of this circuit. A_V = ____**1**____. When you increased the value of V_{CC}, did you notice a significant change in the amount of voltage gain? ____**No**____. If so, describe the amount and direction of change; if not, describe which readings did change. ________.

Lab Projects pages 395–397

PART 22 Summary

Complete the following review questions, indicating the appropriate response by placing a check in the box next to the correct answer.

1. The output ac waveform from a common-emitter amplifier is shifted ________ degrees relative to the input ac waveform.
   - ☐ 0
   - ☑ 180
   - ☐ 90
   - ☐ 360

2. The voltage gain of a common-emitter amplifier can be greater than 1.
   - ☑ True
   - ☐ False

3. Increasing the value of the collector resistor in a common-emitter amplifier ________ the voltage gain of the circuit.
   - ☐ has no effect on
   - ☑ increases
   - ☐ reduces

4. For the common-emitter circuit, the dc voltage readings show that the base-emitter junction
   - ☑ is forward biased
   - ☐ has no bias
   - ☐ is reverse biased
   - ☐ is negatively biased

5. As you decrease the amount of ac signal applied at V_{in} for a common-emitter amplifier, a corresponding decrease in V_{out} indicates that gain of the circuit is also decreasing.
   - ☐ True
   - ☑ False

6. You can identify a common-emitter amplifier by noting that
   - ☐ the emitter is connected to the negative terminal of the power source
   - ☐ the input and output terminals are both connected to the emitter
   - ☑ neither the signal input nor the signal output is connected to the emitter
   - ☐ none of these is true

7. The output ac waveform from a common-collector amplifier is shifted ________ degrees relative to the input ac waveform.
   - ☑ 0
   - ☐ 180
   - ☐ 90
   - ☐ 360

8. A common-collector amplifier is also known as a voltage follower because the output voltage "follows" the input voltage.
   - ☑ True
   - ☐ False

9. Increasing the value of V_{CC} for a common-collector amplifier
   - ☐ increases the voltage gain
   - ☑ has little, if any, effect on the voltage gain

Lab Projects pages 399–400

10. You can identify an NPN common-collector amplifier by noting that
    - ☑ the collector is connected directly to the positive terminal of the power source
    - ☐ the collector is grounded, or connected to circuit common
    - ☐ the output is taken from the collector
    - ☐ none of these is true

Lab Projects pages 399–400

PART 23

BJT AMPLIFIER CLASSES OF OPERATION

PROJECT 71 BJT Class A Amplifier

SPECIAL NOTE:

The following formula will be helpful:

$$A_V = V_{out}/V_{in}$$

where:

A_V is the voltage gain of an amplifier
V_{out} is the signal voltage level at the output of the amplifier (usually peak-to-peak)
V_{in} is the signal voltage level at the input of the amplifier (usually peak-to-peak)

PROCEDURE

1. Connect the circuit as shown in Figure 71-1.

2. Set the dc source for a V_{CC} of +9 V. Make sure the signal source (function generator in the sine-wave mode or an audio oscillator) is disconnected from the circuit or set for 0 V output. Consider the negative side of the source as common, and measure the circuit's dc voltages cited in the Observation section.

OBSERVATION V_E (voltage from emitter to common) = **0.88** V_{dc}.
V_C (voltage from collector to common) = **4.9** V_{dc}.
V_B (voltage from base to common) = **1.58** V_{dc}.

CONCLUSION Use the data from Observation step 2 to determine: V_{CE} (voltage between collector and emitter) = **4.38** V_{dc}. V_{BE} (voltage between the base and emitter) = **0.7** V_{dc}. For Class A operation of a BJT amplifier, V_{CE} should be close to (*V_{CC}, 1/2 V_{CC}, zero volts, −0.7 V*) **1/2 V_{CC}**, so the transistor is (*conducting, not conducting*) **conducting** while no signal is applied at V_{in}. Confirm whether the transistor is conducting or not conducting by determining the voltage across the collector resistor (R_4) and dividing by the resistor value to determine the amount of collector current: I_C = **1.9** mA.

3. Connect the function generator (sine-wave mode) or audio oscillator, and use the oscilloscope to set the input signal to 1 kHz at 0.5 V peak-to-peak. If the signal you find at V_{out} is clipped on either or both alternations, **reduce the input signal level until the clipping action is no longer observed**.

Lab Projects pages 403–405

4. Leave one channel of the oscilloscope connected to V_{in} and connect the second channel of the oscilloscope to V_{out}. If the signal you find at V_{out} is not a clean sinusoidal waveform, reduce the V_{in} signal level until the waveform at the collector is a clean sine wave. Sketch the waveforms and determine the readings specified in the Observation section.

OBSERVATION V_{in} Waveform:

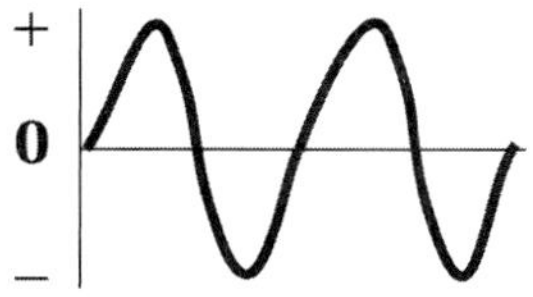

V_{out} Waveform:

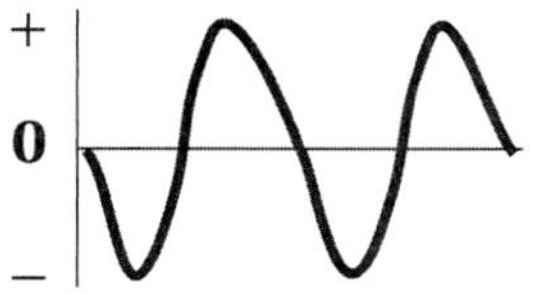

V_{in} = **0.5** $V_{P\text{-}P}$; V_{out} = **2.07** $V_{P\text{-}P}$.

CONCLUSION Comparing the phases of the waveforms at V_{in} and V_{out}, you can say they are (*in phase, 180° out of phase*) **180° out of phase**. What is the voltage gain of this amplifier? **−4**.

5. Slowly decrease the voltage level of the signal to one-half the amount used in step 4. Sketch the waveforms and determine the readings specified in the Observation section.

OBSERVATION V_{in} Waveform:

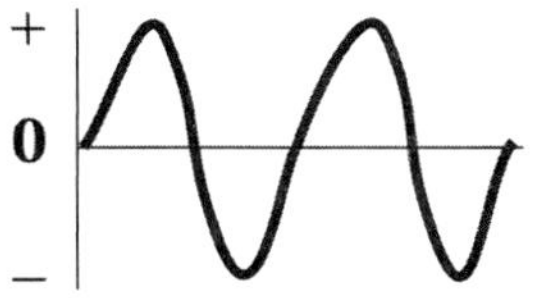

V_{out} Waveform:

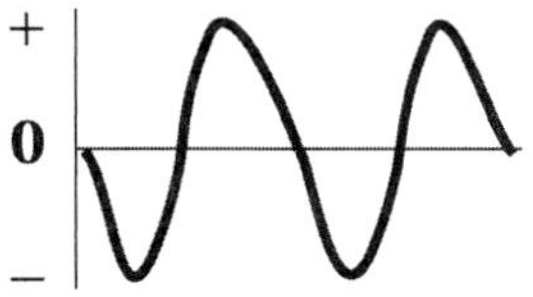

V_{in} = **0.25** $V_{P\text{-}P}$; V_{out} = **1** $V_{P\text{-}P}$.

CONCLUSION The distortion in the output waveform of this amplifier is a sign that the input signal is too large for this circuit design. (*True, False*) **False**.

PROJECT 72 BJT Class B Amplifier

PROCEDURE

1. Connect the circuit as shown in Figure 72-1.

2. Set the dc source for a V_{CC} of +9 V. Connect the signal source (function generator in the sine-wave mode or an audio oscillator) to V_{in}, but make sure its output is at zero volts. Consider the negative side of the source as common, and measure the dc voltages cited in the Observation section.

OBSERVATION V_E (voltage from emitter to common) = **0** V_{dc}.
V_C (voltage from collector to common) = **9** V_{dc}.
V_B (voltage from base to common) = **0** V_{dc}.

CONCLUSION For Class B operation of a BJT amplifier, V_{CE} should be close to (*V_{CC}, 1/2 V_{CC}, zero volts, −0.7 V*) **V_{CC}**. This means the transistor is (*conducting, not conducting*) **not conducting** while no signal is applied at V_{in}. Confirm whether the transistor is conducting or not conducting by determining the voltage across the collector resistor (R_2) and dividing by the resistor value to determine the amount of collector current: I_C = **0** mA.

3. Connect one channel of the oscilloscope to V_{in} and connect the second channel of the oscilloscope to V_{out}.

4. Adjust the signal source for an input signal of 1 kHz at 1.27 $V_{P\text{-}P}$ (0.45 V_{rms}). If the signal you find at V_{out} is clipped ("flattened") on **both** half cycles, **reduce the input signal level until the clipping occurs on just one of the half-cycles**. Sketch the waveforms and determine the readings specified in the Observation section.

OBSERVATION V_{in} Waveform:

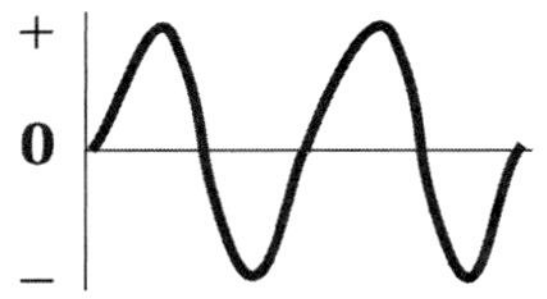

V_{out} Waveform:

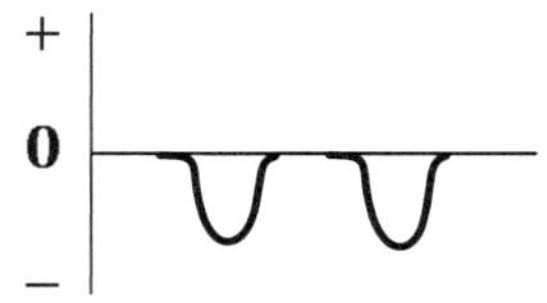

V_{in} = **1.27*** $V_{P\text{-}P}$; V_{out} = **6.8*** $V_{P\text{-}P}$.

***Results may vary.**

CONCLUSION Comparing the phases of the waveforms at V_{in} and V_{out}, you can say they are (*in phase, 180° out of phase*) **180° out of phase**. The distortion noted on one alternation of the output waveform of this amplifier is a sign that something is wrong. (*True, False*) **False**.

Lab Projects pages 407–409

5. Connect a 47-kΩ resistor between the base of the transistor and V_{CC}. Sketch the waveform for V_{out}.

OBSERVATION V_{out} Waveform:

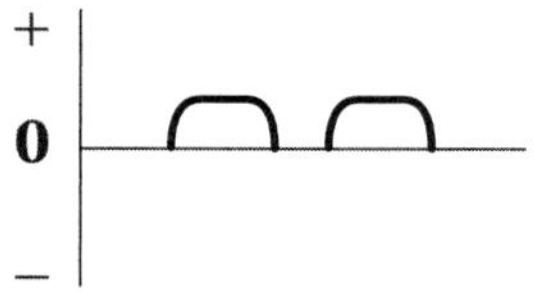

CONCLUSION Adding a resistor between the base and V_{CC} gave this amplifier some (*forward, reverse*) ____**forward**____ emitter-base bias that it didn't have before. This means the amplifier is no longer operating as a Class B amplifier. (*True, False*) ____**True**____.

PROJECT 73 BJT Class C Amplifier

PROCEDURE

1. Connect the circuit as shown in Figure 73-1.

2. Set the dc source for a V_{CC} of +9 V. Connect the signal source (function generator in its sine-wave mode or the audio oscillator) to V_{in}, but make sure its output is at zero volts. Consider the negative side of the source as common, and measure the dc voltages required in the Observation section.

OBSERVATION V_E (voltage from emitter to common) = **0** V_{dc}.
V_C (voltage from collector to common) = **9** V_{dc}.
V_B (voltage from base to common) = **–1.5** V_{dc}.

CONCLUSION For Class C operation of a BJT amplifier, V_{CE} for an NPN transistor should be (*equal to V_{CC}, 1/2 V_{CC}, zero volts, less than zero volts*) **less than zero volts**, so the transistor is (*conducting, not conducting*) **not conducting** while no signal is applied at V_{in}. Use the formula $I_C = (V_{CC} - V_C)/R_C$ to determine the actual amount of collector current when no signal is applied. I_C = **0** mA. When there is no signal applied to this amplifier, it can be said that the emitter-base junction is (*forward, reverse*) **reverse** biased.

3. Connect one channel of the oscilloscope to V_{in}, and connect the second channel of the oscilloscope to V_{out}.

4. Adjust the signal source for an input signal of 1 kHz at 4.2 $V_{P\text{-}P}$ (1.48 V_{rms}). If the signal you find at V_{out} is clipped on **both** alternations, **reduce the input signal level until the clipping occurs on just one of the alternations**. Sketch the waveforms and determine the readings specified in the Observation section.

OBSERVATION V_{in} Waveform:

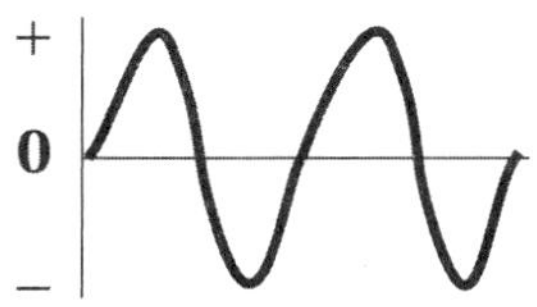

V_{out} Waveform:

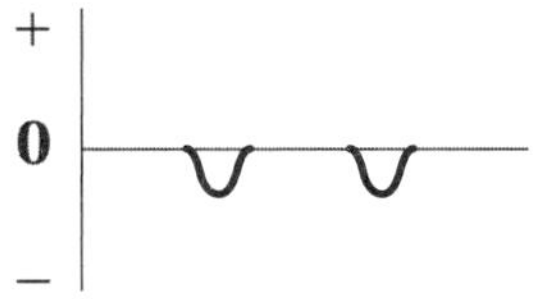

V_{in} = **4.2*** $V_{P\text{-}P}$; V_{out} = **3.9*** $V_{P\text{-}P}$.

***Results may vary.**

CONCLUSION Comparing the phases of the waveforms at V_{in} and V_{out}, you can say they are (*in phase, 180° out of phase*) **180° out of phase**. The distortion noted on one

Lab Projects pages 411–413

alternation of the output waveform of this amplifier indicates conduction during (*more than, equal to, less than*) **less than** 180° of the input waveform. This type of distortion for a Class C amplifier is a sure sign that something is wrong. (*True, False*) **False**.

5. Connect a 47-kΩ resistor between the base of the transistor and V_{CC}. Sketch the waveform for V_{out}.

OBSERVATION V_{out} Waveform:

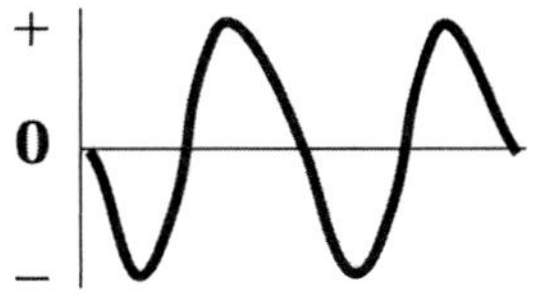

6. Remove the 47-kΩ resistor you added in step 5 and replace it with a 10-kΩ resistor (between the base of the transistor and V_{CC}). Sketch the waveform for V_{out}.

OBSERVATION V_{out} Waveform:

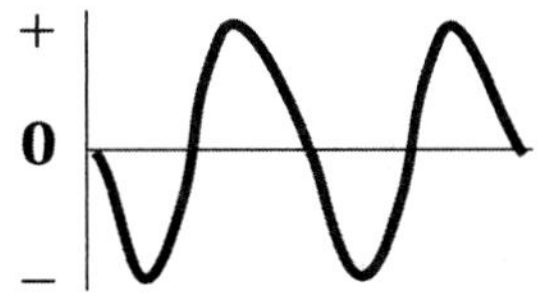

CONCLUSION As you decrease the amount of resistance between the base and V_{CC} in this circuit, the amount of (*forward, reverse*) **reverse** bias at the emitter-base junction becomes less, and the transistor begins conducting over a (*greater, lesser*) **greater** portion of the input ac waveform.

PART 23 Summary

Complete the following review questions, indicating the appropriate response by placing a check in the box next to the correct answer.

1. A Class A amplifier actually amplifies ________ of an ac sine wave that is applied to the input.
   - ☑ all 360°
   - ☐ about 180°
   - ☐ less than 180°
   - ☐ none

2. The base-emitter junction of a Class A common-emitter BJT amplifier is ________ biased.
   - ☑ forward
   - ☐ reverse
   - ☐ zero

3. The output waveform of a Class A common-emitter BJT amplifier is ________ compared to its input waveform.
   - ☐ distorted
   - ☐ shifted 90°
   - ☑ shifted 180°

4. A Class A amplifier is ____________ when there is no signal applied at the input.
   - ☑ conducting
   - ☐ nonconducting

5. The base-emitter junction of a Class B common-emitter BJT amplifier is ________ biased.
   - ☐ forward
   - ☐ reverse
   - ☑ zero

6. A Class B amplifier actually amplifies ________ of an ac sine wave that is applied to the input.
   - ☐ all 360°
   - ☑ about 180°
   - ☐ much less than 180°
   - ☐ none

7. A Class B amplifier is ____________ when there is no signal applied at the input.
   - ☐ conducting
   - ☑ nonconducting

8. A Class C amplifier actually amplifies ________ of an ac sine wave that is applied to the input.
   - ☐ all 360°
   - ☐ about 180°
   - ☑ much less than 180°
   - ☐ none

9. The base-emitter junction of a Class C common-emitter BJT amplifier is ________ biased.
   - ☐ forward
   - ☑ reverse
   - ☐ zero

Lab Projects pages 415–416

10. Of all the classes of amplifier operation, the Class C amplifier is noted for
- ☐ the least amount of signal distortion and the highest amount of efficiency
- ☐ the least amount of signal distortion but the least amount of efficiency
- ☑ the greatest amount of signal distortion but the highest amount of efficiency
- ☐ the greatest amount of signal distortion and the least amount of efficiency

Lab Projects pages 415–416

PART 24

JFET CHARACTERISTICS AND AMPLIFIERS

PROJECT 74 Common-Source JFET Amplifier

PROCEDURE

1. Connect the circuit as shown in Figure 74-1. Set the dc source for a V_{DD} of +12 V. Connect the signal source (audio oscillator or function generator in the sine-wave mode) to V_{in}, but make sure its output is set for zero volts output.

2. Measure the dc voltages with respect to circuit common as required in the Observation section.

OBSERVATION V_S (voltage from source to common) = ____1.6____ V_{dc}.
V_G (voltage from gate to common) = ____0____ V_{dc}.
V_D (voltage from drain to common) = ____4.5____ V_{dc}.

CONCLUSION Use the formula, $V_{DS} = V_D - V_S$, to determine the voltage drop between the source and drain. V_{DS} = ____2.9____. Based on the voltages you observed in this step, do you have good reason to suppose this is a Class A amplifier? ____yes____. Explain your response.
____Conducting with no signal input____.

3. Connect one channel of the oscilloscope to V_{in}, and connect the second channel of the oscilloscope to V_{out}. Adjust the signal source at V_{in} for a signal of 1 kHz at 0.5 $V_{P\text{-}P}$. If the signal you find at V_{out} is distorted on either or both alternations, reduce the input signal level until the distortion is no longer observed. Sketch the waveforms and determine the readings specified in the Observation section.

OBSERVATION V_{in} Waveform:

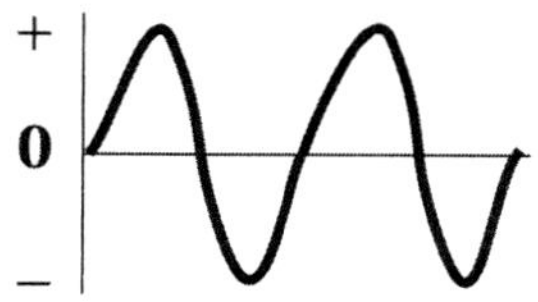

V_{out} Waveform:

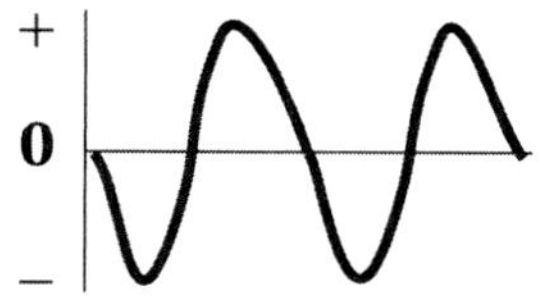

V_{in} = ____0.34____ $V_{P\text{-}P}$.
V_{out} = ____3.18____ $V_{P\text{-}P}$.

 Lab Projects pages 419–421

CONCLUSION The oscilloscope traces indicate that the output voltage from a common-source amplifier is (*in phase, 180° out of phase*) **180° out of phase** with its input waveform. Use the readings you found for V_{in} and V_{out} to determine the voltage gain (A_V) of this circuit. A_V = **9.35**. (Recall that $A_V = V_{out}/V_{in}$.) You have seen in this circuit that the voltage gain of a common-source amplifier can be greater than 1. (*True, False*) **True**.

4. Decrease the setting of the dc source for a V_{DD} of +9 V. If either peak of the waveform at V_{out} is clipped, reduce the input signal level until the clipping effect is no longer apparent. Gather the information requested in the Observation section.

OBSERVATION V_{in} Waveform:

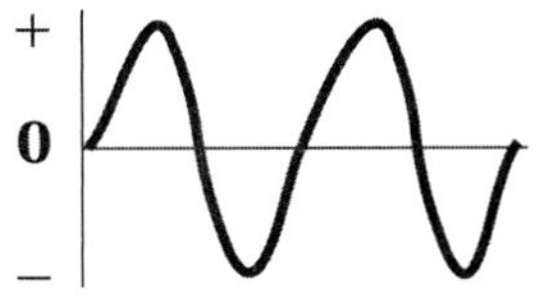

V_{out} Waveform:

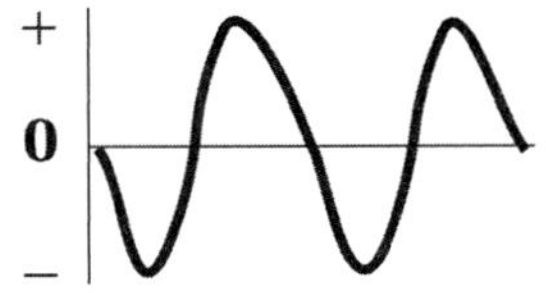

V_{in} = **339** m$V_{P\text{-}P}$.
V_{out} = **357** m$V_{P\text{-}P}$.

CONCLUSION Use the readings for V_{in} and V_{out} to determine the voltage gain of this circuit. A_V = **1.05**. Decreasing the value of V_{DD} increases the voltage gain. (*True, False*) **False**.

Lab Projects pages 419–421

PROJECT 75 Power FETs

PROCEDURE

1. Connect the circuit shown in Figure 75-1.
2. Apply the 12-V supply to the circuit (V_{DD}).
3. Adjust the potentiometer until the gate-source voltage $V_{GS} = 0$ V. Measure the drain-source voltage (V_{DS}) and the voltage across R_3 (V_{R_3}). Increase V_{GS} in steps indicated in the Observation section. Calculate the LED voltage (V_{D_1}), the drain current (I_D), $R_{DS(on)}$, and note the brightness.*

OBSERVATION

$V_{GS} = 0$ V.
$V_{DS} =$ **12 V**.
$V_{R_3} =$ **0 V**.
$V_{D_1} =$ **0 V**.
$I_D =$ **0 mA**.
$R_{DS(on)} =$ **open**.
Brightness **off**.

$V_{GS} = 2$ V.
$V_{DS} =$ ______.
$V_{R_3} =$ ______.
$V_{D_1} =$ ______.
$I_D =$ ______.
$R_{DS(on)} =$ ______.
Brightness ______.

$V_{GS} = 4$ V.
$V_{DS} =$ ______.
$V_{R_3} =$ ______.
$V_{D_1} =$ ______.
$I_D =$ ______.
$R_{DS(on)} =$ ______.
Brightness ______.

$V_{GS} = 8$ V.
$V_{DS} =$ ______.
$V_{R_3} =$ ______.
$V_{D_1} =$ ______.
$I_D =$ ______.
$R_{DS(on)} =$ ______.
Brightness ______.

$V_{GS} = 12$ V.
$V_{DS} =$ ______.
$V_{R_3} =$ ______.
$V_{D_1} =$ ______.
$I_D =$ ______.
$R_{DS(on)} =$ ______.
Brightness ______.

* **Answers will vary. Once the threshold voltage is exceeded the LED will be on and will increase in brightness as V_{GS} increases.**

CONCLUSION Complete the graph in Figure 75-1, which shows how I_D increases with V_{GS}. The brightness of the LED seems to correlate better with the amount of (V_{GS}, V_{DS}) **V_{DS}** than with the amount of (V_{GS}, V_{DS}) **V_{GS}**.

Lab Projects pages 423–429

From the component data sheets: The threshold voltage for the power FET is **2** V. The maximum drain current I_D for the power FET is **5.6**. The maximum continuous LED current is **32 mA (for this circuit).** Explain why the resistor R_3 is required in the circuit. **Since the LED voltage drop is above 1.5 V, the rest of the voltage is dropped across the MOSFET and R_3. R_3 limits the current.**

As V_{GS} increases, $R_{DS(\text{on})}$ *(increases, decreases, stays the same)* **decreases** ________.

4. Connect the circuit shown in Figure 75-2.
5. Apply the 12-V supply to the circuit (V_{DD}).
6. Adjust the potentiometer until the gate-source voltage V_{GS} = 0 V. Measure the drain-source voltage. Increase V_{GS} in steps indicated in the Observation section. Calculate the motor voltage V_M and note the motor rotation, direction, and speed.*

OBSERVATION

V_{GS} = 0 V. Motor rotating (*Yes/No*) **No**.
V_{DS} = **12 V**. Direction (*cw/ccw*) **—**.
V_M = **0 V**. Speed (*slow, fast*) **—**.

V_{GS} = 2 V. Motor rotating (*Yes/No*) ________.
V_{DS} = ________. Direction (*cw/ccw*) ________.
V_M = ________. Speed (*slow, fast*) ________.

V_{GS} = 4 V. Motor rotating (*Yes/No*) ________.
V_{DS} = ________. Direction (*cw/ccw*) ________.
V_M = ________. Speed (*slow, fast*) ________.

V_{GS} = 8 V. Motor rotating (*Yes/No*) ________.
V_{DS} = ________. Direction (*cw/ccw*) ________.
V_M = ________. Speed (*slow, fast*) ________.

V_{GS} = 12 V. Motor rotating (*Yes/No*) ________.
V_{DS} = ________. Direction (*cw/ccw*) ________.
V_M = ________. Speed (*slow, fast*) ________.

* **Answers will vary. Once the threshold voltage is exceeded, the motor will rotate.**

CONCLUSION

The motor speed seems to correlate better with the amount of (V_{GS}, V_{DS}) **V_{DS}** than with the amount of (V_{GS}, V_{DS}) **V_{GS}**. The voltage rating for the motor is **9 to 18 V**. The current rating for the motor is **(see data sheet)**.

Explain why no series resistor is required for the drain-source circuit. **The motor is designed to operate with 12 V applied to it.**

Lab Projects pages 423–429

Did the motor direction change as V_{GS} was increased (*Yes/No*)? **No**. How can the direction of motor rotation be changed? **Reverse the motor wire connections.**

Try your solution. Did your solution work (*Yes/No*)? **Yes**. Now, which direction is the motor turning (*cw/ccw*) **If the motor was turning CW, now it should be turning CCW. If the motor was turning CCW, now it should be turning CW.**

7. Connect the circuit shown in Figure 75-3.

8. Apply the 12-V supply to the circuit (V_{SS}).

9. Adjust the potentiometer until the gate-source voltage V_{SG} = 0 V. Measure the source-drain voltage (V_{SD}) and the voltage across R_3 (V_{R_3}). Increase V_{SG} in steps indicated in the Observation section. Calculate the LED voltage (V_{D_1}), the drain current (I_D), $R_{DS(on)}$, and note the brightness.*

OBSERVATION

V_{SG} = 0 V.
V_{SD} = **12 V**.
V_{R_3} = **0 V**.
V_{D_1} = **0 V**.
I_D = **0 mA**.
$R_{DS(on)}$ = **open**.
Brightness **off**.

V_{SG} = 2 V.
V_{SD} = ______.
V_{R_3} = ______.
V_{D_1} = ______.
I_D = ______.
$R_{DS(on)}$ = ______.
Brightness ______.

V_{SG} = 4 V.
V_{SD} = ______.
V_{R_3} = ______.
V_{D_1} = ______.
I_D = ______.
$R_{DS(on)}$ = ______.
Brightness ______.

V_{SG} = 8 V.
V_{SD} = ______.
V_{R_3} = ______.
V_{D_1} = ______.
I_D = ______.
$R_{DS(on)}$ = ______.
Brightness ______.

V_{SG} = 12 V.
V_{SD} = ______.
V_{R_3} = ______.
V_{D_1} = ______.
I_D = ______.
$R_{DS(on)}$ = ______.
Brightness ______.

* **Answers will vary. Once the threshold voltage is exceeded, the LED will be on. The LED increases in brightness as V_{SG} increases.**

Lab Projects pages 423–429

CONCLUSION Complete the graph in Figure 75-3, which shows how I_D increases with V_{SG}. The brightness of the LED seems to correlate better with the amount of (V_{SG}, V_{SD}) **V_{SD}** than with the amount of (V_{SG}, V_{SD}) **V_{SG}**.

From the component data sheets: The threshold voltage for the power FET is **−2** V. The maximum drain current I_D for the power FET is **12** A. The maximum continuous LED current is **32 mA (for this circuit)**. Explain why the resistor R_3 is required in the circuit. **The resistor R_3 limits the current flow through the LED**

______.

As V_{SG} increases, $R_{DS(on)}$ (*increases, decreases, stays the same*) **decreases** ______.

10. Connect the circuit shown in Figure 75-4.

11. Apply the 12-V supply to the circuit (V_{SS}).

12. Adjust the potentiometer until the gate-source voltage V_{SG} = 0 V. Measure the source-drain voltage (V_{SD}). Increase V_{SG} in steps indicated in the Observation section. Calculate the motor voltage V_M and note the motor rotation, direction, and speed.*

OBSERVATION

V_{SG} = 0 V. Motor rotating (*Yes/No*) **No**.
V_{SD} = **12 V**. Direction (*cw/ccw*) **—**.
V_M = **0 V**. Speed (*slow, fast*) **—**.

V_{SG} = 2 V. Motor rotating (*Yes/No*) ______.
V_{SD} = ______. Direction (*cw/ccw*) ______.
V_M = ______. Speed (*slow, fast*) ______.

V_{SG} = 4 V. Motor rotating (*Yes/No*) ______.
V_{SD} = ______. Direction (*cw/ccw*) ______.
V_M = ______. Speed (*slow, fast*) ______.

V_{SG} = 8 V. Motor rotating (*Yes/No*) ______.
V_{SD} = ______. Direction (*cw/ccw*) ______.
V_M = ______. Speed (*slow, fast*) ______.

V_{SG} = 12 V. Motor rotating (*Yes/No*) ______.
V_{SD} = ______. Direction (*cw/ccw*) ______.
V_M = ______. Speed (*slow, fast*) ______.

* **Answers will vary. Once the threshold voltage is exceeded, the motor will start moving.**

CONCLUSION

The motor speed seems to correlate better with the amount of (V_{SG}, V_{SD}) **V_{SD}** than with the amount of (V_{SG}, V_{SD}) **V_{SG}**. The voltage rating for the motor is **9 to 18 V**. The current rating for the motor is **(see data sheet)**.

Explain why no series resistor is required for the source-drain circuit. **The motor is designed to operate with 12 V applied**

Did the motor direction change as V_{SG} was increased (*Yes/No*)? **No**. How can the direction of motor rotation be changed? **Reverse the motor wire connections**

Try your solution. Did your solution work (*Yes/No*)? **Yes**. Now, which direction is the motor turning (*cw/ccw*) **If the motor was turning CW, now it should be CCW. If the motor was turning CCW, now it should be turning CW.**

Lab Projects pages 423–429

Data Sheet

SR-65S Super High Speed Electric Motor
(273-0256) Specifications Faxback Doc. # 19832

Type: .... Heavy Duty Series/Carbon Brush

Mounting Screws: .... 2.5 mm

Outer Diameter of Motor: .... 27.7 mm or 1.091 Inch

Shaft Diameter: .... 2.3 mm or 0.0905 Inch

Motor Extension: .... 10.0 mm or 0.394 Inch

Length of Shaft (case to end): .... 15.7 mm

Shaft to Base: .... 51.0 mm

Weight (approx.): .... 49.0 gm

At 9 Volts:

Speed, No Load: .... 11,500 RPM

Current Drain at No Load: .... 0.28 - 0.350 mA

Speed at Maximum Efficiency: .... 9,200 RPM

Current Drain at Max. Efficiency: .... 1.86 A

Stall Torque: .... 180 G-cm

At 12 Volts:

Speed, No Load: .... 17,500 RPM

Current Drain at No Load: .... 0.29 - 0.360 mA

Speed at Maximum Efficiency: .... 13,200 RPM

Current Drain at Max. Efficiency: .... 1.9 A

Stall Torque: .... 240 G-cm

At 18 Volts:

Speed, No Load: .... 24,000 RPM

Current Drain at No Load: .... 0.32 - 0.40 A

Speed at Maximum Efficiency: .... 18,000 RPM

Current Drain at No Load: .... 1.98 A

Stall Torque: .... 350 G-cm

Mounting holes are tapped for M2.6 screws on shaft end on 16 mm centers.

Specifications are typical; individual units might vary. Specifications are subject to change and improvement without notice.

(IR/EB 4/16/97)

PART 24 Story Behind the Numbers

Answers will vary. Data and graph are provided to give a perspective of data table and graph content.

Table 1

V_{DD} = 24 V data

Measurements				Calculations		
V_{GG}	V_{GS}	V_{DD}	V_{DS}	I_G	I_D	I_S
V	V	V	V	mA	mA	mA
0	0	24	8	0	16	16
−0.5	−0.5	24	11.75	0	12.25	12.25
−1	−1	24	15	0	9	9
−1.5	−1.5	24	17.75	0	6.25	6.25
−2	−2	24	20	0	4	4
−2.5	−2.5	24	21.75	0	2.25	2.25
−3	−3	24	23	0	1	1
−3.5	−3.5	24	23.75	0	0.25	0.25
−4	−4	24	24	0	0	0
				0	0	0
				0	0	0
				0	0	0

Lab Projects pages 431–437

Table 2

$V_{GG} = 0$ V data

Measurements				Calculations		
V_{GG}	V_{GS}	V_{DD}	V_{DS}	I_G	I_D	I_S
V	V	V	V	mA	mA	mA
0	0	0	0	0	0	0
0	0	1	0.2	0	0.8	0.8
0	0	2	0.4	0	1.6	1.6
0	0	4	0.8	0	3.2	3.2
0	0	6	1.2	0	4.8	4.8
0	0	8	1.6	0	6.4	6.4
0	0	10	2	0	8	8
0	0	12	2.4	0	9.6	9.6
0	0	14	2.8	0	11.2	11.2
0	0	16	3.2	0	12.8	12.8
0	0	18	3.6	0	14.4	14.4
0	0	20	4	0	16	16
0	0	22	6	0	16	16
0	0	24	8	0	16	16

Table 3

$V_{GG} = -1$ V data

Measurements				Calculations		
V_{GG}	V_{GS}	V_{DD}	V_{DS}	I_G	I_D	I_S
V	V	V	V	mA	mA	mA
−1	−1	0	0	0	0	0
−1	−1	1	0.25	0	0.75	0.75
−1	−1	2	0.5	0	1.5	1.5
−1	−1	4	1	0	3	3
−1	−1	6	1.5	0	4.5	4.5
−1	−1	8	2	0	6	6
−1	−1	10	2.5	0	7.5	7.5
−1	−1	12	3	0	9	9
−1	−1	14	5	0	9	9
−1	−1	16	7	0	9	9
−1	−1	18	9	0	9	9
−1	−1	20	11	0	9	9
−1	−1	22	13	0	9	9
−1	−1	24	15	0	9	9

Table 4

$V_{GG} = -2$ V data

Measurements				Calculations		
V_{GG}	V_{GS}	V_{DD}	V_{DS}	I_G	I_D	I_S
V	V	V	V	mA	mA	mA
−2	−2	0	0	0	0	0
−2	−2	1	0.33	0	0.67	0.67
−2	−2	2	0.67	0	1.33	1.33
−2	−2	4	1.33	0	2.67	2.67
−2	−2	6	2	0	4	4
−2	−2	8	4	0	4	4
−2	−2	10	6	0	4	4
−2	−2	12	8	0	4	4
−2	−2	14	10	0	4	4
−2	−2	16	12	0	4	4
−2	−2	18	14	0	4	4
−2	−2	20	16	0	4	4
−2	−2	22	18	0	4	4
−2	−2	24	20	0	4	4

Graph 1

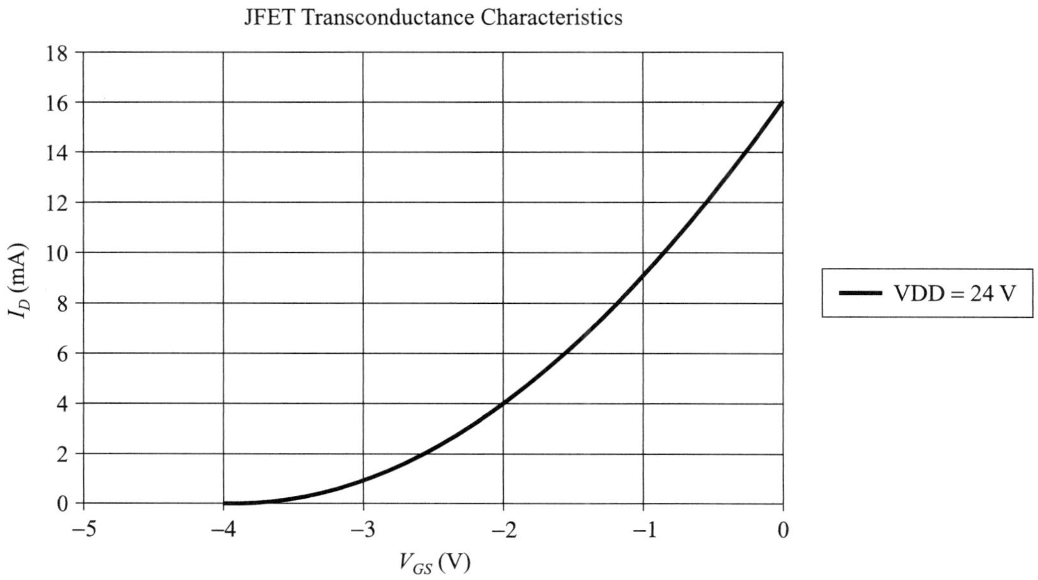

Lab Projects pages 431–437

Graph 2

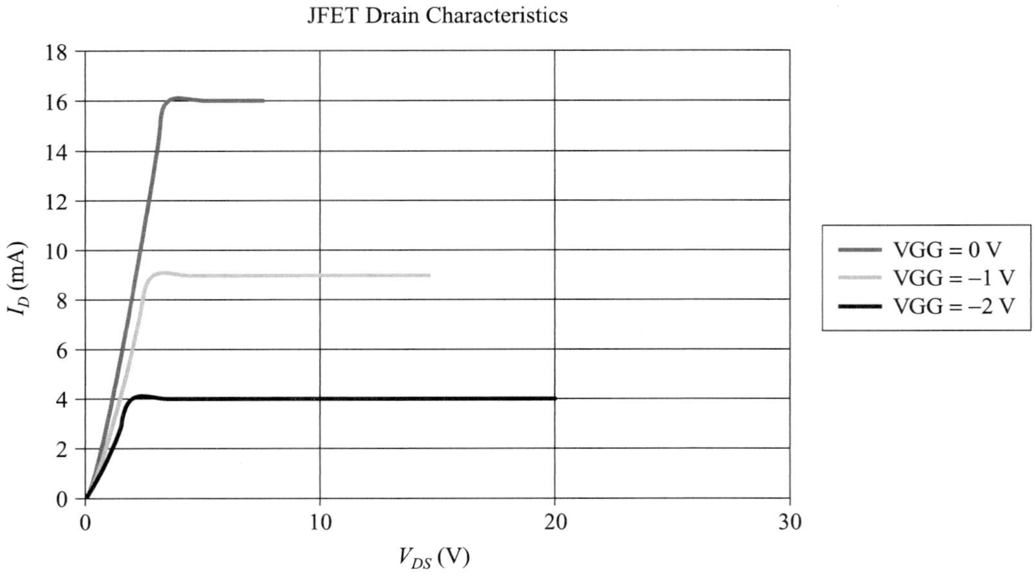

Answers to Analysis Questions

1. The maximum amount of drain current flows through the channel region of a JFET when V_{GS} is at 0 V (*True/False*) **True**.

2. The maximum amount of drain current is called (*pinch-off current, leakage current, drain-source saturation current I_{DSS}*) **drain-source saturation current (I_{DSS})**.

3. The gate-source junction of the n-channel JFET for V_{GG} = 0 V (*has no bias, is reverse biased, is forward biased*) **has no bias**.

4. The gate-source junction of the n-channel JFET for V_{GG} = −1 V (*has no bias, is reverse biased, is forward biased*) **is reverse biased**.

5. An n-channel JFET is (*an enhancement-mode, a depletion-mode*) **a depletion-mode** device.

6. For the n-channel JFET, as the gate source voltage V_{GS} increases in the negative direction, the amount of drain current I_D (*increases, decreases*) **decreases**.

7. The gate current for the n-channel JFET is greater than 10 μA (*True/False*) **False**.

Tri-Power Supply

To produce both positive and negative voltages from a tri-power supply where each output is independent, you will connect the power supply as shown in Figure 75-5.

Lab Projects pages 431–437

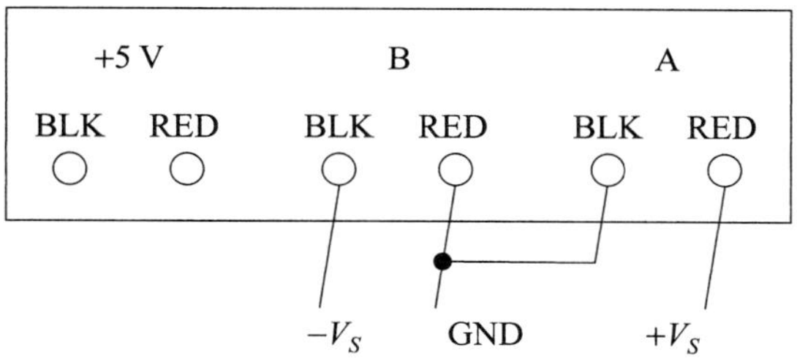

FIGURE 75-5 Tri-power supply (independent mode)

The black terminal of output A is connected to the red terminal of output B. These two terminals will be the ground for the power supply and your circuit board.

- Output A (red terminal) will be a positive voltage with respect to the circuit ground.
- Output B (black terminal) will be a negative voltage with respect to the circuit ground.

Lab Projects pages 431–437

PART 24 Summary

Complete the following review questions, indicating the appropriate response by placing a check in the box next to the correct answer.

1. When a JFET is properly used, the drain current is proportional to the gate current.
   - ☐ True
   - ☑ False
2. The source-gate junction of a JFET requires little bias current and signal current because it is
   - ☑ normally reverse biased
   - ☐ normally forward biased
   - ☐ normally not used
   - ☐ made of a nonconductive metal oxide film
3. A JFET should always be biased for
   - ☑ depletion-mode operation
   - ☐ enhancement-mode operation
   - ☐ sine-wave mode operation
4. For an n-channel JFET, making the gate more negative with respect to the source
   - ☐ causes drain current to increase
   - ☑ causes drain current to decrease
   - ☐ has no significant effect on drain current
5. For a common-source amplifier that uses an n-channel JFET, making the gate more negative with respect to the source causes the drain voltage to increase in the ________________ direction.
   - ☑ positive
   - ☐ negative
6. The main difference between an n-channel JFET amplifier and a p-channel JFET amplifier is
   - ☐ the n-channel version inverts the incoming signal, whereas the corresponding p-channel version does not
   - ☑ the polarity of the bias voltages
   - ☐ the n-channel version operates in the depletion mode, whereas the corresponding p-channel version operates in the enhancement mode
7. A common-drain JFET amplifier
   - ☐ is impractical
   - ☐ has a very low input impedance
   - ☑ is sometimes called a voltage follower
   - ☐ inverts the input signal

Lab Projects pages 439–440

8. When an n-channel JFET is operating as a Class B amplifier, you would expect to find
   - [x] a relatively large negative voltage on the gate
   - [ ] a relatively small negative voltage on the gate
   - [ ] a relatively large positive voltage on the gate

9. You can distinguish a common-gate JFET amplifier by noting that
   - [ ] the input is applied at the gate and the output is taken from the gate
   - [x] the input is applied to the drain and the output is taken from the source
   - [ ] the input is applied to the gate and the output is taken from the source
   - [ ] none of these

10. Inserting a resistor in series with the source terminal of an n-channel JFET can make the source more positive than the gate. This principle is the basis for
    - [ ] preventing the JFET from overheating
    - [ ] increasing the current through the drain terminal
    - [x] establishing self-bias
    - [ ] establishing Class C operation

Lab Projects pages 439–440

PART 25

OPERATIONAL AMPLIFIERS

PROJECT 76 Inverting Op-Amp Circuit

SPECIAL NOTE:

You will need the following formula to complete the work:

$$A_V = -(R_2/R_1)$$

where:

A_V is the voltage gain
R_2 is the value of the feedback resistor
R_1 is the value of the input resistor

PROCEDURE

1. Connect the circuit exactly as shown in Figure 76-1. Make sure the positive terminal of the ±15-Vdc supply is connected to pin 7 of the 741 IC, the negative terminal is connected to pin 4, and the common terminal is connected to the common line as shown in the figure. Be sure you take all meter and oscilloscope readings with respect to the common line of the circuit (and not the −15-V connection of the power source). Measure and record the dc voltages requested in the Observation section.

OBSERVATION DC voltage from pin 7 to common = **+15** V.
DC voltage from pin 4 to common = **−15** V.

2. Connect the function generator (sine-wave mode) or audio oscillator to V_{in}. Connect one channel of the oscilloscope to V_{in}, and connect the second channel of the oscilloscope to V_{out}. Adjust the signal source for an input signal of 1 kHz at 1 $V_{P\text{-}P}$. Sketch the waveforms and determine the readings specified in the Observation section.

OBSERVATION Peak-to-peak voltage from V_{in} to common = **1** V.
Peak-to-peak voltage from V_{out} to common = **4.7** V.

V_{in} Waveform:

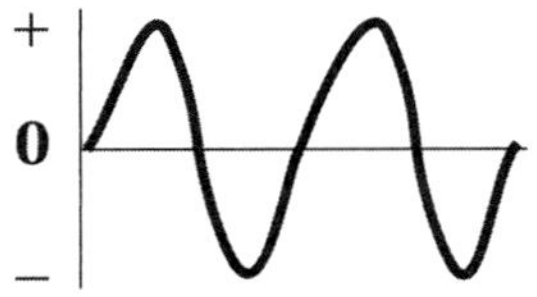

Lab Projects pages 443–445

V_{out} Waveform:

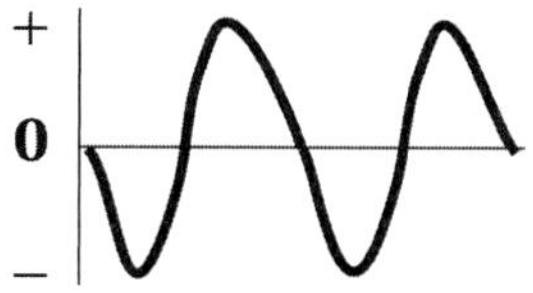

CONCLUSION What is the calculated voltage gain of the circuit shown in Figure 76-1? ___**–4.7**___. What is the actual voltage gain of the circuit as determined by the measured values of V_{in} and V_{out}? ___**–4.7**___. The output waveform is shifted (*0°, 90°, 180°*) ___**180°**___ relative to the input waveform.

3. Increase the level of V_{in} to 2 $V_{P\text{-}P}$; measure and record the values of V_{in} and V_{out}.

OBSERVATION Peak-to-peak voltage from V_{in} to common = ___**2**___ V.

Peak-to-peak voltage from V_{out} to common = ___**9.4**___ V.

CONCLUSION What is the actual voltage gain of the circuit as determined by the measured values of V_{in} and V_{out} in Procedure step 3? ___**≈ –4.7**___. Does increasing the value of V_{in} have any significant effect upon the value of V_{out}? ___**Yes**___. Does increasing the value of V_{in} have any significant effect upon the voltage gain of the circuit? ___**No**___.

4. Replace R_2 with a resistor having a value of 100 kΩ.

5. Adjust the signal source for 1 $V_{P\text{-}P}$. Sketch the waveforms and determine the readings specified in the Observation section.

OBSERVATION Peak-to-peak voltage from V_{in} to common = ___**1**___ V.

Peak-to-peak voltage from V_{out} to common = ___**10**___ V.

V_{in} Waveform:

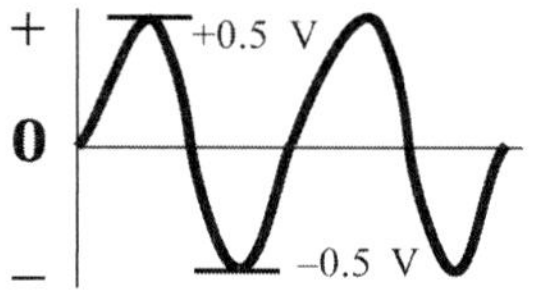

V_{out} Waveform:

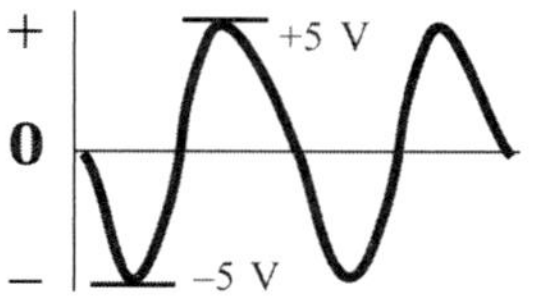

CONCLUSION What is the calculated voltage gain of the circuit when R_2 = 100 kΩ? ___**–10**___. What is the actual voltage gain of the circuit as determined by the measured values of V_{in} and V_{out}? ___**–10**___. Does increasing the value of R_2 have any significant effect upon the voltage gain of the circuit? ___**Yes**___.

Lab Projects pages 443–445

PROJECT 77 Noninverting Op-Amp Circuit

SPECIAL NOTE:

The following formula can be helpful:

$$A_V = (R_2/R_1) + 1$$

where:

A_V is the voltage gain of a noninverting op-amp
R_2 is the value of the feedback resistor
R_1 is the value of the input resistor

PROCEDURE

1. Connect the circuit exactly as shown in Figure 77-1. Make sure the positive terminal of the ±15-Vdc supply is connected to pin 7 of the 741 IC, the negative terminal is connected to pin 4, and the common terminal is connected to the common line as shown in the figure. Be sure you take all meter and oscilloscope readings with respect to the common line of the circuit (and not the −15-V connection of the power source). Measure and record the dc voltages requested in the Observation section.

OBSERVATION DC voltage from pin 7 to common = **+15** V.
DC voltage from pin 4 to common = **−15** V.

2. Connect the function generator (sine-wave mode) or audio oscillator to V_{in}. Connect one channel of the oscilloscope to V_{in}, and connect the second channel of the oscilloscope to V_{out}. Adjust the signal source for an input signal of 1 kHz at 1 $V_{P\text{-}P}$. Sketch the waveforms and determine the readings specified in the Observation section.

OBSERVATION Peak-to-peak voltage from V_{in} to common = **1** V.
Peak-to-peak voltage from V_{out} to common = **5.7** V.

V_{in} Waveform:

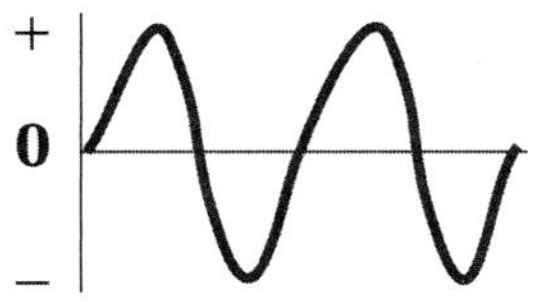

V_{out} Waveform:

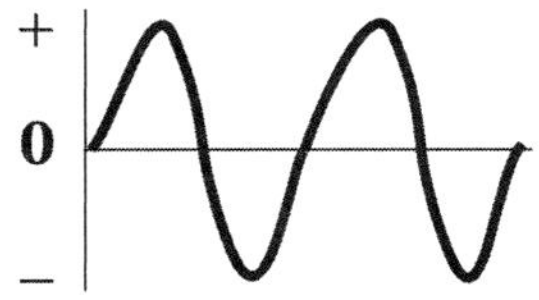

CONCLUSION What is the calculated voltage gain of the circuit shown in Figure 77-1? **5.7**.
What is the actual voltage gain of the circuit as determined by the measured values of

Lab Projects pages 447–450

V_{in} and V_{out}? ___**5.7**___. The output waveform is shifted (*0°*, *90°*, *180°*) ___**0°**___ relative to the input waveform.

3. Replace R_2 with a resistor having a value of 10 kΩ.
4. Leave the signal source at 1 $V_{P\text{-}P}$. Measure and record the data specified in the Observation section.

OBSERVATION Peak-to-peak voltage from V_{in} to common = ___**1**___ V.

Peak-to-peak voltage from V_{out} to common = ___**2**___ V.

CONCLUSION What is the calculated voltage gain of the circuit when R_2 = 10 kΩ? ___**1**___. What is the actual voltage gain of the circuit as determined by the measured values of V_{in} and V_{out}? ___**2**___. Explain why this particular circuit might be called a voltage follower. **Short out R_2 to get a voltage follower**

____

____.

5. Replace R_2 with a resistor having a value of 1 MΩ.
6. Leave the signal source at 1 $V_{P\text{-}P}$. Sketch the waveforms and determine the readings specified in the Observation section.

OBSERVATION Peak-to-peak voltage from V_{in} to common = ___**1**___ V.

Peak-to-peak voltage from V_{out} to common = ___**28**___ V.

V_{in} Waveform:

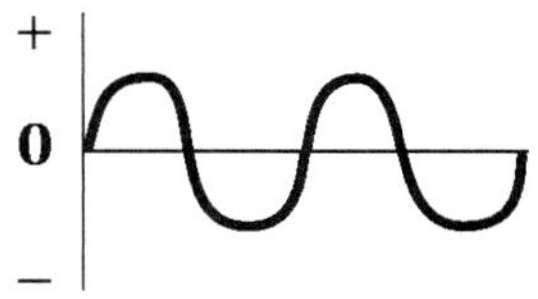

V_{out} Waveform:

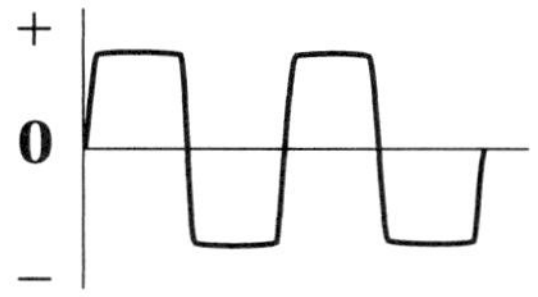

CONCLUSION What is the calculated voltage gain of the circuit when R_2 = 1 MΩ? ___**101**___.

Explain why the output waveform is distorted. **The output peaks cannot exceed the op-amp output saturation voltages (less than the supply voltage levels)**

____.

7. Connect the circuit exactly as shown in Figure 77-2.
8. Connect the function generator to V_{in}. Set the function generator to 1 $V_{P\text{-}P}$ and 1 kHz sine wave. Connect scope channel 1 to V_{in} and scope channel 2 to V_{out}. Sketch the V_{in} and V_{out} waveforms.

Lab Projects pages 447–450

OBSERVATION V_{in} Waveform:

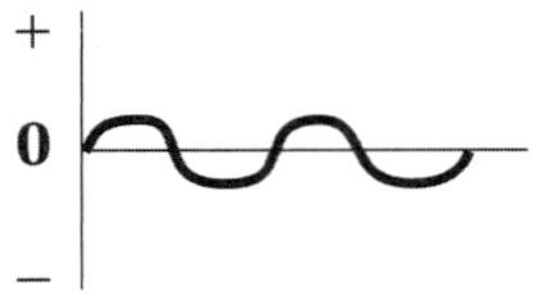

V_{out} Waveform:

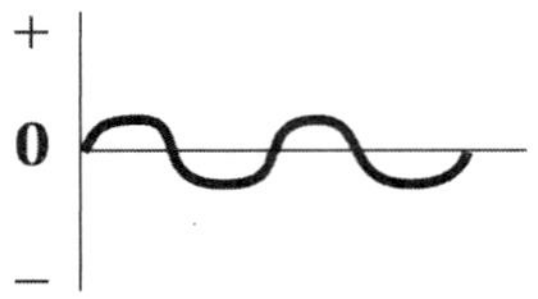

CONCLUSION What is the calculated voltage gain of the circuit? ____**1**____. What is the actual voltage gain of the circuit as determined by the measured values of V_{in} and V_{out}?

____**1**____. Explain why this circuit might be called a unity gain analog buffer. ____ **The circuit has a high input impedance and a low output impedance with a gain of 1**____.

Lab Projects pages 447–450

PROJECT 78 Op-Amp Schmitt-Trigger Circuit

PROCEDURE

1. Connect the circuit exactly as shown in Figure 78-1. Make sure the positive terminal of the ±15-Vdc supply is connected to pin 7 of the 741 IC, the negative terminal is connected to pin 4, and the common terminal is connected to the common line as shown in the figure. (Be sure you take all meter and oscilloscope readings with respect to the common line of the circuit.)

2. Measure and record the dc voltages requested in the Observation section.

OBSERVATION DC voltage from pin 7 to common = **+15** V.

DC voltage from pin 4 to common = **–15** V.

CONCLUSION Explain why having an ac input waveform makes it necessary to have both positive and negative supply voltages for an op-amp circuit. **So the op-amp can respond to both input peaks**

3. Connect the function generator (sine-wave mode) or audio oscillator to V_{in}. Connect one channel of the oscilloscope to V_{in} and connect the second channel of the oscilloscope to V_{out}. Adjust the signal source for an input signal of 1 kHz at 1 $V_{P\text{-}P}$.

4. Sketch the waveforms and determine the readings specified in the Observation section.

OBSERVATION Peak-to-peak voltage from V_{in} to common = **1** V.

Peak-to-peak voltage from V_{out} to common = **1** V.

V_{in} Waveform:

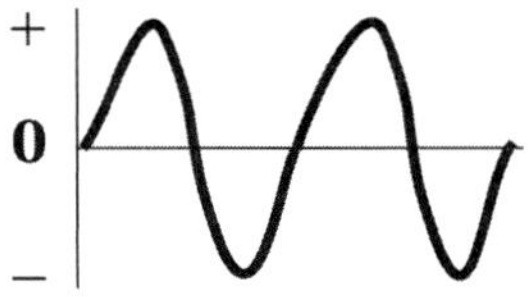

V_{out} Waveform:

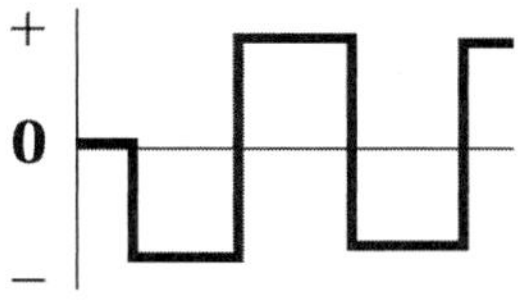

CONCLUSION Calculate the following values for the Schmitt-trigger circuit in Figure 78-1: *UTP* (calculated) = **+0.5** V. *LTP* (calculated) = **–0.5** V. Hysteresis = **1** V. If you superimpose oscilloscope waveforms for V_{in} and V_{out}, you should be able to determine the actual values of: *UTP* (measured) = **+0.5** V. *LTP* (measured) = **–0.5** V. Hysteresis = **1** V. The polarity of the output waveform is shifted

Lab Projects pages 451–453

(*0°, 90°, 180°*) **180°** relative to the input waveform. This is an example of a nonlinear amplifier circuit. (*True, False*) **True**.

5. Adjust the signal source for an input signal 0.1 $V_{P\text{-}P}$. Sketch the waveforms and determine the readings specified in the Observation section.

OBSERVATION Peak-to-peak voltage from V_{in} to common = **0.1** V.

Peak-to-peak voltage from V_{out} to common = **0** V.

V_{in} Waveform:

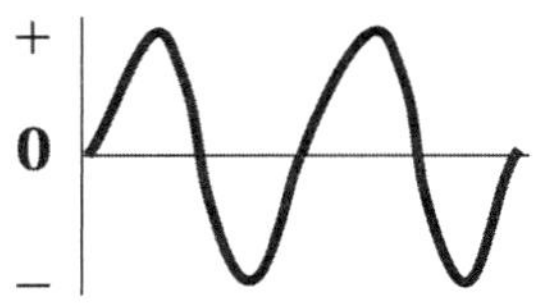

V_{out} Waveform:

CONCLUSION Describe how the waveforms are different from those in Procedure step 4.
No output

How do you account for the differences?
Input is less than UTP and LTP

Lab Projects pages 451–453

PART 25 Summary

Complete the following review questions, indicating the appropriate response by placing a check in the box next to the correct answer.

1. For an inverting op-amp
   - ☐ the input and feedback resistors are both connected to the noninverting input
   - ☐ the input resistor is connected to the inverting input and the feedback resistor is connected to the noninverting input
   - ☐ the input resistor is connected to the noninverting input and the feedback resistor is connected to the inverting input
   - ☑ the input and feedback resistors are both connected to the inverting input

2. The voltage gain of an inverting op-amp is determined by
   - ☑ dividing the amount of feedback resistance by the amount of input resistance
   - ☐ dividing the amount of input resistance by the amount of feedback resistance
   - ☐ dividing the amount of input voltage by the value of the input resistance
   - ☐ multiplying the amount of input voltage by the amount of output voltage

3. An inverting op-amp circuit can never be used as a voltage follower because
   - ☐ the output voltage is always larger than the input voltage
   - ☑ the output waveform is always out of phase with the input waveform
   - ☐ the output is nonlinear

4. What is the voltage gain of an inverting op-amp circuit when the input resistance equals the feedback resistance?
   - ☐ 0
   - ☐ +1
   - ☑ −1
   - ☐ Cannot be determined without knowing the resistor values

5. For a noninverting op-amp
   - ☐ the input is applied to the noninverting input and the feedback resistor is connected to the noninverting input
   - ☐ the input is applied to the inverting input and the feedback resistor is connected to the noninverting input
   - ☑ the input is applied to the noninverting input and the feedback resistor is connected to the inverting input
   - ☐ the input is applied to the inverting input and the feedback resistor is connected to the inverting input

6. The voltage gain of a noninverting op-amp
   - ☐ can be determined by dividing the amount of feedback resistance by the amount of input resistance
   - ☑ can be determined by dividing the amount of output voltage by the amount of input voltage
   - ☐ can be determined by dividing the amount of input voltage by the value of the feedback resistance
   - ☐ is always 1

Lab Projects pages 455–456

7. What is the voltage gain of a noninverting amplifier when the feedback resistance is zero ohms?

- [ ] 0
- [x] 1
- [ ] 2
- [ ] Cannot be determined

8. As long as the output of an op-amp remains undistorted, the amount of input voltage has little to do with the amount of voltage gain for the circuit.

- [x] True
- [ ] False

9. One of the main purposes of a Schmitt-trigger circuit is to transform a sine waveform into a rectangular waveform.

- [x] True
- [ ] False

10. The fact that the input signal for an op-amp Schmitt-trigger circuit goes to the inverting input accounts for

- [ ] the flattening of the output waveform
- [ ] the very high voltage gain of the circuit
- [x] the 180° phase shift of the waveform
- [ ] none of these

Lab Projects pages 455–456

PART 26

OSCILLATORS AND MULTIVIBRATORS

PROJECT 79 Wien-Bridge Oscillator Circuit

SPECIAL NOTE:

You will need the following formula for calculating the operating frequency of a Wien-bridge oscillator:

$$f_r = 1/2\pi RC$$

where:

f_r is the operating frequency of the circuit
R is the value assigned to both resistors in the lead-lag network
C is the value assigned to both capacitors in the lead-lag network

PROCEDURE

1. Connect the circuit shown in Figure 79-1.

CONCLUSION List the components that make up the lead-lag network. **C_1, R_2, C_2, R_1**

This op-amp is connected as (*an inverting, a noninverting*) **a noninverting** amplifier.

2. Connect the oscilloscope to V_{out} and adjust potentiometer R_3 until you see a maximum level of stable oscillation.

3. Sketch two complete cycles of the output waveform in the space provided in the Observation section.

OBSERVATION Output Waveform:

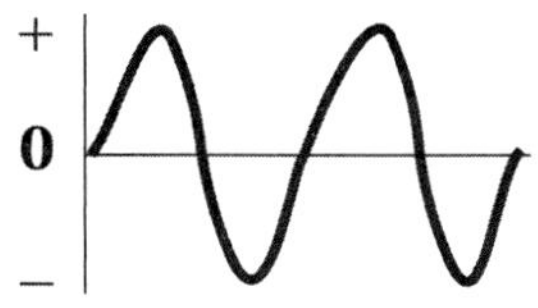

CONCLUSION Describe the form of the output waveform as sinusoidal, rectangular, sawtooth, or triangular: **sinusoidal**.

4. Measure the peak-to-peak amplitude of the waveform at V_{out}.

OBSERVATION V_{out} (peak-to-peak) = **12** V.

CONCLUSION The amplitude of the output waveform is mainly determined by (*the gain of the amplifier, the values of the components in the lead-lag circuit*) **the gain of the amplifier**.

5. Measure the period of the waveform at V_{out}.

OBSERVATION *Period* = **0.63** ms.

CONCLUSION Calculate the operating frequency of this circuit according to the formula for the operating frequency of a Wien-bridge oscillator. **1.59** kHz. Use the formula, $f = 1/period$, to determine the actual operating frequency. **1.58** kHz. Account for the difference, if any, between the calculated and measured frequency. **Tolerances of R_5 and C_5**.

6. Replace resistors R_1 and R_2 in the circuit with 2.7-kΩ resistors. Adjust potentiometer R_3, if necessary, to get a maximum level of stable oscillation.

CONCLUSION According to the formula for the operating frequency of a Wien-bridge oscillator, increasing the value of the lead-lag resistors should (*increase, decrease, have no effect on*) **decrease** the operating frequency.

7. Measure the period of the waveform at the output.

OBSERVATION *Period* = **1.7** ms.

CONCLUSION The actual operating frequency of the circuit is now **590** Hz.

Lab Projects pages 459–461

PROJECT 80 555 Astable Multivibrator

SPECIAL NOTE:

You will need the following formula for calculating the operating frequency of this astable multivibrator:

$$f = 1/0.69C(R_A + 2R_B)$$

where:

f is the output frequency
C is the value of the timing capacitor
R_A and R_B are the values of the timing resistors

PROCEDURE

1. Connect the circuit exactly as shown in Figure 80-1.

CONCLUSION Which resistor in the formula for operating frequency corresponds to resistor R_1 in this circuit? **R_A**. Which corresponds to resistor R_2? **R_B**.

2. Connect the oscilloscope to V_{out} and sketch two complete cycles of the output waveform in the space provided in the Observation section.

OBSERVATION Output Waveform:

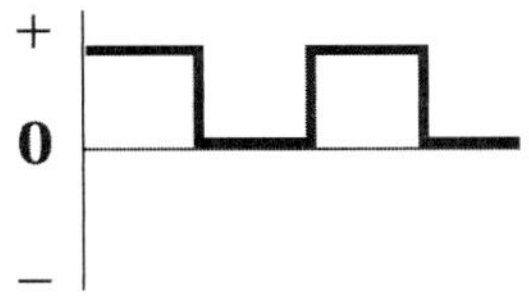

CONCLUSION Describe the form of the output waveform as sinusoidal, rectangular, sawtooth, or triangular: **rectangular**.

3. Measure the peak-to-peak amplitude and period of the waveform at V_{out}.

OBSERVATION V_{out} (peak-to-peak) = **11.6** V.
Period = **1.4** ms.

CONCLUSION Calculate the operating frequency of this circuit according to the formula for the operating frequency of 555-type astable multivibrators. **0.721** kHz. Use the formula, $f = 1/period$, to determine the actual operating frequency. **0.714** kHz. Account for the difference, if any, between the calculated and measured frequency. **Tolerance of components**

4. Replace resistor R_2 with a 10-kΩ resistor.

CONCLUSION According to the formula for the operating frequency of a 555-type astable multivibrator, decreasing the value of either resistor should (*increase, decrease, have no effect on*) **increase** the operating frequency.

5. Measure the peak-to-peak amplitude and period of the waveform at V_{out}.

OBSERVATION V_{out} (peak-to-peak) = **11.6** V.
Period = **0.145** ms.

CONCLUSION The actual operating frequency of the circuit is now **6,896** Hz.

Lab Projects pages 463–464

Part 26 Summary

Complete the following review questions, indicating the appropriate response by placing a check in the box next to the correct answer.

1. At the frequency of oscillation, the lead-lag *RC* network in a Wien-bridge oscillator produces an overall phase shift of ________ degrees.
   - ☑ 0
   - ☐ 90
   - ☐ 180
   - ☐ 270

2. In an op-amp version of a Wien-bridge oscillator, the lead-lag network is connected to the ________ input of the op-amp.
   - ☐ inverting
   - ☑ noninverting
   - ☐ voltage offset

3. In a Wien-bridge oscillator, decreasing the value of either or both resistors in the lead-lag network
   - ☑ increases the operating frequency
   - ☐ decreases the operating frequency
   - ☐ has no effect on the operating frequency

4. In a Wien-bridge oscillator, increasing the value of either or both capacitors in the lead-lag network
   - ☐ increases the operating frequency
   - ☑ decreases the operating frequency
   - ☐ has no effect on the operating frequency

5. In a 555 astable multivibrator, the charge path for the timing capacitor is through
   - ☐ just one timing resistor
   - ☑ both timing resistors
   - ☐ neither timing resistor

6. In a 555 astable multivibrator, the discharge path for the timing capacitor is through
   - ☑ just one timing resistor
   - ☐ both timing resistors
   - ☐ neither timing resistor

7. In a 555 astable multivibrator, decreasing the value of either or both of the timing resistors
   - ☑ increases the operating frequency
   - ☐ decreases the operating frequency
   - ☐ has no effect on the operating frequency

8. For a 555-type astable multivibrator, the discharge time (time the waveform is near zero volts) is always greater than the charge time (time the waveform is near the value of the positive supply voltage).
   - ☐ True
   - ☑ False

Lab Projects pages 465–466

9. In a 555 astable multivibrator, increasing the value of the timing capacitor
   - ☐ increases the operating frequency
   - ☑ decreases the operating frequency
   - ☐ has no effect on the operating frequency

10. According to the formulas for the operating frequency of Wien-bridge oscillators and 555-type astable multivibrators, increasing the amount of supply voltage should
   - ☐ increase the operating frequency
   - ☐ decrease the operating frequency
   - ☑ have no effect on the operating frequency

Lab Projects pages 465–466

PART 27

SCR OPERATION

PROJECT 81 SCRs in DC Circuits

SPECIAL NOTE:

Make all meter readings with respect to the common connections for the circuit.

PROCEDURE

1. Connect the circuit exactly as shown in Figure 81-1. Make sure the 6-Vdc power is turned **off**. Make sure there is **no connection** between points *A* and *B* of the circuit.

2. Turn on the dc power supply and make sure it is set for 6 V.

   NOTE ➤ The lamp should not be on at this time.

   Measure the voltages at the anode and at the gate of the SCR. Record the data in the Observation section.

   OBSERVATION Anode voltage = ______**6**______ V.

   Gate voltage = ______**0**______ V.

   The lamp is (*on, off*) ___**off**___.

   CONCLUSION The cathode-anode terminals of the SCR in this circuit are (*forward, reverse, not*) ______**forward**______ biased. The cathode-gate terminals are (*forward, reverse, not*) ______**not**______ biased. Calculate the voltage across the lamp by subtracting the anode voltage (recorded in the Observation section) from the dc supply voltage: Lamp (load) voltage = ___**0**___ V. Under the conditions observed here, the SCR must be in its (*on, off*) ___**off**___ state. This means there is (*current, no current*) ______**no current**______ flowing through the SCR and through the load.

3. With 6 Vdc still applied to the circuit, make a jumper-wire connection between points *A* and *B*. Provide the data required in the Observation section.

   OBSERVATION The lamp is (*on, off*) ___**on**___.

   Anode voltage = ______**1**______ V.

   Gate voltage = ______**0.7**______ V.

CONCLUSION The cathode-anode terminals of the SCR in this circuit are (*forward, reverse, not*) ___**forward**___ biased. The cathode-gate terminals are (*forward, reverse, not*) ___**forward**___ biased. Calculate the voltage across the lamp: Lamp (load) voltage = ___**5**___ V. Under the conditions observed here, the SCR is in its (*on, off*) ___**on**___ state. This means there is (*current, no current*) ___**current**___ flowing through the SCR and through the load.

4. With 6 Vdc still applied to the circuit, remove your jumper-wire connection between points *A* and *B*. Provide the data required in the Observation section.

OBSERVATION The lamp is (*on, off*) ___**on**___.

Anode voltage = ___**1**___ V.

Gate voltage = ___**0**___ V.

CONCLUSION The cathode-anode terminals of the SCR in this circuit are (*forward, reverse, not*) ___**forward**___ biased. The cathode-gate terminals are (*forward, reverse, not*) ___**not**___ biased.

5. Turn off the dc power supply. Monitor the dc power supply output voltage with your voltmeter and wait for the voltage to drop all the way to zero.

6. Make sure there is **no connection** between points *A* and *B* of the circuit, and turn on the dc power supply. Provide the data required in the Observation section.

OBSERVATION The lamp is (*on, off*) ___**off**___.

CONCLUSION Explain what is necessary to get the SCR conducting once you have restored dc power to the circuit. ___**Forward bias the gate**___
____
______________________________.

Lab Projects pages 469–471

PROJECT 82 SCRs in AC Circuits

SPECIAL NOTE:

Make all oscilloscope readings with respect to the common connections for the circuit.

PROCEDURE

1. Connect the circuit exactly as shown in Figure 82-1. Connect one channel of the oscilloscope to the 6.3-Vac source, and connect the second channel of the oscilloscope to the anode of the SCR. Make sure there is **no connection** between points *A* and *B* of the circuit.

2. With ac power applied to the circuit, note the state of the lamp and the waveforms. Record your findings in the Observation section.

OBSERVATION The lamp is (*on, off*) **off**.

V_{ac} Source Waveform:

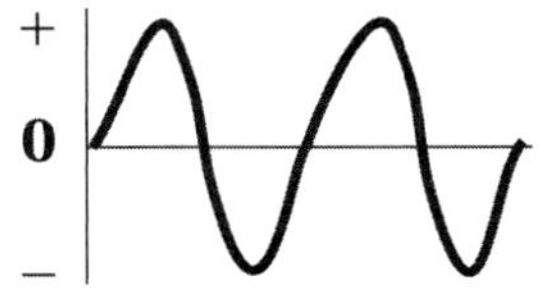

SCR Anode Waveform:

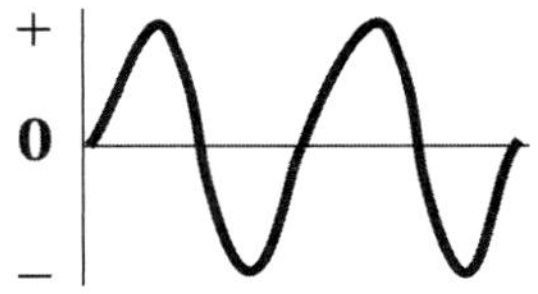

V_{ac} voltage = **17.8** $V_{P\text{-}P}$.

SCR anode voltage = **17.8** $V_{P\text{-}P}$.

CONCLUSION The SCR is evidently (*conducting on both half-cycles, conducting on every other half-cycle, not conducting at all*) **not conducting at all**.

Explain your answer. **SCR is off**.

In this particular operating state, the SCR is acting as (*an open, a closed*) **an open** switch.

3. Connect a jumper wire between points *A* and *B* in the circuit. With ac power still applied to the circuit, note the state of the lamp and the waveforms appearing on the oscilloscope. Record your findings in the Observation section.

OBSERVATION The lamp is (*on, off*) **on**.

V_{ac} Source Waveform:

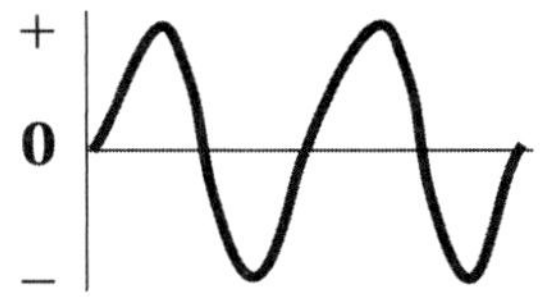

SCR Anode Waveform:

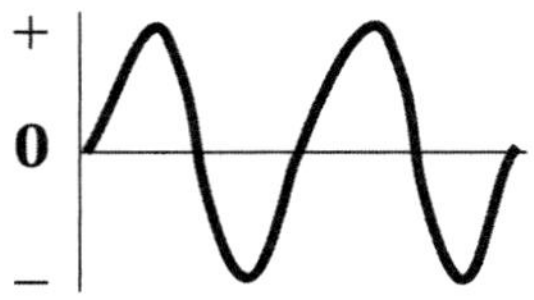

V_{ac} voltage = **17.8** $V_{P\text{-}P}$.

SCR anode voltage = **9.6** $V_{P\text{-}P}$.

CONCLUSION The SCR is evidently (*conducting on both half-cycles, conducting on every other half-cycle, not conducting at all*) **conducting on every other half-cycle**.

Explain your answer. **Acting as a rectifier**.

In this particular operating state, the SCR is acting as a (*full-wave, half-wave*) **half-wave** rectifier.

4. Without interrupting the source of ac power, remove the jumper wire you connected between points *A* and *B* in the circuit. Note the circuit's response and record your findings in the Observation section.

OBSERVATION The lamp is (*on, off*) **off**.

V_{ac} Source Waveform:

SCR Anode Waveform:

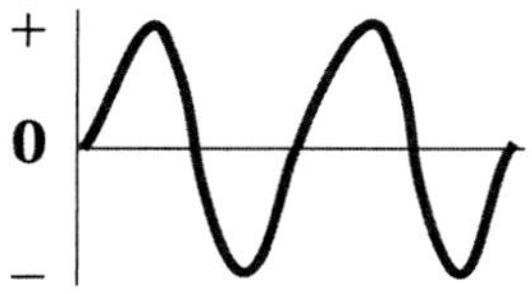

V_{ac} voltage = **12.8** $V_{P\text{-}P}$.

SCR anode voltage = **12.8** $V_{P\text{-}P}$.

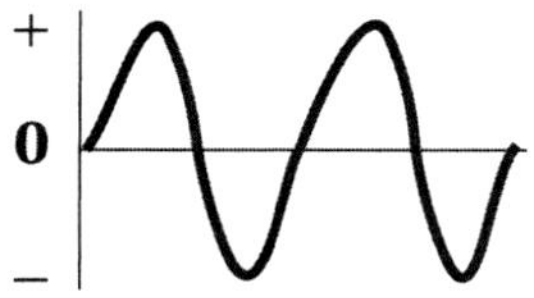

CONCLUSION The SCR is evidently (*conducting on both half-cycles, conducting on every other half-cycle, not conducting at all*) **not conducting at all**.

Explain your answer. **SCR is off**

_______________.

Lab Projects pages 473–475

PART 27 Summary

Complete the following review questions, indicating the appropriate response by placing a check in the box next to the correct answer.

1. In order to begin conduction, an SCR
   - ☐ must be forward biased at the anode, but the gate bias makes no difference
   - ☑ must be forward biased at the anode and the gate at the same time
   - ☐ must be forward biased at the gate, but the anode-cathode bias makes no difference

2. In order to sustain conduction of an SCR,
   - ☑ it must remain forward biased at the anode, but the gate bias makes no difference
   - ☐ it must remain forward biased at both the anode and the gate
   - ☐ it must be forward biased at the gate; however, the anode-cathode bias makes no difference

3. In order to stop conduction of an SCR,
   - ☐ the gate must be at zero volts or reverse biased; the anode bias makes no difference
   - ☑ the anode must be at zero volts or reverse biased; the gate bias makes no difference
   - ☐ the gate and the anode must be reverse biased at the same time

4. In an ac circuit, conduction of the SCR automatically stops when the ac waveform reverses the anode voltage from positive to negative.
   - ☑ True
   - ☐ False

5. For an SCR operating in a dc circuit,
   - ☐ the load current equals the source current minus the SCR current; and the load voltage equals the voltage drop across the SCR
   - ☑ the load current equals the SCR current; and the load voltage equals the source voltage minus the SCR voltage
   - ☐ the load current equals about one-half the SCR current; and the load voltage equals about one-half the source voltage

6. For an SCR that is switched on and forward biased in an ac circuit, the load voltage is equal to the instantaneous values of the source voltage minus the forward voltage drop across the SCR.
   - ☑ True
   - ☐ False

7. If an SCR operating in an ac circuit could be gated on 90° into its forward-biasing half-cycle (instead of 0°),
   - ☐ the load would have more power applied to it
   - ☑ the load would have less power applied to it
   - ☐ there would be no effect on load power

8. At best, an SCR operating from an ac source can only conduct on alternate half-cycles (180° of the full ac waveform).
   - ☑ True
   - ☐ False

Lab Projects pages 477–478

9. An SCR that controls an incandescent lamp in a dc circuit allows the lamp to burn at full brightness because
   - ☐ nearly all dc power is being applied to the lamp
   - ☐ the SCR is acting as a closed switch
   - ☐ there is only a small forward voltage drop across the SCR
   - ☑ all of the above are true

10. An SCR that controls an incandescent lamp in an ac circuit does not allow the lamp to burn at full brightness because
   - ☐ dc power is more energetic than ac power
   - ☑ the SCR cannot allow power to be applied for more than 180° of the ac waveform
   - ☐ the forward voltage drop of an SCR is relatively large
   - ☐ all of the above are true

Lab Projects pages 477–478

PART 28

FIBER-OPTIC SYSTEM CHARACTERISTICS

Story Behind the Numbers

Table 1 Short Cable Data (Circuit1)

V_S (V)	V_D (V)	ID (mA)	V_{out} (V)	I_E (mA)	CTR
0	0	0	0	0	
1	1	0	0	0	
2	1.2	0.8	0.008	0.08	0.1
4	1.2	2.8	0.028	0.28	0.1
6	1.2	4.8	0.048	0.48	0.1
8	1.2	6.8	0.068	0.68	0.1
10	1.2	8.8	0.088	0.88	0.1
12	1.2	10.8	0.108	1.08	0.1
14	1.2	12.8	0.128	1.28	0.1
16	1.2	14.8	0.148	1.48	0.1
18	1.2	16.8	0.168	1.68	0.1
20	1.2	18.8	0.188	1.88	0.1

Lab Projects pages 481–484

Table 2 Long Cable Data (Circuits1)

V_S (V)	V_D (V)	ID (mA)	V_{out} (V)	I_E (mA)	CTR
0	0	0	0	0	
1	1	0	0	0	
2	1.2	0.8	0.004	0.04	0.05
4	1.2	2.8	0.014	0.14	0.05
6	1.2	4.8	0.024	0.24	0.05
8	1.2	6.8	0.034	0.34	0.05
10	1.2	8.8	0.044	0.44	0.05
12	1.2	10.8	0.054	0.54	0.05
14	1.2	12.8	0.064	0.64	0.05
16	1.2	14.8	0.074	0.74	0.05
18	1.2	16.8	0.084	0.84	0.05
20	1.2	18.8	0.094	0.94	0.05

Table 3 Long Cable Data (Circuits2)

V_S (V)	V_D (V)	ID (mA)	V_{out} (V)	I_C (mA)	CTR
0	0	0	5	0	
1	1	0	5	0	
2	1.2	0.8	4.992	0.08	0.1
4	1.2	2.8	4.972	0.28	0.1
6	1.2	4.8	4.952	0.48	0.1
8	1.2	6.8	4.932	0.68	0.1
10	1.2	8.8	4.912	0.88	0.1
12	1.2	10.8	4.892	1.08	0.1
14	1.2	12.8	4.872	1.28	0.1
16	1.2	14.8	4.852	1.48	0.1
18	1.2	16.8	4.832	1.68	0.1
20	1.2	18.8	4.812	1.88	0.1

Lab Projects pages 481–484

Graph 1

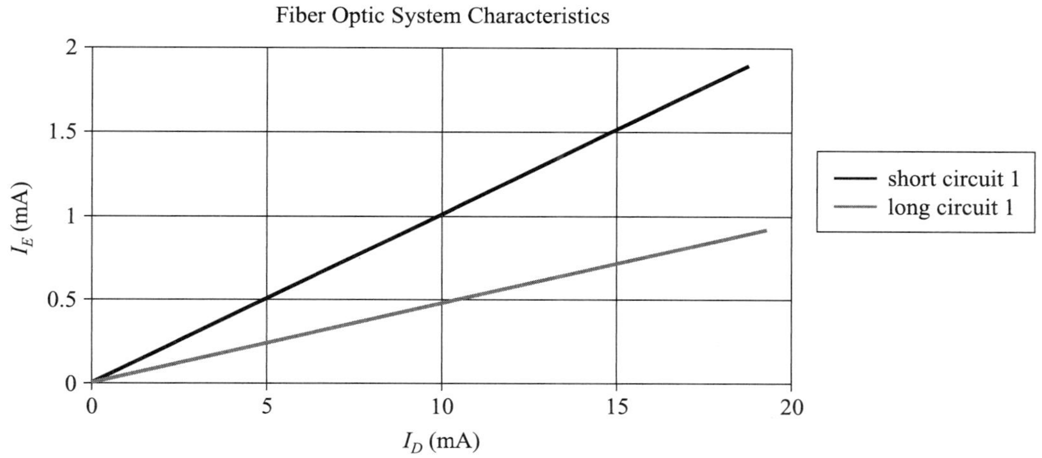

Graph 2

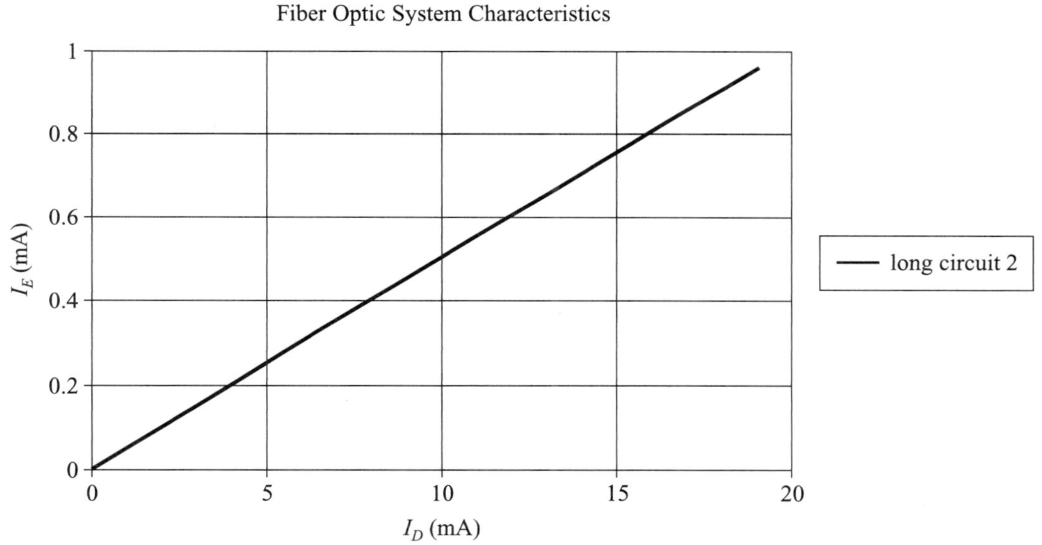

Answers to Analysis Questions

1. a. No
   b. Greater than
   c. Equal to
   d. Greater than. For a, b, and d, the longer fiber-optic cable will have the greater optic power loss (power loss equals the cable length times the cable loss parameter). For c, the photo-emitter circuit did not change in steps 2 and 3.

Lab Projects pages 481–484

2. a. No
   b. Equal to
   c. Equal to
   d. Equal to. For a, the output voltage in step 3 is measured at the photodetector emitter. The output for step 4 is measured at the photodetector collector. For b and d, the cable length is identical in steps 3 and 4. For c, the photoemitter circuit did not change.

Lab Projects pages 481–484